W9-BUY-228

Conference on the Introductory Physics Course

On the Occasion of the Retirement of Robert Resnick

Jack Wilson, Editor

Rensselaer Polytechnic Institute
National Science Foundation

John Wiley & Sons, Inc.
New York • Chichester • Brisbane • Toronto • Singapore

ACQUISITIONS EDITOR Stuart Johnson
MARKETING MANAGER Catherine Faduska
PRODUCTION EDITOR Deborah Herbert
MANUFACTURING MANAGER Mark Cirillo

This book was set in Times Roman by J.H. Cunningham and
printed and bound by Braun-Brumfield, Inc.. The cover was printed by
Braun-Brumfield, Inc..

Recognizing the importance of preserving what has been written, it is a
policy of John Wiley & Sons, Inc. to have books of enduring value published
in the United States printed on acid-free paper, and we exert our best
efforts to that end.

Library of Congress Cataloging in Publication Data:
Conference on the Introductory Physics Course (1993 : Rensselaer
Polytechnic Institute)
Conference on the Introductory Physics Course on the occasion of
the retirement of Robert Resnick / Jack Wilson, editor.
p. cm.
Includes bibliographical references.
ISBN 0–471–15557–8 (cloth : alk. paper)
1. Physics--Study and teaching--Congresses. I. Resnick, Robert,
1923- . II. Wilson, Jack (Jack M.) III. Title.
QC30.C639 1993
530′.071′173--dc20 96-19489
CIP

Printed in the United States of America

10 9 8 7 6 5 4 3 2 1

Table of Contents

PREFACE

Jack M. Wilson

The Conference on the Introductory Physics Course was held on May 20-23, 1993 at Rensselaer Polytechnic Institute in Troy, NY. It was particularly appropriate that this conference be held on the occasion of the retirement of Robert Resnick who, with David Halliday, wrote the book that has defined the introductory course for so many students for so many years. As one student once said after meeting Bob Resnick at a national meeting: "I thought that he invented Physics." Well, Bob and Dave did not really invent physics, but they certainly have explained it first to thousands of introductory students. Over the years since the first introduction of the text, Bob has been a tireless advocate for continuous improvement in every aspect of physics education. He has given generously of his time and talents in support of students and faculty both locally and nationally. David Wallach provided a humorous chronicle of how Resnick and Halliday changed with the times with a chronology of illustrations from the book!

The combination of Bob Resnick's popularity and the timely nature of the conference drew 357 representatives of universities, four year liberal arts colleges, two year colleges, and high schools (with the majority from the universities) to the conference. Delegations came from Canada, Mexico, China, Japan, Germany, and Australia. Given the ferment in the introductory university physics course, there was a lively dialog on the issues. Among the projects represented at the conference was the APS/AAPT Introductory University Physics Project, the M.U.P.P.E.T. project, Workshop Physics, Tools for Scientific Thinking, RealTime Physics, CUPLE: the Comprehensive United Physics Learning Environment, the New E&M, The Physics InfoMall, and the American Institute of Physics: Physics Academic Software Project.

We particularly wanted to consider how that course would change in the light of the issues surrounding computing, cognition, communication, community, and contemporary content. In 1988 it was noted that if one could bring James Clerk Maxwell back from the late 19th century into an introductory physics classroom of 1988, he would recognize nearly everything he saw. Yet, he would recognize almost nothing at any research meeting. This disparity between the "state of the art" in science and the "state of the art" in teaching prevailed in many disciplines. Content in each discipline needed to be updated to reflect modern issues and understanding. In addition, we needed to examine the introductory course in the light of what we are learning about the way students learn. Change, in physics, comes quickly, is readily communicated, and is widely appreciated. Change, in physics teaching, is long in development, is poorly communicated, and is often met with resistance.

Computers have become an integral part of doing physics but we are only now coming to grips with exactly what their role should be in teaching physics. A new area of physics has emerged in the past few years "computational physics" to join theoretical physics and experimental physics. The computational physicist may play a role analogous to that of the instrument builder of earlier this century. The importance and centrality of the computational techniques are recognized by the new journal, Computers in Physics, and by extensive coverage at meetings including the Conference on Computers in Physics Instruction held in Raleigh, North Carolina in August 1988.

The AIP/APS/ASPT sponsored Introductory University Physics Project (IUPP) and Maryland University Project in Physics and Educational Technology (M.U.P.P.E.T.) were each studying the physics curriculum to determine how it might change to reflect physics as it is done today. It was not enough simply to catalog the failings of the present system, we also explored the characteristics such a course should have. Through surveys of the faculty, M.U.P.P.E.T. listed some of the physics skills we wanted our students to acquire, including quantification skills, analytic skills, scales and estimation, approximation skills, numerical skills, intuition, and the ability to break down a large problem into component parts.

A survey of physics texts done by Gordon Aubrecht (Ohio State) for the M.U.P.P.E.T. project in the late 1980's found that 95% of the content of all texts was universal and about 95% of the content was pre-1935. The current texts are enormous, with over 1,000 problems; yet those problems emphasize only a very limited class of skills. Over 90% relate to analytical skills, only a small number relate to the very important estimation skills, and almost none expect the students to use numerical approaches.

Against this backdrop, the Conference on the Introductory Physics Course brought together the key individuals with expertise in each area. After Leon Lederman got the ball rolling with a delightfully humorous evening discussion of "Introductory Physics from K through 21," Bob Resnick put us all back on task with his keynote on "Retrospective and Prospective." Arnold Arons pointed out the pitfalls and potholes waiting for us on the way to curriculum reform. Undaunted, I took up the hopelessly optimistic cause with "Some Possible Futures for the Introductory Course."

Once we had exhausted the past and the future, we turned to the present and asked Lillian McDermott to tell us "How Research Can Guide Us." Thus charged, we took up the issues of computing, video, laboratories, and content reform.

When a group of like minded individuals gets together and begins to plan the future, it is often instructive to ask an outsider to do an independent observation of the community and it's culture. Fortunately, Shiela Tobias has devoted much of her professional career to doing just that for the physics community. Her town-meeting session on "Moving the Mountain: How Do We Get the Physics Community to Change," was a challenge to each of the exponents for change.

As usual at Physics meetings, the hallway conversations, poster sessions, and workshops were among the high points of the meetings. We have included what we can from these sessions in these proceedings to help give the reader the flavor of the meeting. Of course there is no way that any proceedings volume can capture the hands-on experiences of the dozen workshops, many of which made extensive use of the Rensselaer multimedia computing classrooms.

In the end, I came away with my sense of optimism intact. It is a time for change in physics education. Just as the publication of the Halliday and Resnick text launched a new era in 1958, the new formats for physics teaching discussed at this conference was likely to lead to new learning environments for students. Whether it is RealTime Physics, Workshop Physics, the CUPLE Physics Studio, or a yet to be articulated version for the future, it will be an exciting time to be a physics teacher.

In the end, this conference could never have happened with the support, advice, encouragement, and hard work of many persons. Without Robert Resnick, this conference would have lacked the focus and purpose. He inspired me first as an author, then as a mentor, later a friend, and always an inspiration. To Bob students have always come first, and the substance of this conference echoes his dedication.

The support provided by the National Science Foundation (DUE-9350168) and John Wiley & Sons went far beyond the substantial financial support as both made intellectual contributions to the program development and John Wiley & Sons took on the arduous task of creating this proceedings volume.

My thanks go out to the Conference Committee that helped us to put together the program and recruit the speakers. Also, the Proceedings could not have been completed if Stuart Johnson of John Wiley & Sons and Jack Hilton Cunningham, Desktop Publisher, had not stepped in to find a way to produce the volume within very tight financial constraints.

I must particularly single out Linda Kramarchyk, Assistant to the Dean, who really did the communication, logistics, and administrative work of the conference. She has an uncanny ability to translate my vague suggestions into workable arrangements, all while sorting out and prioritizing all the tasks that had to be completed. As usual she was magnificent.

Part I

The Invited Speaker's Papers

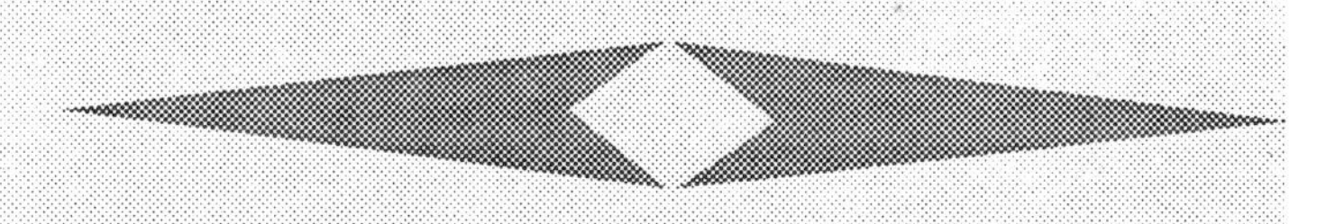

RETROSPECTIVE AND PROSPECTIVE

Robert Resnick, Department of Physics
Rensselaer Polytechnic Institute

Introduction

One member of the planning committee for this conference recommended that the committee simply pick the speakers and let each talk about whatever he or she chooses. That must have been Don Ivey, and — as opening speaker — I can feel free to take his advice. I choose to present a smorgasbord of dishes to you. There will be a bit each of the past, the present, and the futuristic dishes. However, before the conference is over we will have sampled everything and will have absorbed a rather full meal.

So, let me begin selecting dishes. I start with a textbook *Physics, Part I*, Resnick & Halliday and *Part II*, Halliday & Resnick published first in 1960. Why did it succeed? I'm often asked that. It's really only in retrospect that one can approach this question objectively. I might like identifying with the mood of Lord Rutherford, who wrote in a letter late in life as follows: "I've just been reading some of my early papers, and you know, when I'd finished, I said to myself, " Rutherford, my boy, you used to be a damned clever fellow". But instead of believing how clever we were, I'm inclined to say that we authors (Dave and I) somehow unconsciously represented all the needs for change that were then 'in the air' (so to speak) and that we lucked out on the timing. A little personal history helps here.

The book had its origin in my being lecturer in the large physics course for engineers at the University of Pittsburgh. I tried to present the physics with a clarity and a depth I felt was lacking in the books we then used. I was giving perhaps the best lectures of my career both to the undergraduate engineers in general physics and to the graduate physics students in an atomic and nuclear physics course that I also taught then and the students were really turned on. I soon developed a kind of religious fervor about what was wrong with the teaching of physics at all levels, and that led me straight into AAPT meetings and conferences.

It was at a summer conference where I found a co-religionist in Walter Michels, soon to provide the AAPT with one of the most dynamic presidencies it has had, and was persuaded that I should "put my money where my mouth was" (so to speak)...and write an introductory physics textbook. Back home at Pitt. was a talented chairman, David Halliday, whom I informed of my decision. When I found that he was considering writing an atomic physics book to follow his pioneering nuclear one, it didn't take us long to decide on a joint venture...first we would do the general physics book together and then, sometime later, we would do the atomic physics one. (We never did do *that* one, by the way.) We were given a contract by the way, without a single word of manuscript written.

Shortly after the writing started I was enticed away from Pitt. to Rensselaer. This was a decisive move in guaranteeing the success of the book, in my opinion. Every science and engineering student at Rensselaer took a 4-semester introductory physics course that involved the entire physics faculty in one way or another. I was successful in getting foundation and institute support for a complete revision of that course along the lines of my reform proposals. So, in the context of designing new laboratory experiments, new lecture demonstrations, new audio-visual techniques and so forth, the text notes were steadily prepared by the authors, produced, and tested under fire. Although the notes were being used at Pitt. also, we had agreed that the Rensselaer trials would be the control. The student body there was a cut or two above what I had encountered earlier in ability and interest, and the Hawthorne effect was at work...the consciousness of taking part in a major reform program raising the participation and enthusiasm of everyone.

Well, we went through three years of intensive classroom trials at Rensselaer. These trials perfected the textbook. They set the level above the then existing national norm; but the national tide was clearly coming along, though at the time critics felt our level was much too high to be realistic. The trials guaranteed the quality of the material... allowing attention to detail, polishing the physics, and perfecting the goals...a quality that would give the book staying power when used and re-used. The trials enabled us to use our planned expansive writing to clear up all sorts of pedagogic hurdles and student difficulties and to build up the number and quality of examples, figures, tables, questions, problems and other teaching aids. In short, we were able to produce a book that was in tune not only with rising standards and developing trends in physics but also with a need then emerging from the rapid growth of mass education of the sixties, namely, less dependence on instructors and a greater dependence on self-study by students. A textbook was becoming its own study guide.

To be sure the trials could have led to confusion and even disaster if the authors didn't have the judgment to reject a good deal of well-intentioned advice, to take into account contradictory criticisms, and to stick to and to know how to carry out their initial goals. It is a fact that not one of the many external reviews commissioned by the publishers was favorable . . . reviewers really did not see much merit in the book when they tested it against the needs they perceived. Indeed, we learned later that it was only on the advice of Robert Sproull, who was then a Wiley consultant and already a successful physics writer in his own right, that the firm decided to ignore these reviews and to go ahead and publish the book as Dave Halliday and I wanted it.

We knew when we were done that the book would work here at RPI and perhaps at Cal Tech or MIT or Berkeley — we didn't know how fast interest in physics and student standards would rise nationally in that immediate post-Sputnik environment, so that our 'special' book would become the country's norm a few years later. Coinciding with this was a big infusion of younger physicists into academic faculties who themselves yearned to modernize the subject matter, to raise the level and to put in their own enthusiasms, allowing us to use vector algebra (imagine!), to sprinkle modern physics in throughout, to push conceptual items like thought questions and small print sections on special topics or on the history and philosophy of physics, and to give all sorts of references for further reading. Somehow, Dave and I — consciously or unconsciously — represented that whole new wave of change. And it didn't hurt that one of us was a theorist and the other an experimentalist (See Figure 1).

1. **Intensive Classroom Testing — by a large staff on a large and very able student body for three years**
2. **The Hawthorne Effect — text development part of a conscious major reform program**
3. **Growth of Mass College Education — textbook becomes its own study guide**
4. **Rising Post-Sputnik Level of Physics Course — a deeper treatment of fewer topics**
5. **Infusion of Younger Physicsts Into Academic Faculties — leads to modernization of subject matter and greater stress on conceptualization**
6. **The Authors — Intuitively reflecting the needed changes; theorist and experimentalist**

Figure 1. Some Factors in the Success of the Halliday-Resnick Text of the '60's

I remember Jerrold Zacharias saying to me — "You guys spoiled the whole thing. By doing such a great job on bringing out the very best in the old way of doing things you set physics back by a generation, not allowing us to do it really right." And I remember my debt to authors of other books from which I had taught earlier, both for teaching me what went over well but also provoking me on what didn't; I didn't dare look at other books, in preparing my own, for fear of copying. When I told this to Mark Zemansky (what a sweetheart of a guy!) he said "There's nothing wrong with copying, you've just got to be very careful that you don't copy mistakes."

Well, there are plenty of analogies and contrasts today to the early history I just recited; but I leave that to other speakers. However to give you some food for thought, I want to read to you words written 34 years ago, from the Preface to *Physics, Part I*, by Resnick and Halliday. Time permits only selected items, but reading it all would illustrate how little our goals or ideas have changed.

"The most frequent criticisms made in varying degrees of textbooks used in such a course are these: (a) the content is encyclopedic in that topics are not treated with sufficient depth, the discussions are largely descriptive rather than explanatory and analytical, and too many topics are surveyed; (b) the content is not sufficiently "modern," and applications are drawn mostly from past engineering practice rather than from contemporary physics; (c) the organization of the material is too compartmentalized to reveal the essential unity of physics and its principles; (d) the approach is highly deductive and does not stress sufficiently the connection between theory and experiment.

> "In writing this textbook we have been cognizant of these criticisms and have given much thought to ways of meeting them. We have considered the possibility of reorganizing the subject matter. The adoption of an atomic approach from the beginning or a structure built around energy in its various aspects suggest themselves. (And here's what Zach objected to.) We have concluded that our goals can best be achieved by modifying the selection and treatment of topics within the traditional organization. To shuffle freely the cards of subject matter content or to abandon entirely a sequence which represents the growth of physical thought invites both a failure to appreciate the Newtonian and Maxwellian syntheses of classical physics and a superficial understanding of modern physics. A solid underpinning of classical physics is essential to build superstructure of contemporary physics in our opinion.

"To illustrate how we hope to achieve our goals within this framework, we present here the principal features of our book." And then we get specific.

Well, that's enough for the first edition of the book. Keep in mind, however, that as we taught from it here at RPI we had an absolutely first rate battery of lecture demonstrations used for student motivation and interest and for concrete experimental display; that we created our own and showed other films in modern physics; that we promoted TV pick-up of small scale lecture demonstrations; that we developed the use of the overhead projector in imaginative ways. We did use the technology of the day for pedagogic purposes just as today the microprocessor and computer are being used (something you'll hear from Jack Wilson, I'm sure). And we stressed and assigned thought questions, discussed them in recitations and put them on quizzes — conceptualization was not ignored. And the assigned recitation problems were, for a long time, the best in the business, so to speak. We did not ignore the issues I that expect subsequent speakers to address in today's context.

New Approaches

I have given earlier talks in which I present some detail on programs that are experimenting today with new ways of approaching the introductory physics course. Such programs are MUPPET (Maryland University Project in Physics and Educational Technology); the Overview-Case Study Method or Spiral Method of Teaching Physics; the Karlsruhe Approach to Physics; the Introductory University Physics Program; the Colgate Introductory Physics Program, and others. You will hear about some of these later on in the conference. There's lots of ferment out there but certain common issues arise in these programs as to what's wrong with today's courses. Please note how similar all this is to the Preface of R&H written in 1959 (See Figure 2).

- **Modern Physics — excluded or poorly presented**
- **Problems — unrealistic, oversimplified, ignore cognitive insights**
- **No Unity — too much material or not properly organized**
- **Computer/Computational Physics — not used to resolve above criticisms**

Figure 2: Common Criticisms

These are that modern physics is either not sufficiently represented in our introductory course or else not presented in an understandable way; that the problems treated in our courses are simple and unrealistic, and not treated with an awareness of cognitive issues (student misconceptions), or the manner in which physicists actually solve problems; that the unity of physics is not revealed in our course because we either treat too much material or organize the concepts improperly; and that we are not taking advantage of what the computer and computational physics has to offer in resolving these issues.

But there are still other issues and concerns to put on the table here, many relevant to the ones I've already mentioned and some that tend to moderate or contradict .

Less May Be More

First, there's the theme popularized by Phillip Morrison that "less may be more", which I endorse. That is, rather than "cover" all of physics we should "uncover" it, that we can get a reasonable depth of understanding only by avoiding encyclopedic outlines.

It would help us at the college level if high school courses did not imitate the college course. In 1960, consistent with this theme, we (R&H) eliminated from our textbook a whole host of topics typically in earlier courses. I read you a partial list from our 1959 preface. "Many topics are treated in greater depth than has been customary heretofore, and much contemporary material has been woven into the body of the text. To permit this greater depth and inclusion of contemporary material, we have omitted entirely or treated only indirectly much traditional material, such as simple machines, surface tension, viscosity, calorimetry, change of state, humidity, pumps, practical engines, musical scales, architectural acoustics, electrochemistry, thermoelectricity, motors, alternating-current circuits, electronics, lens aberrations, color photometry, and others." One part of the justification to professors was that those topics were treated in the high-school course. It wasn't long before the high schools followed suit and eliminated most of those same topics. This was not a service to high school students either, many of whom did not go on to college or to a college physics course. At the high school level the advanced placement course is the one that might properly resemble the college one. The regular one year junior year or senior year high school physics course, however, should be a survey course that applies physics to the everyday world, one that could be considered a terminal course in physics. In my opinion such a course would even have been better for the college-bound physics major. Perhaps the newly proposed NSTA program for spreading physics over all four high school years is the correct way to go.

However, all this had the effect of building up the pressure on authors to incrementally add this or that topic and, with the coincident and more important growth of modern physics topics, we're nearly back to an encyclopedic textbook today.

Compared to thirty years ago, however, the current treatment of topics *is* deeper; there is more of a story line and a greater stress on unifying principles. Moreover, simply comparing the lengths of current and much earlier textbooks is very misleading. Many of the extra pages in textbooks come from the growth of *pedagogic* features such as thought questions, exercises, problems, worked-out sample problems, hints on problem solving, chapter summaries, photographs and figures and even free-standing essays on applications of physics to our everyday world. All of these features, as well as a much more expansive treatment of individual topics, flow from the growth of mass education in the colleges during that period and from the increased awareness of cognitive issues in student learning and the need to motivate students. These required that the textbook be — more than before — the in-depth teacher, the workbook, and the study guide. So we *have* made much progress, indeed, in the recent past. However, the *number* of topics included in texts has in fact grown just about back to where it was before 1960, even as the course time being allotted to introductory physics is decreasing.

Keep in mind that textbooks are expected to provide material for courses of significantly different lengths and level, for instructors having different tastes and emphases, and for different audiences of students. I believe that it is incumbent on *professors* to exercise their judgment to cut a coherent path through the subject matter. There are many such paths that can very significantly reduce topical coverage, while at the same time include a significant coverage of modern physics and reveal unifying principles and themes. A great many do just that: at some schools fluids, thermodynamics and relativity are omitted from the course which emphasizes Newtonian mechanics, Maxwellian electromagnetism, Waves and Quantum Physics and their applications. Of classical topics, some omit geometric optics or electrical circuits, or confine that to labs; they might not treat a.c. circuits, or polarization, or rigid body statics, or magnetic properties of matter, and so forth. Or they vary the depth of treatment of the included topics. Professors *do* plan their courses; they are not mindless followers of text. Indeed, the texts reflect *their* collective demands and desires — through the strong feedback they give publisher's representatives, the letters they write to authors, the countless manuscript reviews of texts in process of birth, and so forth. There have been plenty of short books available; they simply were not much used. Books that start relatively short soon become longer. The most widely adopted texts are the longer ones. *One result, however, is that we are tempted to cover too much in our course.*

What we need is a campaign to promote "less is more". Perhaps we should have a regular feature in our teaching journals that show how various teachers choose to accomplish that. Don Holcomb, for example, at one point developed criteria for organizing such a physics course with concrete examples of inclusion and exclusion. We need widespread participation in such exercises and publicity for them. It should become acceptable and even fashionable to do less to achieve more (See Figure 3).

- **"Uncover" physics rather than "cover" it**
- **High school should not imitate college**
- **Professors should cut a coherent path thru the sub ject matter in texts — many valid paths exist**
- **We need a publicity campaign to promote "less is more"**
- **Textbook $\neq$ Course**

Figure 3. "Less may be More"

If one is worried about so-called gaps or the potential lack of coherence in a 'less may be more' selection of topics from a long textbook, many options exist on how to address this. Amongst them is the preparation and availability of a large number of shorter textbooks, each following its own selection and emphasis and presenting alternative models. Many of these are on the way, including one by Halliday, Resnick and Barbara Levi, quite apart from those planned by the experimental programs, such as IUPP, that I mentioned earlier.

The Course and the Textbook

As a final comment on this textbook issue, I must state that although in one sense a textbook is more than the course, in that it contains more items of content than will be considered, in another sense the physics course is very much more than the textbook. The textbook cannot take on the burden of doing everything nor is it equipped to do so. Consider, for example, the desire to show that physics is an experimental and observational science; that the students need an overview of it all and not only details; that the excitement and vitality of physics should be conveyed; that scaling, estimation, and approximation skills of physicists should be revealed; that physics as a process needs more emphasis; that modeling of physical systems is important, that student-professor interaction is crucial in eradicating misconceptions and the like. To help meet such objectives, physics courses cry out for variety and quality in the laboratory, in the examinations we give, in the teaching methods we use, in the active (not passive) role that lectures can play, in the use of computers, and in the spectrum of physicists who are the ultimate teachers. We need more emphasis and attention put on improving these aspects than we are now giving them. If 'less may be more' is my first call to action, then this is my second one. It's a real cop-out to clamor about improving the physics course by talking only about subject matter content and not to pay serious attention to these other vital areas. As Tom Rossing has noted, we need to pay much more attention to *how* we teach, and *who* does the teaching and *to whom*, and not just give attention to *what* we teach (See Figure 4).

- **Need to show that physics is an experimental and observational science**
- **Students need an overview of the subject**
- **Excitement and vitality of physics should be conveyed**
- **Scaling, estimation, and approximation skills need to be revealed**
- **Modeling of physical systems important**
- **Student-professor interaction crucial in eradicating misconceptions**
- **Need for 'soaking' time**

Figure 4. Textbook ≠ Course.

Surely we need to focus more attention on the individual student and we must allow for 'soaking time' for concepts to sink in. It has been generally understood in the past that the purpose of teaching a course is not to teach the student but to teach the teacher. Perhaps, then, requiring senior students to teach freshmen or sophomores, for example, isn't a bad idea at all. In fact that's just what I did as an undergraduate at the Johns Hopkins University. But, of course, we must go further than that. Moreover, you can't teach something unknown in terms of something else unknown — so, e.g., teaching quantum mechanics in terms of statistical ideas or distribution functions, or imaginary numbers — or requiring knowledge of computer languages to teach physics topics — must allow the time to learn the unknowns assumed known. And there are constraints, like the high cost of new technology, the expense of teaching small numbers of students, the needs expressed by the users of our courses — students sent to us by engineering departments, for example, and the like. So, in all that we do, let's be realistic in aiming at the ideal.

Changes In Student Body

Another real issue — one that poses at least as serious a problem as the ones we've already discussed — is the overall change in the preparation, outlook, and even the composition of the incoming student body in the U.S. (See Figure 5). Students come to college today with very different experiences than even 30 years ago (e.g. television) and with much less hands-on experience in technology or applied science (rural to urban living). Moreover, they are not as willing or able to work at learning. There are clear signs of a decline in literacy. In spite of better calculus and computer education, there's a decline in knowledge of geometry and space perception. The job market and general sociological scene puts greater emphasis now on training rather than education, on the relevance and application of physics to the perceived early employment field of the student rather than on physics per se. And the increasing specialization of knowledge in engineering and science curricula is pressuring physics professors to cater even more to user's perceived needs rather than being forced to accept the alternative of having the physics course credits reduced. Not unrelated to this has been the pressure on young physics faculty for research grants, tenure, and specialization in their teaching, leading to less concern for and participation on their part in the general introductory courses. And, finally, the changing demographic situation in the U.S., where now women constitute over half the enrolled college population and where by the year 2000 the percentage of college age persons who are minority (black and Hispanic) members will be about 35%, requires that we meet the challenge and seize the opportunity to appeal to student constituencies not traditionally sources of physics students or physics majors.

- **Decline in literacy**
- **Decline in knowledge of geometry**
- **Training stressed rather than education**
- **Relevance of physics to other disciplines stressed, rather than physics per se Less participation by young physics faculty in the introductory course**
- **Must appeal to student constituencieds not traditionally source of physics students**
- **Not as willing or able to work on learning**

Figure 5. Changes in Study Body

Today we have pre-entry summer programs for admitted but poorly prepared students, X-sections (X for extra time) in the introductory course, campus-wide tutorial services, learning centers and writing center, review sessions, student study guides, instructors' manuals, manuals giving solutions to textbook problems for faculty — all reflecting these new conditions. And almost unconsciously we've been demanding less and less of students on our tests and quizzes. Cliff Swartz, of the State University of New York at Stony Brook, has developed a one-semester pre-course — an impedance matching prerequisite to introductory physics — just to get the students today to the level of preparation that we once assumed they had attained. This procedure is spreading to more and more colleges.

For the sake of the student, for the country, and for physics itself we cannot take a "sink or swim" attitude towards our students. Nor must we give up reproducing our own kind. We are at once elitists and democratic. This is the complimentary principle at work, two sides of the same coin. As we lower the entry barriers to physics and increase the stress on pedagogy and application in some cases we must keep the standards of preparation and level of performance and sophistication as high as ever at the same time in other cases. Surely, an honors course in physics or one for dedicated physics majors cannot use the same model as a course for the poorly prepared and/or unmotivated student who's "required to take it" — not if it's to be successful.

Happily though, there's room for many models — even in the calculus-based physics course for engineering and science students. Lengths are different, levels are different, our audiences are different, and our faculties are different. We *need* a large diversity of courses.

My attitude is to "stir the pot" of curriculum reforms aimed at improving *all* these courses. That in itself will heighten the performance of the teacher and the interest and participation of the student. And what's found to be effective and appropriate will be stolen by others — authors included — for their courses. In this way the introductory courses will steadily evolve and perhaps a significant mutation will arise; quantity of change can lead to a qualitatively new course. In other words — evolution, not revolution (See Figure 6).

- **Elitist and Democratic**
- **Need for Diversity**
- **'Stir the Pot' of Curriculum Reform**

Figure 6

Classical *versus* Modern

Finally, I'd like to comment on the theme I've heard many express of classical *versus* modern physics. That probably arose because of the resistance of some to dropping favorite classical topics when pressed to add "modern" topics and has been influenced also by the traditional location of "modern" topics at the *end* of textbooks. By modern I mean topics that require an understanding of relativistic or quantum concepts. The application of *classical* physics to our modern world (such as space travel) I call contemporary physics, rather than modern physics. Hopefully, we're all contemporary. *My point is that classical and modern physics should not be put in opposition to one another* (See Figure 7).

- **Unifying principles equally applicable**
- **Topics of relevance and application in both areas**
- **Modern physics an extension of the classical-continuity**
- **Threads of modern physics can be weaved thru the fabric of classical physics**
- **Appeal to experiment everywhere to show the need for new ideas**

Figure 7. Classical versus *Modern*

For one thing, there are unifying principles in physics that are equally applicable to modern and classical physics. I refer to conservation laws, field concepts, wave concepts, symmetry ideas, analogies of method and the like. Indeed, these should be the themes we stress, using illustrations from both classical and modern areas. There needs to be more emphasis on process and unity than on content per se.

For another, even those who want applied courses and relevant topics can find much of relevance and application in modern physics — be it lasers, semiconductors, superconductors, nuclear energy or you name it. Here the "need to know" will motivate otherwise disinterested students to study modern physics.

Another argument against the inclusion of modern physics is the valid observation that we grow up in a macroscopic classical world in which the concepts of relativity and quantum physics are remote, sometimes counterintuitive, and implausible as starting points. This surely affects *how* and where we should present modern physics topics (and perhaps how deeply we should go into them in a first course) but not really whether or not we should present them. Our approach too often has been to make relativity and quantum physics appear shocking, revolutionary, "gee-whiz", and anti-establishment, — a tremendous contradiction to so-called common sense — whereas the best pedagogy demands that we stress the plausibility of modern physics as an extension of classical physics into the relativistic (high speed) and quantum (small dimension) domains: We should aim at a smooth transition and a natural acceptance of new enlarged views of the world. Indeed, we can reverse the process showing that the classical physics emerges from relativity when we let $c \to \infty$ and from quantum mechanics when we let $h \to o$. You might say that I want continuity as well as unity. Not that this is easy. It probably is no harder than overcoming Aristotelian ideas in teaching Newtonian mechanics — something we *do* try to do — but, if harder, not that much harder.

Indeed, all through classical physics we can foreshadow these modern ideas: in the Doppler effect we can easily introduce the notion of the absence of a reference medium for light and that the relative motion...of the observer and the observed...is what counts; in gravitation I've found it easy to get students to accept the principle of equivalence; in kinetic theory the quantum idea emerges from discussions of specific heats and there are many other places where the underlying discreteness can be revealed. The principle of superposition, boundary conditions, standing waves and ideas of resonance in classical physics are carried on analogously into quantum physics. We can even demonstrate "frustrated internal reflection" easily with paraffin blocks and microwaves, thereby anticipating barrier penetration and tunneling in quantum mechanics. And — anticipating the later need — we can modify our treatment of classical statistical physics with its distribution functions and the like.

In fact, one can weave threads of modern physics all thru the fabric of classical physics. And we can appeal to experiment everywhere showing the need for the new ideas.

One More Particular Plea

The time has come to treat special relativity in the introductory course not as modern physics but as classical physics instead. Not only should a significant treatment be in every course but it should come very much earlier than the typical current placement near the end. It can very properly come just after classical mechanics and the study of mechanical waves, for example.

To begin with, unless relativity is in our standard course the non-physics major student will never in any other course be exposed to this profoundly significant subject. Moreover, the mathematical preparation needed is minimal and students typically are fascinated by the subject. And all my teaching experience shows me that the student really does not understand classical mechanics — and has much less appreciation of electromagnetism — without being exposed to relativity. Relativity reveals dramatically the role of experiment, the importance of reference frames, the breakdown and limitations of Newtonian mechanics, the central role of conservation principles, the invariance of physical laws, [the meaning of length and time], and the very process of physics itself. It can reveal the simplification of otherwise complex problems by the proper choice of reference frame, and the equivalence of relativity and electromagnetism. And examples from nuclear and high energy physics provide meaningful applications of the subject. In our fourth edition of the text 'PHYSICS', Dave Halliday and I, along with Ken Krane, have adopted this very procedure (See Figure 8).

- **Select fewer topics in order to get more depth. "Less is More"**
- **Put emphasis on *how* to teach and *who* teaches *to whom*, not merely on *what* to teach. "Textbook ≠ Course"**
- **Aim for diversity of courses, not a concensus course. "Stir the Pot of Reform"**
- **Classical and modern physics need not and should not stand in opposition to each other. "Continuity and Unity"**

Figure 8. Summary

Conclusion

There are many other important aspects of the introductory course that I've barely touched on, if at all, in my presentation. The role, content and structure of the laboratory is certainly one. The role of lecture demonstrations and audio visual aids, the whole area of testing, and the out-of-course physics related activities are others. And I've mostly ignored the non-calculus, liberal arts introductory physics course. Nor have I faced the issue of the role of computers in all aspects of the course including interactive learning. You can see that I've taken a "less may be more" approach in designing this talk.

However, I've filled the table enough already and I leave it up to others to serve up their dishes to you.

We all love our subject — physics, or natural philosophy. And we want others to share in our love of the subject. So we are here to argue and to compete on how best to make our subject loved. Each generation must learn all over again what the previous one did and advance from there.

A final comment might be in order, however. There's lots of activity and ferment out there. We are now beginning to see a broadening of and an increase in members of the physics profession who are participating in a reformulation of introductory physics. That's all to the good.

IMPROVEMENT OF PHYSICS TEACHING IN THE HEYDAY OF THE 1960'S

Arnold B. Arons, Department of Physics
University of Washington, Seattle

Historical Preliminary

The year 1960 saw the appearance of the first edition of "Physics for Students of Science and Engineering" by Resnick & Halliday and also the genesis of the Commission on College Physics. Bob Resnick played an obvious role in the first event and a significant, albeit less well known, role in the second. To have some perspective toward the events of the '60's, it is useful to be aware of at least a few elements of prior history.

Before World War II, introductory physics texts for scientists and engineers were, for the most part, very much like the algebra based texts of the present day without the twentieth century physics that is now included. They were only slightly different from widely used high school texts such as those by Millikan and Gale and by Black and Davis. In 1944, Francis Sears copyrighted the three volume, calculus-based "Principles of Physics" that he had developed for his two-year introductory physics course at M.I.T. and had it produced for use at M.I.T. by a small printing company, called Addison-Wesley, whose plant was located nearby. The revised edition of 1946 became available to the physics community at large as physicists (such as I myself) were returning to academic work after years of war research. Filling a critical need for highly improved instructional materials, it spread at a velocity only slightly less than that of light and proved to be the making of the Addison-Wesley Publishing Company.

Francis Sears had generated a discontinuous shift to the teaching of introductory physics at calculus-based level. The three volume "Principles of Physics" had a level and content, a mixture of phenomenology and mathematical sophistication, that could just about be assimilated by the more able students in a full two-year sequence. (In the subsequent years up to 1960, several competing texts of similar level and content, such as those by Margenau and Murphy and by Shortley and Williams, for example, emerged but did not achieve anything like the circulation of Sears.) Many of the mid-level students, especially those heading for engineering, were left behind in understanding of the physics and were relying on partial credit from memorization of problem solving procedures.

Murmurs of concern in the teaching community about deleterious effects of increasing pace of coverage and level of abstraction led, in 1957, to the convening by AAPT of a national conference at Carleton College. A key recommendation in the report of the conference reads, "Physics, as a body of knowledge is now far too extensive to receive adequate general coverage in an introductory course. ...physics teachers must reduce drastically the number of topics discussed. ...A more critical and parsimonious selection of content would permit a pace that encourages both reflection on the part of the student and a proper regard for depth and intellectual rigor." [1] Under the influence of some minority pleading, the report also renders lip service to the infusion of at least some historical, philosophical, and humanistic insights into introductory physics for all students, not just for non-science majors. The extent to which these recommendations ultimately influenced nation-wide practice should be obvious to any observer regardless of his or her age.

The first edition of Resnick & Halliday appeared in 1960, and, for the present audience, I need not enlarge on its tremendous influence and its meteoric history except to emphasize that the intent of all the texts of that period was for use in a full two-year introductory course. There began, however, the process of squeezing down of the time span.

(Let me remark parenthetically that, since that time, the calculus based texts that have emerged in competition with each other have become increasingly massive, more colorful, more highly mathematicized, less phenomenological, and more inclusive. The content is very much greater than the majority of students can absorb with any significant degree of understanding in two years while volumes of subject matter are being squeezed into one year, apparently on the premise that, if the material is passed by the student at sufficiently high velocity, the Lorentz contraction will shorten it to the point at which it drops into the hole that is the student mind. This is a problem for the present generation of physics teachers; allow me to return to the past.)

In the post war years, several themes dominated discourse on physics teaching in APS and AAPT. Sparked by James Bryant Conant and other leaders of wartime research (especially those participating in the development of nuclear energy and the atomic bomb), was the concern with wider scientific literacy in the general public. Although this concern had been articulately expressed in committee reports on physics teaching since at least the beginning of the century, little had been seriously done in curriculum development along such lines until the late 1940's and early 1950's. Another widely prevalent theme was that of modernization of the introductory physics course. Still another theme was that of the quality of high school physics teaching, and summer institutes for high school teachers were getting under way on a small scale with private (mostly industrial) sources of support in the mid 1950's.

Alan Waterman, as first director of NSF (which had gotten started in 1950), had been very conservative about initiating foundation activity in science education. He was fearful of the difficulties and policy ramifications that this area entailed and obviously wanted to get NSF procedures well settled and widely respected in hard science first. By the mid 1950's, however, Cold War concerns about threatening shortages of scientists and engineers, made the climate ripe for moving into education. Jerrold Zacharias, with his uncanny sense of timing and his ubiquity and high prestige in science advisory circles, went to Waterman at this juncture with a proposal to develop a new high school physics course. Waterman was ready for a proposal from this echelon of the scientific community, and Zach convened the Physical Science Study Committee in Cambridge in September 1956. It is to be noted that he enlisted leading external figures such as Edward Purcell, Philip Morrison, and I.I. Rabi in the enterprise as well as individuals such as Francis Friedman within M.I.T. itself. (More on this subject later.)

Drafting of "PSSC Physics" was well under way in 1957 when Sputnick, the first Russian satellite, went into orbit months before a successful American launching. President Eisenhower directed that there be vigorous action in science education, and the hey-day of science curriculum activity began. A pilot version of "PSSC Physics" ran in eight selected high schools in 1957-58, in 300 schools in 1958-59, and in 600 schools in 1959-60. The hard cover edition, published by D.C. Heath emerged in 1960. "Area meetings," supported all over the country from headquarters at Cambridge until around 1965, convened local groups of PSSC-using high school teachers several times a year in college or university settings under the auspices of a physics faculty member. I myself ran such meetings at Amherst College from 1958 through 1968 to my great profit and pleasure in personal friendships formed and retained.

Over these and following years, running through the 1960's, there also proliferated the extensive NSF-supported high school curriculum developments in all the other sciences, the elementary and middle school curriculum developments, and the extensive program of summer institutes, first for high school teachers and then including earlier teaching levels as well. By the end of the '60's, there had been added in the physical science domain "Project Physics," "Introductory Physical Science (IPS)" (for ninth grade level), "Physical Science II (PSII)," and elementary school curricula such as "Elementary Science Study (ESS)" "Science Curriculum Improvement Study (SCIS)," and "Science, A Process Approach (SAPA)."

Advent of the Commission on College Physics

By 1959, Zacharias felt that the high school physics curriculum was well in hand. Convinced that "PSSC Physics" (with, in his view, its incomparably lucid text presentations, compelling new laboratory experiments, and vivid instructional films, combined with the ongoing re-education of high school teachers, achieved by running the teachers through all of "PSSC Physics" in six or eight intensive weeks of summer institutes) would be bringing into the colleges and universities masses of students in complete control of introductory physics at PSSC level, made a proposal to NSF to start on a college course that would efficiently follow on the high school one.

When the then ruling hierarchy of the AAPT got wind of Zach's move, they quickly (apparently in fear that Zach would preempt, and put his own imprimatur on, college level physics) sent NSF what amounted to a counter proposal. NSF very reasonably responded with the message that the physics community better get its act together and arrive at some consensus as to what should be done at college level. With funding left over from PSSC, Zach was commissioned to run a conference based on wide representation from different types of colleges and universities. This Conference met three times, first at M.I.T. on 17 - 19 December 1959, second at Washington University on 25 - 27 February 1960, and third at the University of Minnesota on 5 - 7 May 1960. Some seventy three physicists, (coming from sixteen states, from twenty universities, ten colleges, four institutes of technology, four industrial laboratories, and NSF and AIP) attended.
Being a participant myself, I have some personal recollections of what transpired. The meetings involved wide ranging discussion of the level and content of introductory courses, the expected influence of PSSC Physics, implications of the report of the Carleton Conference ,[1] the desirability of increased attention to scientific literacy, the desirability of more twentieth century physics in the introductory course. (In connection with the latter, and fully consistent with PSSC's dedication to rooting all conceptual development in directly observed phenomena, Zach's heart was set on introducing students to quantum phenomena by making universally available a simple and easily performed version of the Stern-Gerlach experiment, demonstrating space quantization with an atomic beam. His personal attachment to atomic beam phenomena was, of course, quite understandable. Unfortunately, a sufficiently reliable and easily performed demonstration is still to be devised.)

Behind the hours of public discussion of what should be done in the way of curriculum improvement at college level was a substantial amount of less public politicking. There were nocturnal gatherings of groups with various shared agendas or interests in curriculum development, discussing strategies to be adopted and mapping desired outcomes of the Conference. Not having been personally attached to any of these interest groups, I am, unfortunately, unable to give an account of what actually transpired within them. However, by the time of the Minneapolis meeting, it was clear that the Conference would not give its blessing to any one course development proposal or throw its support to any one group.

Nevertheless a strong consensus was reached as to what might be done. As Walter Michels's first report of the Commission on College Physics diplomatically puts it, "It was clear that the job ahead was too important , too varied, too demanding to be made the sole responsibility of any one organization or group. Further, it was recognized that every academic institution must accept its share of the task, as must every organization of physicists. From these considerations there gradually emerged the concept of an ad hoc commission, made up of representative individuals who would pledge themselves to a continuing effort in the coordination of work being done throughout the country. [i.e., that the Commission] should act as a 'nerve center' in the improvement of physics instruction...]"[2]

A proposal was submitted to NSF, support was granted quickly, and the Commission came into existence. It became the prototype for similar organizations in the fields of chemistry, biology, earth sciences, and engineering. NSF support for the commissions ceased in the early '70's, and the only descendant organization that has survived to this day is the biological sciences' BSCS which managed continuity through a transition to other than federal sources of support.

The Conference recommended a 17-member commission, three of whom were to be ex-officio (namely the president and president elect of AAPT and the Director of AIP). The initial members, appointed by the Conference, were: Sandy Brown (MIT), Fran Friedman (MIT), Gerald Holton (Harvard), Bob Hulsizer (Illinois), Elmer Hutchisson (AIP), Charles Kittel (Berkeley), Walter Michels (Bryn Mawr, Chairman), Phil Morrison (Cornell), Vince Parker (Louisiana State, AAPT), Melba Phillips (Washington University), Ed Purcell (Harvard), Bob Resnick (RPI), Mat Sands (Cal. Tech.), Francis Sears (Dartmouth), Frank Verbrugge

(Minnesota, AAPT), and Jerrold Zacharias (MIT). Seven of the Commissioners were appointed for 4-year terms and seven for 2-year terms.

In subsequent years, some of these individuals left the Commission, either through expiration of terms or through resignation because of the pressure of other duties. Among individuals joining the Commission through direct election or ex-officio were: Fay Ajzenberg-Selove (Haverford), Arnold Arons (Amherst), Stan Ballard (Florida), Herman Branson (Howard), Ed Condon (Colorado), Mal Correll (Colorado), Dick Crane (Michigan), Ken Davis (Reed), Ken Ford (Irvine), Tony French (MIT), Ron Geballe (University of Washington), Mort Hammermesh (Minnesota), Len Jossem (Ohio State), Walter Knight (Berkeley), Bill Koch (AIP), Ed Lambe (Stony Brook), Bob Leighton (Cal. Tech.), Bob Little (Texas), Dick Mara (Gettysburg), James Mayo (Morehouse), Alan Portis (Berkeley), Bob Pound (Harvard), Allan Sachs (Columbia), Ralph Sawyer (AIP), "Duke" Sells (Geneseo), Van Williams (AIP), and Betty Wood (Bell Labs.).

The activities of the Commission on College Physics were managed by a succession of temporary executive officers and staff members on leave from their parent institutions. Among these were: Robert B. Bennett, Ronald Blum, Paul Camp, Philip DiLavore, John Fowler, Ben Green, Everett Hafner, Len Jossem, Thomas Joyner, Ed Lambe, Lyle Phillips, John Robson, Peter Roll, Richard Roth, Arnold Strassenburg, and Richard West. (Len Jossem subsequently became a member and chairman of the Commission.)

Without assessing contributions of others, let me say right at this point, as a first hand witness, that Bob Resnick had been a leading figure in the discussions that led to invention of the Commission. His voice was continually one of reason, good sense, and fruitful ideas in the subsequent councils. His leadership and initiative contributed directly to various enterprises subsequently undertaken or catalyzed by the Commission.

Let us note the professional level of the Commission membership. The presence of a large cluster of distinguished senior leaders in research, statesmen of the physics profession, made for more than critical mass. It gave the Commission visibility, credibility, and prestige throughout the physics community; it helped make it respectable for research physicists to devote time and contribute ideas to improvement of physics teaching. Not all members of the Commission participated directly in conferences or curricular projects (although many did), but all of them contributed their wisdom and statesmanship to the councils, and these contributions were not dilatory; they were seminal. Viewing this perspective, let us note that, at the present time, although a few distinguished senior physicists are setting an admirable example by leading and contributing to important educational enterprises, we are not currently beneficiaries of a massive, organized, highly visible contribution such as that provided by the Commission on College Physics in the '60's. Would a revival of such participation aid or hinder our efforts to improve science education at the present time? Would it be cost effective? I do not pretend to have quick or glib answers to these questions, but I offer them for thoughtful consideration and revue in the existing context.

Commission Activities

The activities of the Commission over its active life of about ten years were far too numerous and ramified for detailed review in this paper. Interested readers are referred to references 2 through 5 for such information. I only wish to mention a few instructive highlights and bring out some cautionary aspects. What can history teach us?

The Commission directly supported many conferences over the entire spectrum of physics instruction. Some of these conferences generated fruitful activity and results; others did not. The Commission initiated or catalyzed many activities that were quickly spun off to be implemented by Commission members or other recruits. Such activities spanned the range of defining needed instructional films, laboratory experiments, and demonstrations; discussion and outlining of curricula for various student populations; analyzing and recommending practices in laboratory instruction; initiating and spinning off various publishing ventures, and so forth.

Bob Leighton and Mat Sands, inspired by Commission discussions, returned to Cal Tech and persuaded Richard Feynman to teach the freshman physics course. We all know the result. The "Feynman Lectures on Physics" are still in print as one of the treasures of our profession. Charles Kittel, one of the first members, inspired by Zach's visions of a college course that would continue where PSSC left off, withdrew from the Commission in order to devote his time to chairing the generation of the Berkeley Physics Course, with its well known and valuable sequence of texts. Ed Purcell contributed to the Berkeley series his tour-de-force of presenting electricity and magnetism from a relativistic standpoint. Alan Portis developed the Berkeley Lab experiments.

It is useful to assess, in hindsight, the meaning and results of these efforts. The Feynman "Lectures" never fulfilled their original intent — that of generating a magnificent introductory college course following on PSSC in high school — despite numerous attempts to use them in this context (in the euphoria that always attends the appearance of new materials.) The mismatch between the "Lectures" and the college freshman, even in calculus-based physics courses, was far too great to make the course viable. This unavoidable assessment is neither derogatory nor pejorative. The "Lectures," as I have already remarked, are one of the treasures of our profession — a synthetic overview of physics through the eye of one of our greatest physicists. There is, of course, no better source that graduate students can study to prepare for qualifying examinations. Still more significantly, all of us at faculty level have sharpened our insights, seen previously unsensed elegance and beauty, deepened our understanding of our subject, and improved our teaching with the help of Feynman's unparalleled grasp.

Similar things can be said about the Berkeley course. The sequence is replete with beautiful presentations that are appropriate at second or third year levels, but it is just as mismatched to introductory level as are the Feynman "Lectures." Sitting down and writing curricular materials ex-cathedra, by conjecture as to what is desirable and appropriate, and without understanding the cognitive and intellectual problems of the audience being addressed, is not a fruitful way of generating instructional materials — regardless of the eminence and professional competence of the progenitors. (It should be noted in contrast that the best parts of the excellent elementary school science curricula were generated when the leading authors themselves went out into the classrooms with pilot drafts and returned to re-work them after first hand experience with the children.)

In other activity inspired by discussions in the Commission, Bob Resnick fostered a workshop on generation of simple, inexpensive laboratory and demonstration experiments. He also encouraged, supported, and essentially unleashed the many subsequent contributions made through RPI by Walter Eppenstein and Harry Meiners. Harry was funded to tour the country in his van and penetrate college and university prep rooms to ferret out ideas and designs of exceptionally good, and not too common, demonstration experiments. His monumental two-volume compendium stands to this day as an invaluable teaching resource.

Bob Resnick, by and large, associated himself with and cultivated fruitful and sensible operations. He had a knack for staying away from complex projects. (The mathematically sophisticated will recognize complex projects as the set for which costs are real and results imaginary.) This cannot be said for all of us on the Commission. I, to my lasting regret, associated myself with one of its most complex projects, namely the Seattle summer writing conference of 1965.

This misplaced effort had been generated, as usual, with earnest good intent. Conferences in the early '60's had examined, and outlined suggestions for, curricula for physics majors, non-science majors, and an array of other groups. The curriculum for physics majors, with its intent of preparation of the majority of such students for graduate school and research, came to be labeled the "R Curriculum," where the R, of course, stood for Research. Discussions kept returning to the problem posed by individuals who might (and should) wish to take physics for educational purposes other than preparation for graduate work. The idea emerged of defining an "S Curriculum" for this audience, where the S stood for Synthesis. The hope was even expressed that an inspired and well designed S Curriculum might eventually replace the continually worrisome R and lead to a result that actually implemented recommendations such as those, now fading into temporal distance, of the Carleton Conference.[1] One of the deeply embedded notions was to exploit films and other types of graphic design in more extensive and aggressive ways than had been prevalent up to that time. Eventually the rather specious and self congratulatory summary report of the enterprise was titled "Instruction by Design."

Several of the wise and senior statesmen on the Commission, sensitive to the by then accumulating experience with the generation of mismatched materials, warned us that we would assemble in Seattle and write for each other's approbation rather than for students. I confess that I was afraid of precisely this effect, but I must also confess that I was attracted by the possibility of seeing something of the Pacific Northwest (hitherto beyond my compass), and I naively thought that I might be able to influence at least some of the writing by telling participants about a few of the things I had slowly and painfully begun learning about the learning problems of the learners.

I absolutely and completely failed in the latter respect. The warnings of the elders were right on the mark. Participants wrote either for their own or for their colleagues' approbation. All one had to do was to present this elegant material in his or her own way; the presentations of others not having succeeded because of not having been sufficiently elegant or clear. Not one single participant was about to listen to my descriptions of problems of concept formation or to take such elements into account in their writing. (The only thing I accomplished personally was to take advantage of a newly devised IBM programming language called "'coursewriter" that Ed Adams, an IBM physicist, brought to Seattle. I wrote a small, completely Socratic, dialog on the concept of "weightlessness," based on key word recognition, and I satisfied myself that such dialogs, if generated in sufficient number, might eventually be useful in instruction. This dream has never come to fruition despite some further sporadic attempts.)

At the Seattle Conference, some 18 monographs were written on various aspects of physics. They died in the files, receiving virtually no use or acceptance. They were devastatingly reviewed by experienced reviewers (Mark Zemansky once approached me in confidence and inquired as to how "'so and so" could possibly have written such appalling junk.) They exerted no influence on further writing. Had they been used, they would have been found to be just as mismatched to the intended audience as were the higher quality materials I mentioned earlier, and they did not have anything like the elegance or the beauty of the latter. In other words, not having any Feynmans or Purcells among us, we did not even generate any good mismatched materials. The project was just about as complex as it could be.

I relate this cautionary tale because this is a syndrome that history has not yet helped us escape. I see in progress at the present time costly projects that are at least as complex as the one I have described. Santayana's warning about the necessity of knowing history still goes unheeded. In this context "knowing" history means understanding why certain efforts failed. It is not enough to plunge into exactly similar modes of activity with the unfounded assurance that we will do it successfully because of the elegance of our insights and the purity of our ideals.

Advent of Research on Learning of Physics

I recall several times in Commission meetings when Zacharias would bring his fist down on the table and say that "we should be doing research on learning." Occasionally there would be an echo, but no action was generated. When the term "research" was used in this context, what Zach and others meant was devising ways of generating flawless instructional materials — text, films, and other devices — that would impart full knowledge and error free understanding to the docile and fascinated learner, i.e., research would consist of refining powerful ways of generating effective instructional "delivery systems."

I had no objection whatsoever to improving delivery systems and have no objection at the present time; it is an important activity. When, on one occasion, however, I timidly ventured to suggest that students were having great difficulty comprehending velocity and acceleration and interpreting related diagrams, that they did not assimilate the Law of Inertia and persisted in Aristotelian and medieval modes of thought, that the Third Law was extremely subtle and that students resisted ascribing, to inanimate objects, the capacity of exerting force, that these learning obstacles did not go away with one lucid explanation but persisted with great resistance to displacement, I was told by one of the eminent members of the Commission, for whom I had tremendous respect, that I was describing personal tricks unique to my own style and procedures and that this had nothing to do with research. Other commissioners reacted similarly albeit less articulately.

It was clear that the descriptions I was giving were being viewed not as generalizable insights but as quirks of one individual's personal approach. The Commission, as a body, had no conception of what we currently think of when we refer to research on teaching and learning. I had no choice but to subside. Many of my friends of the present day are likely to be astonished by such timidity on my part. I can only say that I was younger then, and, moreover, as Gerry Holton was to remark subsequently in an analogous situation, I was sitting side by side with some of the giants on whose shoulders we stood.

On one occasion at the Seattle Conference the director of the operation was leading a group discussion on the teaching of kinematics. I ventured to try to describe some of the difficulties I had observed among the students and pointed out that I had had some modest success in generating understanding by asking students to place their hand on the edge of the table and then execute, with their hand, the rectilinear motion presented in a position-time or velocity-time diagram. I pointed out how important it was to have the students simultaneously describe the velocity changes and the accelerations in words. I described what happened when you started a student with a position-time history parallel to the t-axis: Most students would visibly tense their muscles, get ready to move their hand, and then say, in obvious astonishment, "It's standing still!" What I hoped to convey to the group was the possible usefulness of coupling the verbal text presentations with direct kinesthetic experience in order to help generate mastery of abstract concepts. I had already incorporated such assignments in my textbook "Development of Concepts of Physics."[6] The director cut off my discussion, described it to the group as an "Arons gimmick," suggested that we were wasting valuable time, and returned to the discussion of text presentations.

I am delighted that we have come a long way since the summer of '65; we now have the sonic range finder tied to the computer and all the beautiful consequences. The system is even widely used; its impact on learning is documented and accepted; and a big issue is made of the importance of kinesthetic experience. Much as I love and admire this system, however, I can't resist saying that one can still achieve a great deal with the hand on the edge of the table.

Indeed, we have come a long way from the frame of mind that saw research on teaching and learning as exclusively a matter of producing and refining delivery systems. The shift really got under way in the late '60's and early '70's when Bob Karplus, influenced by his exposure to Piagetian techniques and insights, generated the SCIS elementary school curriculum and began investigating abstract thinking among pre-college students. It was greatly accelerated by the seminal paper by McKinnon and Renner,[7] showing that only one third of a group of college students performed successfully on Piagetian tasks on ratio reasoning and control of variables. Since then the enterprise has blossomed. From AAPT sessions on research convening a handful of people in the smallest room available, we have standing-room-only sessions with an audience of hundreds. We have flourishing publication and respected leaders in the field.

The problem now is how to make the insights being acquired achieve deeper and more lasting impact on existing delivery systems and, still more importantly, on sensible and realistic choice of subject matter , on intellectually digestible volume and pace of coverage , and on matching the impedance of different student populations, principally in our introductory courses but also further up the line. A key issue still remains regarding what I call '"honest" presentation. Honest presentation entails showing students how we know, why we believe, why we accept, what is the evidence for, the concepts and theories we throw at them. Deluging them as rapidly as possible in the end results of developmental sequences is not what I am willing to describe as honest; it is vulnerable to John Gardner's trenchant criticism of much of American education to the effect that we hand our students the cut flowers and forbid them to see the growing plants.

I suggest that these are some of the directions in which our own history impels us with respect to physics teaching. Perhaps with a little humility and knowledge of the past we can avoid the condemnation to relive it.

References

1. "Improving the quality and effectiveness of introductory physics courses." Report of a conference sponsored by the AAPT. *Am.J.Phys.* 25, 417 (1957). [Also additional papers by Holton, Michels, and Whaley in the same issue.]
2. "Progress Report of the Commision on College Physics." *Am. J. Phys.* 30, 665-686 (1962).
3. "Progress Report of the Commission on College Physics." *Am. J. Phys.* 32, 398-432, (1964).

4. "Progress Report of the Commission on College Physics." *Am. J. Phys.* 34, 834-894 (1966).
5. "Progress Report of the Commission on College Physics." *Am. J. Phys.* 36, 1035-1080 (1968).
6. Arons, A. B. "Development of Concepts of Physics." Addison-Wesley, Reading, MA. 1965.
7. McKinnon, J. W. and Renner, J. W. "Are Colleges Concerned with Intellectual Development?" *Am J. Phys.* 39, 1047 (1971).

Some Possible Futures for the Introductory Physics Course

Jack M. Wilson, Director, Center for Innovation in Undergraduate Education, Rensselaer Polytechnic Institute

Introduction

As we try to discern the future of the introductory physics course, we will need to look back at where we have been, take stock of where we are, and identify the issues that will affect the course of events. Assuming that a few of the other speakers have adequately examined the past and that most of the speakers are describing the current state of the art, I will try to concentrate on the future. First let us identify the forces shaping our future. I will refer to these as community, communication, cognition, contemporary physics, and computers — the five C's.

Next we will consider how the course will change under the influence of these issues. Remember the old children's rhyme: No more lectures, no more books, no more teachers' dirty looks. This will be our guide to a future where the lecture is changed and de-emphasized, the text book is supplanted by interactive texts with access to rich databases of text, video, simulations, and problem solving tools, and the classroom climate will be changed to an interactive and communicative environment that can accommodate diversity of learning styles, preparation, culture, and interests.

We will then move through the design process for an interactive workshop physics course for large research university environments that must deal with a thousand or more students per semester. I will try to illustrate both the process and the results by describing the efforts at Rensselaer to replace the existing traditional lecture/recitation/lab course with an integrated workshop physics approach.

The Forces Shaping Our Future

There are five forces that we must all take into account as we discuss the past present and the future of the introductory course. I will refer to these as community, communication, cognition, contemporary physics, and computers. These five C's provide a shorthand way of focusing us on the task at hand without forgetting the important issues that are driving the process.

Community

The physics community is changing and it must change. Physics has always been done by a global community, but we have not really communicated that to many students who pass through a physics course on the way to other destinations. Physics has never been, but must become, a community inclusive of women, African-Americans, and Hispanics. The amazing success of incorporating Asians into the physics community needs to repeated with other communities where the problem may be far more intractable. We simply can no longer afford to fill our ranks only from the traditional cultural groups that have populated physics.

In order to become more accessible, physics must learn to accommodate diversity, diversity in cultural backgrounds, diversity in preparation of university students, diversity in interests, diversity in learning styles, and diversity in career paths. This is not so easy to accomplish within the extraordinarily rigid confines of our curriculum and our career path.

Communication

Communication has always been a critical part of physics education. In fact, success in doing physics depends upon ones ability to communicate ones work to others. The physics community has various formal mechanisms for this communication that have been enshrined in the "publish or perish" dictum, the "invited lecture," and the colloquium. In the first part of this century and earlier, this formal mechanism of communication was augmented by an informal system of visiting and letter writing. The former remains popular today, but the latter has declined in importance. These mechanisms have been supplemented (and in some cases nearly supplanted) by less formal more rapid forms of communication by telephone, fax, and electronic network. Today anyone foolish enough to rely on the formal communication mechanisms may be doomed forever to publication in "Physics Yesterday."

In spite of this obvious importance of communication there is little in the average introductory physics course to develop the student's abilities in these areas (unless one includes listening and solving homework problems as communications skills!). The model at most research universities features a teacher centered environment which puts a premium on presentation skills for the teacher, but requires little in the way of communication capability from the students. Most physics courses encourage the model of the "rugged individualist" competing against his or her fellow students to rise to the top of the curve. Physics is a cooperative as well as a competitive science, but the emphasis in the introductory course is on competition. Cooperation implies communication. Practicing cooperative skills requires effective use of communication. Treisman has demonstrated the spectacular success of cooperative learning techniques in encouraging the success of minority students in particular.[1]

The future of the introductory physics course includes far more use of communication skills; written, oral, and electronic. In addition to writing across the curriculum, we need to consider computing across the curriculum, design across the curriculum, and cooperative learning across the curriculum. We also need to explore the way electronic communication can connect students and faculty without regard to geography.

Cognition

We know far more now about the way students learn than we did in the middle part of this century when the existing introductory physics course involved. Today we have had and will have talks from some of the leaders in the research in physics education that is exploring aspects of the cognitive sciences as applied in the domain of physics. The work of Arons,[2] Fuller,[3] Goldberg,[4] McDermot,[5] Thornton,[6] Tobias,[7] and others highlights the successes and the challenges posed to us by the results of research in physics education. In spite of these insights, the typical university physics course is relatively unchanged. We continue to devote time and resources to activities that are minimally effective in stimulating student learning while giving short shrift to those things which are more effective.

We pretend to teach them and they pretend to learn

The most negative characterization of our present educational system is a paraphrase of the joke about the Russian employment contract that says " they pretend to pay us and we pretend to work." Arons, Hestenes[8], and others have challenged that insidious contract by actually asking the students what they have learned, and the physics community is not happy with what they found. Laws,[9] Mazur, and others were so affected by the results that they abandoned the traditional practices in favor of approaches that are demonstrably more effective.

Laws introduction of "Workshop Physics" at Dickinson University was particularly radical. Since lectures have been shown to be so ineffective at stimulating student learning, she simply abandoned the lecture completely in favor of a workshop that combined features of both the laboratory and recitation.

Contemporary Physics

The disparity between the way Physics is done and the way it is taught has driven several of the projects to reexamine the content of the introductory physics course and develop more up to date materials. Current interest might be dated back to editorials by Wilson[10] and Rigden[11] and the formation of the M.U.P.P.E.T. project in 1984 and IUPP in 1987.

The Maryland University Project in Physics and Educational Technology (M.U.P.P.E.T.), was founded by Redish, MacDonald, and Wilson to reexamine the content of the introductory physics course in the light of the advances in computing, cognition, and contemporary physics and to use technology to enable changes in curriculum and pedagogy.[12,13] Gordon Aubrecht, as a visiting scientist on the M.U.P.P.E.T. project, performed a comparative analysis of the present introductory physics course and same course earlier in this century.[14] He also compared the skills found in the present course with a list of desired skills developed by interviewing active research physicists and asking the question: "What skills do you want students to acquire before coming to work on a project with you?" His analysis showed quite clearly that the traditional course does not address many of the desired skills.

The Introductory University Physics Project (IUPP) was founded by AIP, AAPT, and APS and funded by the National Science Foundation to engage a nationwide group of physicists in the question of just what physics needs to be taught. From the beginning they took the approach that less is more, that the course should reflect contemporary ideas in physics, and that there should be a story line to the course. Many of the participants in this conference were inspired by IUPP to create the innovations described here.

The most controversial aspect of IUPP is perhaps the inclusion of ideas from quantum mechanics. Merzbacher,[15] Moore, and Aubrecht[16] have all published materials that have generated productive and energetic controversy on these subjects. Arnold Arons' strong critique of IUPP perhaps typifies both the vehemence about and basis for the controversy.[17]

Both the proponents and opponents of including contemporary physics often bog down in the discussion of including relativity and quantum mechanics. Leaving aside the question of how anyone could consider the ideas of the first three decades of this century to be "modern physics," many people miss the main point that *physics is an uncertain science.* If there is a central difference between the way physics is done and the way physics is taught it is this. Introductory physics is taught as if it was a deterministic science that predicts exact values for physical measurements based upon analytic solutions to differential equations. Physics, as done, requires dealing with analytical *and* numerical solutions to (mainly) non-linear equations that result in probabilities for values. Introductory Physics is NOW a domain of rigid laws and certain outcomes, but in spite of Einstein's objections, God really does play dice with the universe.

Where in introductory physics does the student learn of the restrictions on certainty due to:

1. Statistical (information) limitations.
2. Quantum limitations
3. the fact that all real physical systems are non-linear.

The argument about the inclusion of contemporary physics tends to ignore the fact that the current introductory physics course teaches the student a view of physics that is fundamentally wrong. Physics does not deal with absolute certainties; it deals with probabilities. Much of the time in advanced undergraduate courses is spent dispelling these misconceptions that we built so assiduously in the introductory courses. For the majority of the students for whom this is the terminal course, these misconceptions become the view of physics which they carry throughout their life and career.

Computers

Because I have spoken and written so often on this subject (often in collaboration with Joe Redish who will speak on this subject later), I will shorten my remarks by outlining some of the issues and referring to previous work on this subject. My philosophy over the years has remained consistent and can be summed up by repeating: "Computing has changed the ways physics is done, and it needs to continue to change the way physics is taught."[18,19]

Computing has entered physics learning through a variety of routes. Computers are used in the laboratory for data acquisition, storage, analysis, and visualization.[20,21,22,23] They are also used for pre-laboratory preparation for students.[24] In the lecture they are often used as lecture demonstration devices that allow the lecturer to display simulations and visualize data. Recently they have begun to be used for display and analysis of computer based video.[25] Computational physics projects are often the most popular topics for undergraduate research (not to mention graduate research).[26] Problem solving is done with both spreadsheets[27,28,29] and programming tools.[30] Many of these same tools can be used in homework, individual study, and remediation.

In spite of all of this excellent work, computing remains on the periphery of most introductory courses. For the student and faculty users of computer curricular materials, there is often just too high a threshold of effort. Since many materials do not work together and follow no standards for user interface design, data structures, and hardware, users must invest too much time in learning the interface and getting everything working. What is needed is a comprehensive and unified approach to delivering sophisticated integrated computing tools and curriculum materials to users that allow them to invest less time in learning how to use them and more time in learning physics.

The consortium to develop the Comprehensive Unified Physics Learning Environment (CUPLE) was founded to try to bring together the many talented individuals working in these areas to create an environment that would meet the needs of the typical physics instructor who does not have the time to invest in learning so much about computers.[31] The initial collaborators include many names familiar to the physics community: Wilson, Redish, Priscilla Laws, Ronald Thornton, Peter Signell, Dean Zollman, and Patrick Cooney.

Although CUPLE is ultimately targeted at an environment similar to Priscilla Laws' Workshop Physics course, it is more often worked into traditional courses as laboratory activities, lecture demonstrations, class room experiments, outside of classroom activities, and many other ways.[32]

What Is the Future for the Introductory Course?

Whenever I ponder the future of the introductory physics course, I think of the old children's rhyme *No more lectures, no more books, no more teachers dirty looks!* If this is tempered by the famous Gilbert and Sullivan line: "*What never? ... Well, hardly ever!*" it may just describe the general features of the future introductory physics course.

No more lectures?

I expect that we will see a significant de-emphasis of the lecture as the primary learning environment. Will the lecture disappear? Probably not. But, it will change dramatically. When lecturing is done it will likely be combined with more interactive techniques. Ten or fifteen minute mini-lectures may displace hour long pontifications. These mini-lectures may occur in studio or workshop style environments or in interactive lectures with worksheets or interactive lecture demonstrations. The mini-lectures are likely to be combined with cooperative learning techniques that encourage the students to cooperate rather than compete in the solution of problems.

The team approach to physics implies that evaluation will have to be done in a fundamentally different way. There will have to be less emphasis on individual quizzes and tests and more on student projects that require team work.

The de-emphasis of the lecture and the emphasis on cooperation implies a very different role for the teacher. The availability of information rich environments made possible by the advances in communications and networking and by new learning environments like CUPLE will reduce the reliance upon the instructor as the primary source of information. Rather than a repository of information the teacher will become a guide to information. Rather than a lecturer the teacher will become a mentor. Reducing the reliance upon the lecture will increase the time available for hands on activities, which will be linked much more closely (both in time and content) to the other class activities. Rather than explaining a concept and then telling the class that "next Tuesday you will see this in lab," the instructor may simply tell the student to try it out right then.

No more books?

Clearly there will be books, but these "books" are likely to be quite different from the static texts used for the last few centuries. Hypermedia books can be created by teams of faculty, graduate students, and undergraduate students. These materials can be adapted to local conditions and linked to other materials. Illustrations in these books can be dynamic simulations in which the students can alter the picture and see the physical results. In the example shown below the students can grab the object with the mouse and drag it to other positions and see how the image changes as a result. Notice that this illustration is also linked to a database of String and Sticky Tape experiments (From Ron Edge's book) and the Freier and Anderson Demonstration Handbook.

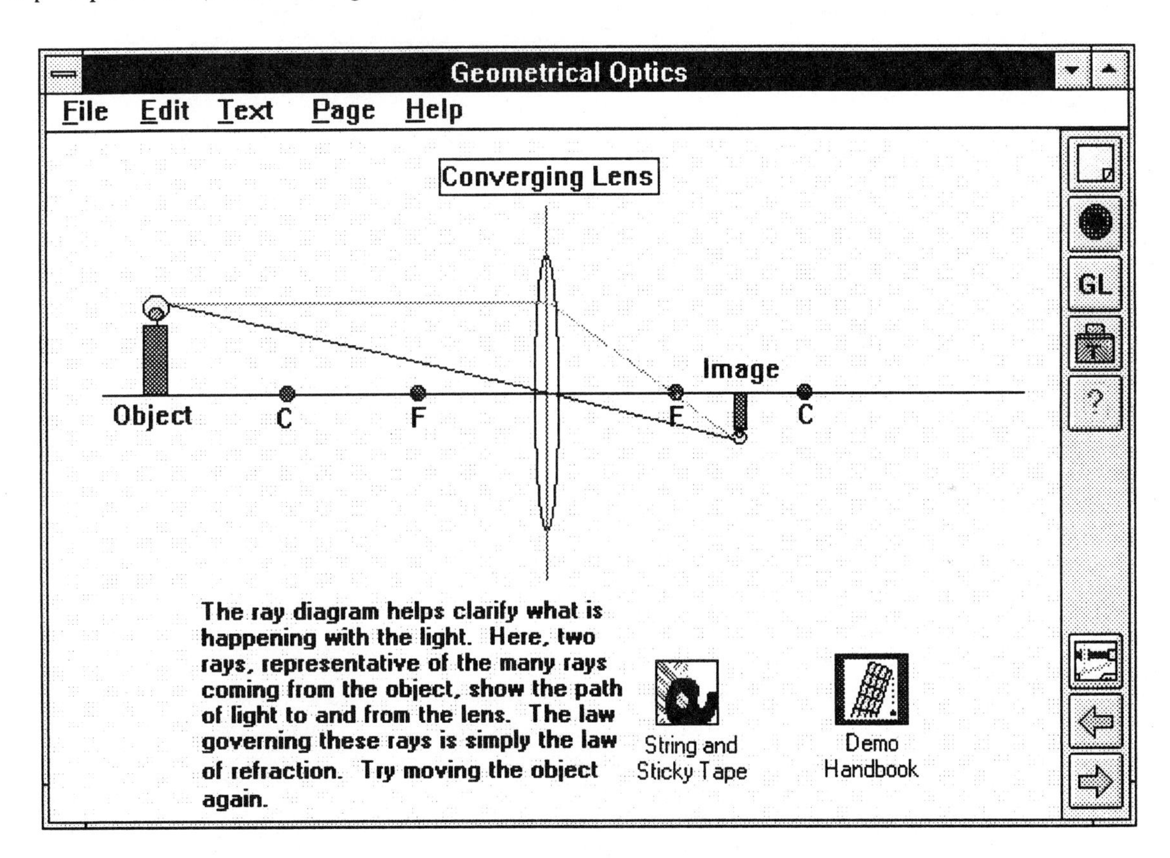

Figure 1.

These hypermedia texts will also link students through the network to huge databases of information. These databases will include full motion video of various physical phenomena and will provide students with the powerful tools to display and analyze these videos. Students will also have live links to experts and other students through a live video window on each computer which they can connect to others in other regions. The first test of a course taught using this technology was conducted through a joint Rensselaer/AT&T/Bell Laboratory interactive distance learning course that linked classrooms and workstations in Troy, Dallas, Chicago, Cincinnati, and Holmdale, NJ with an interactive multimedia course that included workstation video conferencing and application sharing.

The idea of hypermedia interactive texts may seem far fetched, but at Rensselaer much of this is already implemented in the introductory physics course. During the 1992-93 academic year they were tested in classes with over three hundred students. In 1993-94 they will be delivered to all 750 students in the course. We are presently installing an IBM ES9000 as a regional repository for multimedia courseware and for public domain compressed digital video. As of the fall of 1993, this multimedia mega-server was in prototype operation at the Rensselaer Center for Innovation in Undergraduate Education faculty resource room. During the summer of 1993, a team of undergraduate students from Rensselaer, Harvey Mudd College, Lund University in Sweden, Munich University in Germany, and several other U.S. schools has created over 500 megabytes of compressed digital video that can be delivered over the network nearly anywhere.

No more teachers' dirty looks

This is intended to symbolize a much improved classroom climate. There must be a better environment to accommodate the cultural, racial, and gender diversity that we will see in our classrooms. Accommodating this diversity will be a measure of the success of the introductory physics course of the future. We also need to accommodate the diversity of learning styles, interests, and preparation. It is difficult for me to see how we might do this in the context of the traditional course. We have all had the experience of trying to select an approach knowing that it was a compromise that would reach some students, bore others, and leave still others in complete confusion.

The use of cooperative learning techniques, the availability of hypertext learning environments, and the changes in the structure of the classrooms provide an opportunity to allow for this diversity. It is up to us to make that vision a reality.

Introductory Physics Courses at the Research Universities

American universities are presented with unprecedented opportunities for making significant improvements in the quality of their undergraduate programs at a time when they are also faced with daunting challenges on the economic front.[33] For the research universities, in particular, the need to show sustained improvements in educational quality must be satisfied while at the same time making substantial improvements in the economic efficiency of the educational process. The central challenge is to incorporate the powerful tools provided by modern computing, to redesign the course structure to take into account the research on student learning provided by the cognitive scientists, to revise the curriculum to emphasize the skills needed in this changed environment, to develop students communication skills, and to accommodate the growing diversity of the community of learners.

For the research university this challenge has another dimension: economics. This dimension has made it difficult for the major universities to assume leadership positions in the reform efforts. At many major universities physics (and the other introductory courses), have to educate a thousand or more students per year in a production environment that has been refined by years of experience and cannot afford an experimental failure.

How can we incorporate the workshop approach developed for classes numbering tens of students into a research university environment that deals with students numbering in the thousands? There have been some notable efforts to make large lectures more interactive. At Rensselaer, Tufts,[34] and Oregon State,[35] faculty are using interactive lecture demonstrations that lead directly into homework assignments or laboratory type activities. Eric Mazur at Harvard uses hypermedia computing tools to put some of this interaction into his large lectures at Harvard.[36]

American universities have enshrined 18th century educational models in bricks and mortar that provide daunting obstacles to those who would innovate. When Harry McLaughlin proposed to teach Math III at Rensselaer using the cooperative learning approaches similar to those used by Karl Smith at Minnesota, Eric Mazur at Harvard, or Uri Treisman at U.C. Berkeley, he was faced with the nearly impossible task of testing this program in a classic 19th century lecture hall. He had to surmount some major obstacles caused by the inflexible design of the room and it was difficult to see how the program could be institutionalized without increased cost and significant impact on faculty. When curriculum reforms are considered, facility development must be considered in the same breath.

Is it possible for us to design a system that would allow us to use the workshop approach to the introductory physics course that would provide a far better learning environment for our students, that would integrate the class, laboratory, and recitation, and that would allow us to use powerful computer based materials as an ever-present tool for student use in class?

In February of 1993 a meeting of experts convened at Rensselaer to discuss the future of the classroom in the broadest sense. There was remarkable unanimity from the experts on these issues. Representatives from industry, academe, and architectural firms were unanimous in their recommendation that we consider various alternatives that would allow us to respond to the issues above. William Graves, Professor Mathematics and Associate Provost for Information Technology at the University of North Carolina, was particularly eloquent in his call for both curriculum reform and innovative approaches to class room structure.[37] As we have seen, there are significant changes occurring in undergraduate educational programs and few institutions have managed to accommodate change thus far, let alone begin to plan for the coming changes. At the February meeting, the experts were also unanimous on one other issue: flexibility is the key to survive and prosper during this period of rapid change. In order to address each of the issues discussed above, we will need to have facilities that will allow experimentation, that will acts as the research laboratory for educational change, and that will inform the changes that are always ongoing on the rest of campus.

The Experts' Meeting resulted in an evolving strategic plan for radical reform of the introductory mathematics and science courses at Rensselaer. The model that evolved was inspired by Priscilla Law's Workshop Physics model, but was significantly modified to fit the research university environment. Suppose that these courses were reorganized to be delivered in the workshop format with about 60 persons per workshop section. What would be the financial impact?

Prior to the Computing in Calculus reforms accomplished in the last four years, Mathematics, Physics, Chemistry, and Computer Science were all taught in the same format. The model of the introductory course at Rensselaer is one in which there are approximately 750 students in a single course (now Physics, Chemistry, Computer Science, or Introduction to Engineering Analysis). These 750 students are divided into two lecture sections with 375 students per section. There are 25 recitation sections with 30 students each and 30 lab sections with 25 students.

How does the workshop model with 60 students compare economically to the traditional model? Can we afford the "luxury" of this far more personal and effective environment? As we can see below, the economics of this model is quite competitive with the economics of the traditional lecture/recitation/lab format. In each case we assumed that each workshop would be taught by a faculty member with one graduate student available to teach an hour or two per week. In our working model we are also assuming that undergraduate assistants will work in each course, but they are not included in this calculation since their cost is small in comparison to other costs and they are not involved in today's courses. The cost of facilities is also not included since we were making a direct comparison of salaries, the largest factor. Facilities cost for the workshop model should be somewhat less than the traditional model since only one type of facility is required rather than the three rooms required for the traditional model. Preliminary study indicates that room scheduling will be simplified by the newer model.

Clearly there are alternate models to our existing courses that can be economically competitive and more educationally effective.

During the summer of 1993, we completely renovated a classroom for the first offering of the Workshop Calculus course. In the spring of 1994 the first CUPLE based workshop physics course will be conducted. This classroom accommodates 64 students in a comfortable workshop facility. There are 32 work tables with open workspace and a computer workstation, each designed for two students. The tables form three concentric partial ovals with an opening at the front of the room for the teachers worktable and for projection. The workstations are arranged so that when students are working together on an assigned problem, they turn away from the center of the room and focus their attention on their own small group workspace. The instructor is able to see all 32 workstation screens from the center of the oval, and thereby receives direct feedback on how things are going for the students. In the Physics course, the workstation will be running the CUPLE system, have full access to networked multimedia, and include a microcomputer based laboratory system for data acquisition, analysis, and visualization.

	Students/ Section	Sections	Hours per Section	Total hours	Faculty hours	TA hours	Cost
Math. I	931		4			Revenue	$1,241,333
Lecture	116	8	4	32	32	0	$106,667
Recitation	28	33	1	33	0	33	$51,563
Lab	28	33	1	33	0	33	$51,563
							$209,792
Workshop	62	15	5	75	45	30	$196,875
						Savings	$12,917
Physics 1	685		4			Revenue	$913,333
Lecture	343	2	2	4	4	0	$13,333
Recitation	23	30	2	60	24	36	$136,250
Lab	21	33	1.5	49.5	0	49.5	$77,344
							$226,927
Workshop	62	11	5	55	33	22	$144,375
						Savings	$82,552
Chem/Mater 1	727		4			Revenue	$969,333
Lecture	364	2	2	4	4	0	$13,333
Recitation	30	24	2	48	48	0	$160,000
Lab	91	8	1.5	12	10	160	$283,333
							$456,667
Workshop	61	12	5	60	36	24	$157,500
						Savings	$299,167

Table 1.

When the instructor wants to conduct a discussion or give a mini-lecture, he or she is able to ask the students to turn back toward the center of the room. This removes the distraction of having a functioning workstation directly in front of the student during the discussion or lecture period.

This yields a classroom in which multiple foci are possible. Students can work together as teams of two or two teams my work together to form a small group of four. Discussion as a whole is facilitated by the semicircular arrangement of student chairs. Most students can see one another with a minimum of swiveling of chairs.

This type of classroom is friendly even to those instructors who tend toward the traditional style of classroom in which most of the activities are teacher centered rather than student centered. Projection is easily accomplished, and all students have a clear view of both the instructor and any projected materials.

As a facility in which the instructor acts more as a mentor/guide/advisor, the classroom is unequaled. Rather than separating the functions of lecture, recitation and laboratory, the instructor can move freely from lecture mode into discussion, and can assign a computer activity, ask the students to discuss their results with their neighbors, and then ask them to describe the result to the class. Laboratory simply becomes another one of the classroom activities that is mixed in with everything else.

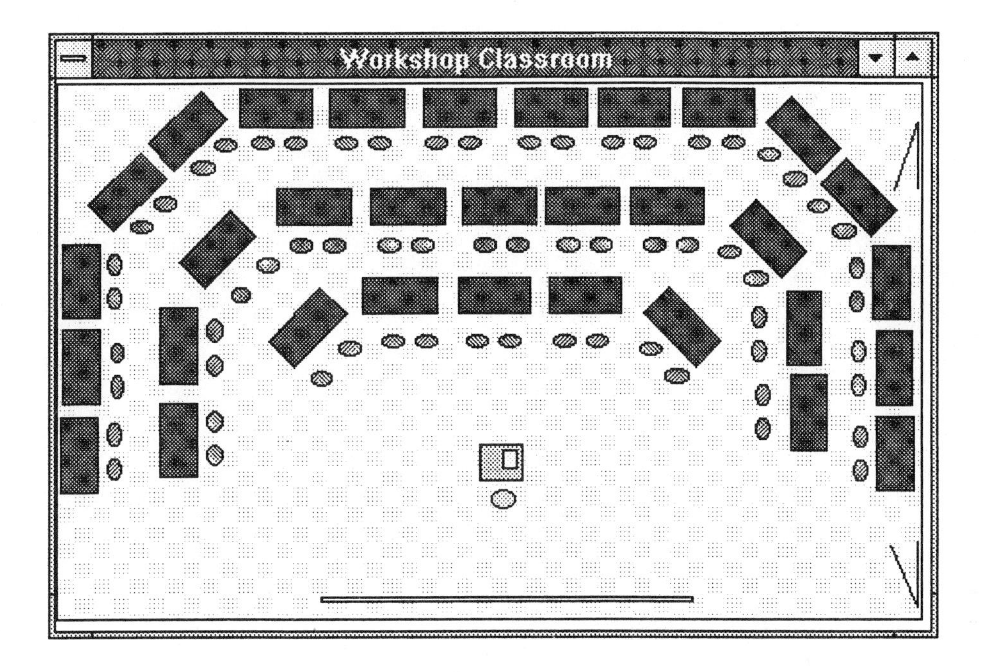

Figure 2

We plan to integrate the lecture, laboratory, and recitation into a single integrated workshop course. This course will use the latest in computing tools and incorporate use of cooperative learning approaches. We will create a powerful link between the lecture materials and the problem solving and hands on laboratories. This is a link that is tenuous at best at this time.

How would this operate if it became the standard approach to introductory physics? The 700 enrolled in the large semesters would be divided into 12 sections of 50-60 persons. We propose to structure the course as a four contact hour and four credit hour course taught in two periods each two hours in length. One room could conceivably serve all sections with a Monday/Wednesday or Tuesday/Thursday format. Since the rooms would be in continuous use for Physics I, there would not need to be hours reserved for setup as is presently the case in our lecture halls.

The most important feature is that each student would experience the Physics 1 environment in a more personal and interactive environment that has been demonstrated to be more effective than the traditional lecture/recitation/lab approach.

Conclusion

The introductory physics classroom of the future may be a very different place than the present day classes modeled after the 18th and 19th century European models. The teacher will be more important than ever, but his or her role will change from more of a lecturer to more of a mentor. For many physicists this role will be more akin to the role of research supervisor.

The classroom will likely be noisier, with far more prediction, discussion, and group activity. Students will be actively involved in the learning process. There will be more provision for "hands-on" experiences.

Computing will be ever present, but the emphasis will be on computing tools for browsing databases of text, video, simulations, and problems and for acquiring, storing, analyzing, and visualizing data.

The classroom will be much friendlier to individual student differences with provision for dealing with both remediation and enrichment on the same subject.

In short the introductory physics classroom of the future will be a much more exciting place to learn, and to teach.

References

1 Phillip Uri Treisman, "Teaching Mathematics to a Changing Population," Report of the Professional Development Program at the University of California, Berkeley (1990). and "Mathematics Achievement Among African American Undergraduates at the University of California, Berkeley: An Evaluation of the Mathematics Workshop Program," *Journal of Negro Education* **59** No. 3 (1990).

2 A.B. Arons, "Student Patterns of Thinking and Reasoning," Parts 1, 2, and 3; *Physics Teacher* 21, 576(1983); 22 21(1984); 22, 88(1984).

3 R.G. Fuller, "LVs, CDs, and New Possibilities," in Proc. Conf. Computers in Phys. Instr. (Reading, MA, Addison Wesley 1989).

4 F. Goldberg and L.C. McDermott, "An Investigation of Student Understanding of the Real Images Formed by a Converging Lens or Concave Mirror," *Am. J. Phys.* **55**(2), 108-119(Feb. 1987).

5 L.C. McDermott, "Research on Conceptual Understanding in Mechanics", Physics Today 37, 24-32(1984).

6 R. Thornton, "Learning Physical Concepts with Real-Time Laboratory Measurement Tools," *Am. J. Phys.* **58**(9), 858(1990).

7 S. Tobias, *They're not Dumb, They're Different!* (Tucson, AZ; Research Corporation) 1990.

8 I.A. Halloun and D. Hestenes, "The Initial Knowledge State of Physics Students," *Am J. Phys.* **53**, 1043-1055 (1985).

9 Priscilla Laws, "Workshop Physics: Learning Introductory Physics by Doing It," *Change Magazine,* (July/Aug 1991).

10 J. Wilson, "What to Teach?", *AAPT Announcer* 14, 91(1984).

11 J. Rigden, "The Current Challenge: Introductory Physics," *Am. J. Phys.* **51** 516 (1983).

12 W. M. MacDonald, E.F. Redish, and J.M. Wilson, "The M.U.P.P.E.T. Manifesto," *Computers in Physics* 1, 23 ff, (July/Aug 1988).

13 J.M. Wilson and E. F. Redish, "Using Computers in Teaching Physics," *Physics Today* **42**(1) 34 (January 1989)

14 G.J. Aubrecht, "Should there be Twentieth Century Physics in Twenty First Century Textbooks?" *AAPT Announcer* **16**(4), 104 (1986).

15 E. Merzbacher, "Can Quantum Mechanics be Taught in a One-Year Calculus Based Introductory Physics Course?", *AAPT Announcer* 17, 74(1987).

16 *Quarks, Quasars, and Quandries,* G. A. Aubrecht ed.; "Proceedings of the Conference on the Teaching of Modern Physics," J. M. Wilson Conf. Dir. (College Park, MD: AAPT, 1987).

17 A. Arons, "Uses of the Past: Are We Condemned to Relive It?", *AAPT Announcer* 18, 82(1988).

18 J.M. Wilson and E. F. Redish, "Using Computers in Teaching Physics," *Physics Today* **42**(1) 34 (January 1989)

19 J.M. Wilson, "Computer Software Has Begun to Change Physics Education," *Computers in Physics* **5**(6), 580(Nov/Dec 1991).

20 R. Thornton, "Learning Physical Concepts with Real-Time Laboratory Measurement Tools," *Am. J. Phys. 58(9), 858(1990).*

21[21] J. M. Wilson "The Impact of Computers on the Physics Laboratory," in *Research on Physics Education,* (Paris: Editions du Centre National de la Recherche Scientifique 1984) 445ff.

22 J. M. Wilson, M. K. Summers, and K. Hirata, "The Microcomputer as a Laboratory Instrument," in *Trends in Phys. Ed.* (Tokyo, Japan: KTK Publ. 1986), 238.

23 J.M. Wilson, "Combining Computer Modeling with Traditional Laboratory Experiences in the Introductory Mechanics Laboratory for Physics Majors," *AAPT Announcer,* **17**(2), 80(May, 1987).

24 J. Wilson, "Experimental Simulation in the Modern Physics Laboratory," *Am. J. Phys.* 48, 701 (1980).

25 J. M. Wilson, "Laboratory Data Acquisition through Computer Controlled Video," *AAPT Announcer* **21**(4), 67(Dec. 1991).

26 E.F. Redish, "The Impact of the Computer on the Physics Curriculum," Proc. Conf. Computers in Phys. Instr. (NY: Am. Inst. Phys. 1988).

27 C. Misner and P. Cooney, *Spreadsheet Physics,* (Addison Wesley 1991)

28 Potter, Peck, and Barkley, *Dynamic Models in Physics,* (N. Simonson 1989)

29 D. Dykstra and R. Fuller, *Wondering about Physics,* (Wiley 1988)

30 W.M. MacDonald, E.F. Redish, and J.M. Wilson., "The M.U.P.P.E.T. Manifesto," *Computers in Physics* **1**(1), 23(July/Aug 1988).

31 J.M. Wilson, E.F. Redish, and C.K. McDaniel, "The Comprehensive Unified Physics Learning Environment: Part I. Background and System Operation," *Computers in Physics* **6**(2), (Mar/Apr 1992). J.M. Wilson, E.F. Redish, and C.K. McDaniel, "The Comprehensive Unified Physics Learning Environment: Part II. Materials," *Computers in Physics* **6**(3), (May/Jun 1992).

32 J.M. Wilson, E.F. Redish, and C.K. McDaniel, "The CUPLE Project: A Hyper- and Multi-media Approach to Restructuring Physics Education," a chapter in *Sociomedia: Multimedia, Hypermedia, and the Social Construction of Knowledge*, E. Barrett, ed., (MIT Press, Boston, MA, 1992).

33 "Time to Prune the Ivy," *Business Week,* (May 24, 1993)

34 R. Thornton, "Learning Physical Concepts with Real-Time Laboratory Measurement Tools," *Am. J. Phys.* **58**(9), 858(1990).

35 D. Sokoloff and R. Thornton, "Interactive Lecture Demonstrations," preprint May 1993.

36 E. Mazur, "A Hypermedia Approach Toward Teaching Physics," *AAPT Announcer* **21**(2), 61 (May 1991).

37 W. Graves, "The Future of the Classroom" Expert's Conference, Troy, NY, 10 February 1993.

How Research Can Guide Us In Improving the Introductory Course

Lillian C. McDermott, Department of Physics
University of Washington, Seattle

Introduction

We have by now a rich source of documented information that indicates that the difference between what is taught and what is learned in introductory physics is often greater than most instructors realize.[1,2] Results from research also suggest that a major reason for this discrepancy is that the way in which physics is traditionally taught does not match the way in which most students learn. This paper presents some of this evidence and illustrates how research can be used to guide the design of more effective instruction.

Instruction in the traditional mode

Instruction in physics has traditionally been based on the instructor's view of the subject and the instructor's perception of the student. Most teachers are eager to transmit not only specific information and skills but also to convey their excitement about physics. Having obtained a particular insight after considerable intellectual effort, they want to save students from going through the same struggle. Instruction often proceeds from the top down, from the general to the particular. Students are not actively engaged in the process of abstraction and generalization. By presenting general principles and showing how to apply them in a few cases, instructors hope to teach students how to do the same in new situations. Recalling how they were inspired by their own experience in introductory physics, many instructors tend to think of students as younger versions of themselves. In actual fact, such a description fits only a very small minority. In the U.S., approximately one in every 30 university students taking introductory physics will major in the subject.

To many students, physics is a collection of facts and formulas. There is a common perception that the key to solving a problem is finding the right equation.[3] Students often do not recognize the critical role of reasoning in physics, nor understand what constitutes a physical explanation.

Despite differences in size, resources and student population, the content and level of the introductory course do not vary much from one institution to another. The teaching of physics has evolved to its present state as the result of the experience and the common heritage of instructors. Accurate and comprehensive texts have helped establish a common standard, but they have also produced a situation that is highly resistant to change.[4] For intellectual and practical reasons, it is unlikely that any radical revolution in physics teaching will succeed. Thus, the response of the Physics Education Group at the University of Washington to the challenge of the introductory course is to try to improve instruction through a process of cumulative, incremental change that is guided by research.

Constraints are imposed on the development of instructional materials for implementation in a large research-oriented department. The end result must be practical for use with many students in an environment in which responsibility for the introductory course is shared by many faculty members. Innovations that require faculty to spend more time on teaching are unlikely to survive. Any increase in instructor-student contact hours will have to be met by teaching assistants. Thus, provision for their proper preparation is crucial. An additional condition that our group set on materials that we produce is that they be sufficiently flexible to be useful in institutions that differ in size and mission from our own.

Modification of the traditional mode

We are developing a set of tutorial materials (tentatively entitled *Tutorials in Introductory Physics*) consistent with the requirements outlined above. This supplementary curriculum is being designed for use in conjunction with the lectures, laboratory and textbook of a typical calculus-based or algebra-based course. The project was motivated by the decision of the Physics Department to improve instruction in introductory calculus-based physics by making the laboratory compulsory for all students and replacing one of the four lectures/week with section of 20 - 25 students. Our group has taken responsibility for implementing these sections. We have instituted required tutorials that involve almost 1000 students each academic quarter. The term *tutorial* is used to emphasize the difference in content and teaching style between these sessions and typical recitation or quiz sections.

The tutorials are not devoted to quantitative problem-solving or a review of the lectures. Their purpose is to promote active learning. During the sessions, the students work together in groups of three or four. The structure is provided by worksheets that guide them through carefully sequenced activities designed to deepen conceptual understanding and to cultivate scientific reasoning skills. The instructor does not lecture or provide answers but engages in Socratic-style dialogue to help students find their own answers to questions posed by the worksheets. There is an emphasis on interpreting symbolic representations of abstract quantities (e.g., vector, flux, field) and on making connections between the formalism (e.g., equations, graphs, diagrams) and the real world. Similar qualitative exercises assigned as homework provide additional practice.

The lectures and tutorials are integrated through a system of pretests and course examinations. Each tutorial is preceded by a ten-minute "pretest" on material that has been treated in lecture by usually not yet in a tutorial. The pretests inform the instructor about the level of student understanding and help identify for the students what they are expected to learn in the tutorials. A significant part of every examination is based on the tutorials.

Generalizations from Research On Learning and Teaching

We approach the task of developing a supplementary curriculum for introductory physics by asking questions such as the following: Is the standard presentation of a topic adequate for students to develop a functional understanding? If not, what gaps need to be filled? Do the various elements of a typical course - example problems in lecture or text, lecture demonstrations, laboratory experiments, computer programs - deepen and extend student understanding? If not, how can these be made more effective?

We begin to address these questions by investigating how well students understand a topic before, during and after traditional instruction. Our methods range from individual demonstration interviews to descriptive studies of student performance in the classroom.[5,6] Careful analysis of the information obtained helps us identify the conceptual and reasoning difficulties that students have with the material. We draw on these insights in designing instructional strategies to address specific difficulties. These are incorporated into teaching sequences that form small coherent portions of curriculum. Development, testing and revisions take place in a continuous cycle on the basis of classroom experience.

The discussion in this section is organized under six generalizations drawn from research on the learning and teaching of physics.[7] Although the illustrations are taken from investigations by our group, the generalizations could be supported with findings by other investigators. Similar conclusions have also been reached by experienced instructors who have probed student understanding in less formal ways in the classroom.[8]

Example from electric circuits

The criterion most often used in physics as a measure of mastery of the subject is performance on standard quantitative problems. As course grades attest, many students who complete an introductory course can solve such problems. However, they are often dependent on memorized formulas and do not develop a functional understanding, i.e., the ability to do the reasoning necessary to apply concepts to new situations.

- **Facility in solving standard quantitative problems is not adequate criterion for functional understanding.** *Questions that require qualitative reasoning and verbal explanation are essential.*

We have been investigating student understanding of electric circuits over a period of many years.[9] One task that has proved particularly effective for eliciting common difficulties is based on three simple circuits consisting of identical bulbs and ideal batteries. (See Figure 1 below.) One circuit has a single bulb; another has two bulbs in series; the third has two bulbs in parallel. Students are asked to rank the five bulbs according to relative brightness and to explain their reasoning. This comparison requires no calculations. A simple qualitative model, in which bulb brightness is related to current or potential difference, is sufficient to determine that A = D = E > B = C.

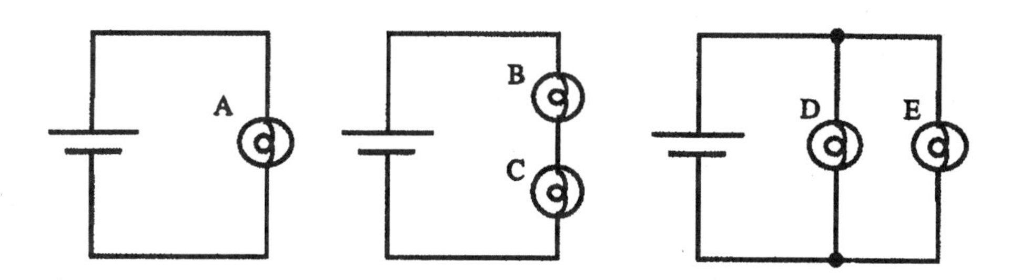

Figure 1. Students were asked to rank by brightness the five identical bulbs in the circuits shown and to explain their reasoning. They were told to assume that the batteries are ideal. The correct response is A = D = E > B = C.

We have administered this task to more than 500 university students. Whether before or after instruction, only about 15% of the students in a calculus-based course give a proper ranking. We have obtained the same results from high school physics teachers and from university faculty who teach other sciences and mathematics.

When this question was first asked on an examination in calculus-based physics, we analyzed the responses in great detail.[10] The students had already solved relatively complicated circuit problems. Nevertheless, almost every possible bulb order appeared. Inappropriate use of formulas was common. For example, many students calculated the equivalent resistance of the series and parallel circuits, substituted the values into the formulas for the power dissipated in a resistor, and associated the results with the brightness of individual bulbs in the series and parallel networks. Technical terms, such as current, potential and energy, were frequently used interchangeably. Students often did not seem to understand how basic concepts were related or how they differed. Some very serious misconceptions were prevalent. Two of the most widespread were the belief that current is "used up" in a circuit and that the battery is a constant current source.

Results from qualitative questions on other topics besides electric circuits have corroborated the generalization that success on standard quantitative problems is not a reliable indicator of functional understanding.[11] Moreover, on certain types of qualitative questions, student performance has been essentially the same: before and after a traditional algebra-based or calculus-based course, with and without a standard laboratory course, and regardless of the reputation of the instructor as a lecturer.[12]

- **A coherent conceptual framework is not typically an outcome of traditional instruction.** *Students need to participate in the process of constructing qualitative models that can help them understand relationships and differences among concepts.*

Our investigation of student understanding of electric circuits revealed that students who had successfully used Ohm's law and Kirchhoff's rules often could not do the qualitative reasoning necessary to predict and explain the behavior of very simple circuits. The nature of their conceptual and reasoning difficulties indicated that they had not integrated the concepts of current, potential difference and resistance into a consistent framework. Their skill in mathematical manipulation may have been sufficient to solve standard problems. However, when presented with a situation for which no procedures had been memorized, many students could not apply the appropriate concepts correctly.

An instructional strategy that has proved effective for helping students synthesize the basic electrical concepts into a coherent structure is to have them develop a qualitative model for a circuit.[13] Students are guided through the process of inductive and deductive reasoning necessary for constructing a model that enables them to predict and explain the relative brightness of bulbs in resistive circuits.[14] The preferred mode is through "hands-on" experience in the laboratory, but model-building can also be accomplished through interactive lectures and tutorials, in which demonstrations replace experiments.

Our experience, and that of others as well, shows that an emphasis on concept development does not detract from performance on quantitative problems. Many students need explicit instruction on problem-solving procedures to develop the requisite skills. However, premature introduction of algorithmic procedure often encourages students to concentrate solely on the formalism, rather than on the underlying physics. We have found that a more effective approach is to lay the foundation necessary for a sound qualitative understanding before proceeding to a mathematical analysis. Although less time is available for numerical problem-solving, examination results indicate that students who have learned in this way often perform better on quantitative problems and much better on qualitative problems than those taught in the traditional manner.[15]

Example from dynamics

The Atwood's machine is often used to illustrate the application of Newton's Laws to a system of two bodies in which the motion of one is affected by the other. (See Figure 2 below.) Our group investigated student understanding of this device in a series of studies.[16,17] We identified some serious conceptual and reasoning difficulties related to the acceleration *a* and the tension *T*, the same variables that students are typically asked to find. Other studies have yielded similar results.[18] In order to obtain more detailed information on the nature and prevalence of specific difficulties, we decided to examine how students think about the motion of simpler compound systems.[19]

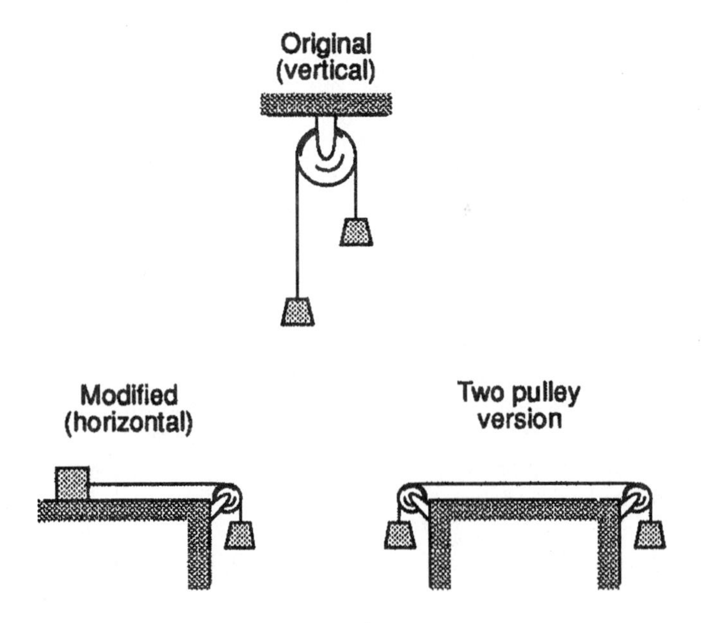

Figure 2. Versions of the Atwood's machine that appear in the standard introductory physics curriculum.

We designed a pretest based on a system of two blocks that move together in a horizontal linear direction under the constraint of a connecting string. The pretest was administered to a calculus-based physics class of about 100 students. The relevant material had been covered in lecture but not yet in a tutorial. The same pretest was also given to about 20 first-year graduate students, who were teaching assistants in introductory physics courses. The results from the two groups support the next generalization.

- **Certain conceptual difficulties are not overcome by traditional instruction.** *Persistent conceptual difficulties must be explicitly addressed by multiple challenges in different context.*

The questions on the pretest referred to the demonstration shown in Figure 3, in which two blocks (Block A and Block B) connected by a string (String #2) are pulled across a table by another string (String #1) attached to Block A. The mass of Block A is less than that of Block B. The students were told to assume that the strings are massless and that there is friction between the blocks and table.

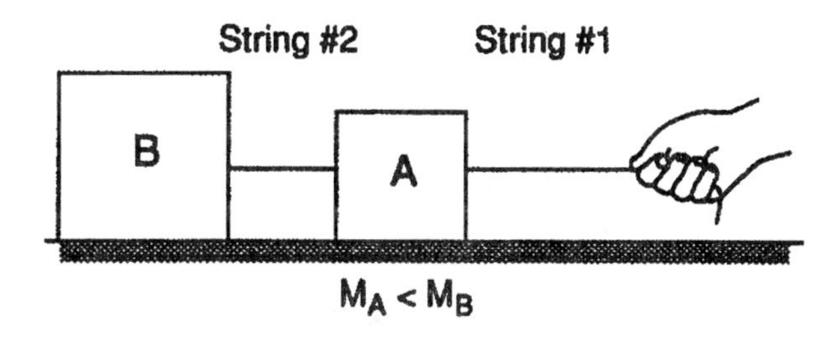

Figure 3. Diagram from pretest used to elicit student difficulties related to the role of the string.

On the first question, the students were asked to compare the acceleration of Block A with that of Block B. Approximately 85% stated correctly that the accelerations were equal. Instead of invoking the kinematical constraint that the separation between the two blocks must remain constant, the remaining 15% tried to use dynamical arguments. They assumed that the forces exerted by the two strings were equal and reasoned that since $M_A < M_B$ and since "F=ma," the acceleration of Block A must be greater. This question has since been administered in three other classes, altogether to about 450 students. Although the four classes were taught by three different instructors, about the same percentage (15%) of students in each class stated that the accelerations were different.

Another question on the pretest asked the students to compare the force exerted by String #1 on Block A with that exerted by String #2 on Block B. Only about 40% realized that the force exerted by String #1 must be greater in order to overcome the frictional force on both blocks and accelerate the entire system. The remaining students claimed that the forces were equal or that String #2 exerted the greater force. Each prediction reflected a different incorrect view of the role of the string.

About 40% of the class used Newton's Second Law to predict that, since the accelerations were equal and $M_B > M_A$, the force exerted by String #2 was greater. These students seemed to believe that the force exerted by each string depended only on the mass of the block to which it was directly attached and which it was pulling forward. The remaining 20% of the students stated that the force exerted by String #1 on Block A and String #2 on Block B were equal. These students seemed to believe that the force exerted by String #1 was transmitted unchanged to String #2. The following comments support this interpretation: "It is the same force," and "the force exerted on String #1 goes through [Block A] onto String #2." The external force applied to the system was not distinguished from the force applied by String #2 on Block B.

An example of how ongoing research guides the design of our instructional materials is the way in which we arrived at the decision to use unequal masses for Block A and Block B. On the initial version of the pretest, we had $M_A = M_B$. In comparing the force exerted by String #1 on Block A with that exerted by String #2 on Block B, about half of the students recognized that the force exerted by String #1 was greater. The most common incorrect response given by the others was that the forces were equal. These students seemed to think of the string either as only pulling forward on one block or as merely transmitting a force applied to one block to the other. Since both ideas lead to the same incorrect conclusion that the forces are equal, it was difficult to identify the conceptual and reasoning difficulties of individual students. We realized that the choice of $M_B > M_A$ would enable us to probe more deeply into student understanding of the role of the string.

Difficulty with the pretest was not limited to the introductory physics students. For $M_B > M_A$, a large percentage of the graduate students also made incorrect comparisons. All knew that the accelerations of the two blocks must be equal, but only 60% realized that the force exerted by String #1 on Block A must be greater than that exerted by String #2 on Block B. The other 40% said that String #2 exerted the larger force. Like the introductory students who made this error, they focused solely on the relative masses of the two blocks. As one graduate student stated, "to cause the same acceleration the force on Block B must be much larger. ... the mass of B is bigger ..." Many of the graduate students who drew free-body diagrams neglected to show the force exerted by String #2 backward on Block A.

The performance of the graduate students on the pretest suggests that study of advanced physics does not necessarily deepen qualitative understanding of Newton's Laws.[20] The nature of the errors that they made corroborates evidence from introductory courses that certain conceptual difficulties are persistent and resist traditional instruction.

Underlying some of the difficulties with the role of the string was a poor understanding of the concept of a system in classical mechanics. Responses to questions on pretest, course examinations and worksheets indicated that students had considerable difficulty in: (1) isolating an appropriate system, (2) identifying correctly all the forces present, (3) discriminating properly between Third Law force pairs, and (4) recognizing that it is the net force on a system that determines the acceleration. Students often could not decide which $\boldsymbol{F}$, which $\boldsymbol{m}$, and which $\boldsymbol{a}$ to associate with which system.

Even when students were able to choose an appropriate system and variables, they often failed to recognize the relationships between interacting systems and to do the reasoning necessary to predict the consequences of these relationships. The performance of students on a question from a course examination supports the generalization below.

- **Growth in reasoning ability does not usually result from traditional instruction.** *Scientific reasoning skills must be expressly cultivated.*

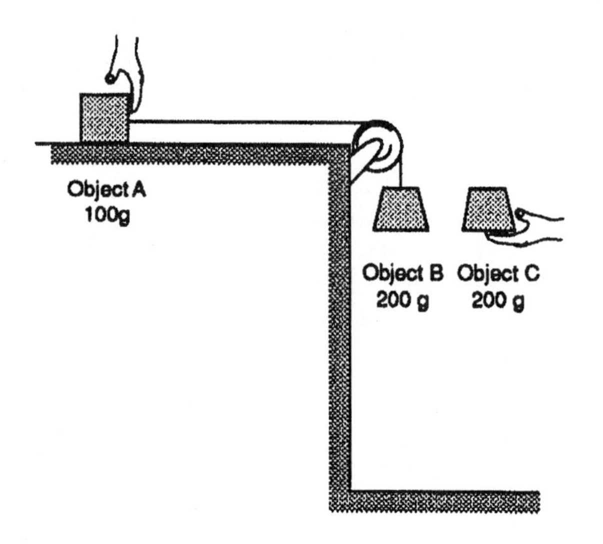

Figure 4. Diagram from a question to access student understanding of the role of the string. (a) Students are asked to compare the motion of Block B with that of an identical object falling freely. (b) Students are asked how the tension would change when the hand holding Block A in place is removed.

The examination question was based on the diagram of the modified Atwood's machine shown in Figure 4. Students were asked how the tension would change when the hand holding the sliding block (Block A) is removed. The mass of Block A is 100 g. The hanging block (Block B) has a mass of 200 g. Also in the diagram, but not involved in this question, is a block of mass 200 g (Block C) that is held at the same height as Block B.

As the system begins to move freely, the tension in the string decreases. However, 75% of the class predicted that the tension would remain the same. The most common incorrect explanation was that removing the hand would affect only Block A and not the string or Block B. The students did not recognize the need to think holistically about the effect of making a local change in a system of interacting object.

- **Connections among concepts, formal representations, and the real world are often lacking after traditional instruction.** *Students need repeated practice in interpreting physics formalism and relating it to the real world.*

Lack of skill in relating the formal representations of physics to a physical system also contributed to the difficulties that students had in analyzing the motion of a compound system.[21] For example, students often failed to make proper correspondence between the system (real or sketched) and the free-body diagrams for the components. As evidence, we present results from a pretest in which students were shown an unbalanced vertical Atwood's machine with the heavier block held higher than the lighter one. (See Figure 5(a) below.) They were asked to predict the motion that would take place when the system was released and to compare the weights of the blocks and the tension at both ends of the string. They were expected to justify their answers with free-body diagrams.

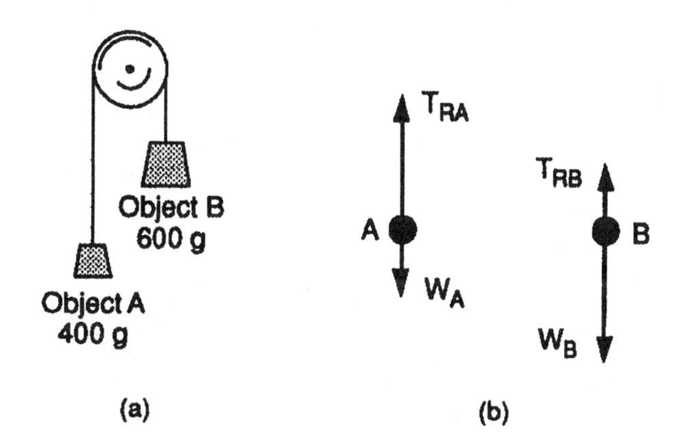

Figure 5. (a) Physical system used in tutorial on original Atwood's machine. (b) Typical incorrect free-body diagram draw by students to represent the forces extended on Block A and on Block B.

Most students predicted that the heavier block would drop and the lighter one would rise. Almost all stated that the weight of the heavier block was greater than the tension and that the weight of the lighter block was less. Many of their free-body diagrams, however, showed different magnitudes for the forces exerted by the string on the two blocks. A common mistake was to show the upward force on the lighter block as greater than the upward force on the heavier block. (See Figure 5(b) above.) As in the case for the pretest already discussed, this error suggests a perception of the string as directly transmitting a force applied to one object (in this case the weight of the block) directly to the other.

- **Teaching by telling is an ineffective mode of instruction for most students.** *Students must be intellectually active to develop a functional understanding.*

The types of difficulties that have been discussed in this section are not usually due to failure of the instructor to present the material correctly and clearly. Explanations by an instructor are rarely sufficient for helping students overcome deep-seated difficulties. Learning how to apply the Laws of Motion to a compound system requires significant mental effort and an extended period of time. The instructor cannot hasten the process by working example problems for students and expecting them to imitate the procedures demonstrated. No matter how lucid the lecture or well-written the text, meaningful learning will not occur unless students are intellectually active. Those who learn successfully from traditional instruction through lecture, textbook and problem-solving do so because they constantly question their own comprehension, confront their difficulties and persist in trying to overcome them. Most students do not bring this degree of intellectual independence to their study of introductory physics.[22]

Development of the Tutorials: *implementation of results research*

We found that almost all of the conceptual and reasoning difficulties evoked by the Atwood's machine were also elicited by the linear motion of two blocks connected by a string. Evidently, many students in a traditional introductory course do not develop sufficient functional understanding of Newtonian mechanics to be able to apply basic concepts to simple physical systems. A major purpose of the tutorials is to promote development of a consistent conceptual framework.

An instructional strategy that we have found effective for engaging students intellectually is to generate a conceptual conflict that they must resolve.[23] The first step is to *elicit* a suspected difficulty by contriving a situation in which students are likely to make a particular error. Once the difficulty has been exposed, the instructor must insist that students *confront* and *resolve* the issue. It is also necessary that the tendency to make a given error be evoked and overcome in a variety of situations. Otherwise, students may memorize special cases. The difficulty may remain latent and, if sufficiently serious, preclude future learning. To be able to integrate counter-intuitive ideas into a coherent framework, students need time to *apply* the same concepts and reasoning in different contexts, to *reflect* upon these experiences and to *generalize* from them.

The primary aim of the strategy outlined above is not to eradicate misconceptions but to help student synthesize related concepts into a coherent structure that will enable them to predict and explain some simple physical phenomena. To achieve this goal, they must have the opportunity to examine the way they think about the world, to compare their ideas with the formal concepts of physics, and to convince themselves of the utility of the physicist's view. Efforts to address conceptual and reasoning difficulties that ignore the conceptions held by students are not likely to be effective.

Example from a tutorial

To illustrate the nature of the instructional materials, we describe and activity that occurs late in a teaching sequence in which students' progress from the study of a very simple two-body system (two blocks in contact) to progressively more complex systems involving strings and other constraints. A pretest on the vertical Atwood's machine precedes the tutorial. The pretest, which has already been described, elicits several misconceptions, including the common belief that the weight of each block acts directly on the other block.

The tutorial begins by having the students draw new free-body diagrams for the same physical system used in the pretest. (See Figure 5.) This time, however, values are given for the weights (e.g., 4N and 4N) and the students are asked to compare the net forces on the two blocks. Many argue on the basis of their diagrams that the magnitudes are the same. Since each block has its own weight acting down on it and the weight of the other block acting up on it, the magnitude of the net force on each block is |6N - 4N|. When asked about the accelerations, the students conclude that they must be different because the masses are different. At this point, a suggestion to consider the effect of the string reminds the students of the kinematical constraint that is imposed.

Recognizing that the acceleration must be equal, the students realize that the net force on each block cannot be the same. As they confront this contradiction, they are asked to examine their free-body diagrams.

Most of the incorrect diagrams show the upward force exerted by the string to be different for each block. Invariably, some of the students recall that in an earlier tutorial they determined that the tension in a massless string must be the same at both ends. As they try to resolve this discrepancy, the students recognize that the weight of one block does not act directly on the other. Instead, the string exerts an equal upward force on each block: on one in the same direction as the motion, and on the other in the opposite direction. The students infer that the tension has a value between the two weights and that the acceleration is less than that of the free fall.

Effectiveness of the tutorial system

Before tutorials were instituted in all regular sections of the calculus-based course, we were able to compare the performance of students in the tutorial system with that of other students by giving the same examination questions to the two groups. The results from two of these questions are summarized below. Both were based on the horizontal Atwood's machine in Fig. 4. This form of the device was not treated in the tutorials.

In the first question, students in two large classes (one with tutorials, the other without) were asked to compare the accelerations, or the time of fall, of Block B and Block C when released simultaneously. They were to consider the situation both with and without friction. With friction present, almost all of the student in both classes predicted correctly that Block C would have the greater acceleration, or that it would strike the floor first. With friction absent, about 70% of the tutorial students made this prediction. In the other class, however, only about 45% of the students gave the correct response. The explanations given by students who had participated in the tutorials indicated that they recognized the presence of a constraint on the motion. In the other class, however, there were few references to the upward force exerted by the string on Block B. When this question was given in another tutorial class to students who had not yet had the relevant tutorial, about 55% responded correctly. Thus it appears that some transfer of reasoning skills had taken place because of participation in the tutorial system.

In the second question, students in three large classes (two with tutorials, one without) were asked how the tension would change when the hand holding Block A was withdrawn. The poor performance of the class without tutorials has already been cited earlier in this paper as an illustration of the failure of students to reason holistically about a system consisting of interacting parts. In the two tutorial classes, there was a much smaller tendency to treat the blocks and string independently. More than 50% of these students (as opposed to 25% in the class without tutorials) recognized that when Block B accelerates downward, the tension is less than the weight. Although the lectures were different, the results from both classes in the tutorial system were essentially the same.

Overall, on these and other questions, the tutorial students were better at drawing free-body diagrams, identifying Third Law force pairs and analyzing dynamical systems qualitatively. Moreover, their responses often included dynamical arguments. In contrast, the justification presented by the other students tended to be limited to algebraic formulas.

In addition to their use in tutorial sections, the materials have also provided a basis for interactive lectures. During the lecture, the instructor asks students for oral or written responses to questions posed on the worksheets. Students are given an opportunity to discuss answers with one another before participating in a discussion led by the instructor.

Preparation of teaching assistants

As has already been noted, the success of the tutorial system is critically dependent on the tutorial instructors. In out Physics Department, as in many others, small sections of large lecture courses are delegated to teaching assistants. Few have had experience in teaching, especially in the inquiry-oriented manner appropriate for the tutorials.

The generalization that teaching by telling is often ineffective has implications for the education of precollege and university teachers. The corollary below applies to both.

- **Most teachers teach as they have been taught.** *Teachers should have the opportunity to learn in the way they are expected to teach.*

This last generalization is based on more than twenty years of experience in courses for preparing precollege teachers to teach physics as a process of inquiry. It is also based on direct observation of graduate students who serve as teaching assistants in these courses.

Whether intended or not, teaching methods are learned by example. If a given topic has been learned by listening to lecture, reading a textbook and working problems, there is a strong temptation to teach it in the same way. It is difficult for successful learners, of whom graduate students are a prime example, to recognize that how they learned physics may not be how most students learn best. Teaching assistants need guidance in learning how to teach physics in a way that promotes student involvement rather than as a body of information that can be memorized.

There is, of course, another compelling reason for devoting special attention to the preparation of teaching assistants. As has been illustrated earlier, even graduate students often have difficulty with topics from basic physics. Most have not thought carefully about the content of the introductory course since taking it. Yet if they are to be expected to teach physics by guiding students through a line of reasoning, rather than by supplying information and answers, they must have a deep understanding of the material.

The preparation of teaching assistants as leaders of the tutorials takes place in a year-long seminar that is offered for credit. At the beginning of these weekly meetings, the TA's take the same pretests as the students in the introductory course. Afterwards they examine the responses given by the students and discuss possible reasons for their errors. The TA's then proceed step-by-step through the tutorial scheduled for that week. They work collaboratively in small groups in the same way that the undergraduates are expected to do during the tutorial sessions. Experienced TA's, who are already familiar with the materials (usually members of the Physics Education Group), help the others by asking questions rather than giving answers. The experienced TA's illustrate by their approach the role that tutorial instructors should take in the tutorial sessions. By the end of the seminar, the TA's not only have a better understanding of the material but also how the tutorial sessions should be conducted to promote student learning. It has been our experience that graduate students often transfer insights on teaching gained in the seminar and tutorials to their other teaching assignments, such as grading examinations and teaching laboratory.

It is also important that faculty who lecture in the courses attend at least part of the seminars. By participating in the discussions of student performance on the pretests, the instructors get direct feedback on how well the material has been understood. In addition, the interaction between the instructors and the teaching assistants helps insure that the lecture and tutorial components of the course will be consistent.

Conclusion

Not very long ago, the physics community considered research in physics education an activity in which instructors thought carefully about the subject matter and worked to produce clear and unambiguous lectures, textbooks and laboratory manuals. Improvement in instruction was often equated with an increase in clarity of presentation. Perhaps the most significant contribution that research, as it is conducted today, can make to the quality of instruction is to underscore the importance of focusing greater attention on what students are actually learning. To improve the match between the curriculum and the needs and abilities of students, we need the type of detailed knowledge about the nature of student understanding that can only be obtained from in-depth studies. Moreover, it is not only learning by students that should be subjected to scholarly inquiry. Instructional strategies designed to address student difficulties need to be tested for effectiveness in the classroom.

To be of significant and timely benefit to large numbers of students, the results of research in physics education should be implemented in curriculum. Authors of textbooks and other instructional materials can draw on insights acquired through research to help students overcome difficulties that have been identified as especially persistent. If this is done, the amount of material will have to be reduced. Although interesting to physicists, information added for enrichment can be overwhelming to students. They often cannot distinguish what is essential from what can be omitted in a first encounter with a new topic.

Both by necessity and tradition, the lecture is likely to continue to be the dominant mode of instruction at colleges and universities. Instructors can incorporate findings from research in physics education into their presentations. Lectures need not always be a passive learning experience.[24] Modifications in the style and content of textbooks and lectures can be heard and textbooks read with little intellectual engagement. Students may respond similarly to study guides, laboratory manuals and computer programs.

Meaningful learning, which connotes the ability to interpret and use knowledge in situations different from those in which it was initially acquired, requires that students be intellectually active. Development of functional understanding cannot take place unless students themselves go through the reasoning involved in the development and application of concepts. They need to be given a great deal of practice in solving multi-step qualitative problems and in explaining their reasoning.[25] To be able to transfer a reasoning skill learned in one context to another, students need multiple opportunities to use that same skill in different contexts. They are unlikely, however, to persevere in this type of intellectual activity unless the course structure, including examinations, reflects this same emphasis. We have found that the tutorial system that has been described is an effective means of encouraging students to make the necessary mental commitment.

Acknowledgments

The tutorial project has been a joint effort by members of the Physics Education Group. Peter S. Shaffer deserves particular recognition for his leadership and many contributions. Special thanks are also due to E.F. Redish (University of Maryland), a member of our group for the 1992-1993 academic year, for his active role in the tutorials. Also deeply appreciated is the support by the National Science Foundation that has enabled us to conduct a coordinated program in research, curriculum development and instruction.

References

1 L. C. McDermott, "What we teach and what is learned - Closing the gap," *Am. J. Phys.* **59**, 301-315 (1991).

2 For two articles with extensive references to the literature on student difficulties in mechanics, see L.C. McDermott, "Research on conceptual understanding in mechanics," *Phys. Today* **37** (7), 24-32 (1984) and I. Halloun and D. Hestenes, "Common sense concepts about motion," *Am. J. Phys.* **53**, 1056-1065 (1985). An extensive listing of research reports on student difficulties with electric circuits appears in the bibliography Students' Alternative Frameworks and Science Education, 3rd edition, edited by H. Pfundt and R. Duit (Institute for Science Education, Kiel, Germany, 1991).

3 H. S. Lin, "Learning physics vs. passing courses," *Phys. Teach.* **20**, 151-157 (1982); D.M. Hammer, "Two approaches to learning physics," *Phys. Teach.* **27**, 664-670 (1989); D. M. Hammer (1993), Ph.D. Dissertation, Science and Mathematics Education (unpublished).

4 An example of a well-known text that has served as a model for many others is R. Resnick, D. Halliday and K. Krane, *Physics* (Wiley, New York, 1992).

5 For examples of the use of individual demonstration interviews in research by the Physics Education Group, see D. E. Trowbridge and L. C. McDermott, "Investigation of student understanding of the concept of velocity in one dimension," *Am. J. Phys.* **48**, 1020-1028 (1980); D. E. Trowbridge and L. C. McDermott, "Investigation of student understanding of the concept of acceleration in one dimension," *Am. J. Phys.* **49**, 242-253 (1981); R. A. Lawson and L. C. McDermott, "Student understanding of the work-energy and impulse-momentum theorems," *Am. J. Phys.* **55**, 811-817 (1987); F. M. Goldberg and L. C. McDermott, "An investigation of student understanding of the real image formed by a converging lens or concave mirror," *Am. J. Phys.* **55**, 108-119 (1987).

6 For an example of a descriptive study by the Physics Education Group, see in addition to Ref. 9, 10, and 16, L. C. McDermott, M. L. Rosenquist and E. H. van Zee, "Student difficulties in connecting graphs and physics: Examples from kinematics," *Am. J. Phys.* **55**, 503-515 (1987).

7 These generalizations are discussed in more detail in L. C. McDermott, "How we teach and how students learn - A mismatch?" *Am. J. Phys.* **61**, 295-298 (1993).

8 See, for example, A. B. Arons, *A Guide to Introductory Physics Teaching* (Wiley, New York, 1990).

9 L. C. McDermott and P. S. Shaffer, "Research as a guide for curriculum development: an example from introductory electricity, Part I: Investigation of student understanding," *Am. J. Phys.* **60**, 994-1003 (1992); Erratum to Part I, *Am. J. Phys.* **61**, 81 (1993); P. S. Shaffer and L. C. McDermott, "Research as a guide for curriculum development: an example from introductory electricity, Part II: Design of instructional strategies," *Am. J. Phys.* **60**, 1003-1013 (1992).

10 L. C McDermott and E. H. van Zee, "Identifying and addressing student difficulties with electric circuits," in *Aspects of Understanding Electricity,* Proceedings of an International Workshop, Ludwigsburg, Germany, edited by R. Duit, W. Jung, and C. v. Rhoneck (Verlag Schmidt & Klaunig, Kiel, Germany, 1984), pp. 39-48.

11 In addition to Ref. 2, 5 and 9, see I. A. Halloun and D. Hestenes, "The initial knowledge state of college physics students," *Am. J. Phys.* **53** 1043-1055 (1985); R. K. Thornton and D. R. Sokoloff, "Learning motion concepts using real-time microcomputer-based laboratory tools," *Am. J. Phys.* **58**, 858-870 (1990); D. Hestenes, M. Wells and G. Swackhamer, "Force concept inventory," *Phys. Teach.* **30**, 141-158 (1992).

12 In addition to the papers on geometrical optics and electric circuits in Refs. 5 and 9, see E. Mazur, "Qualitative vs. quantitative thinking: Are we teaching the right ting?" *Optics and Photonics News* **2**, 38, (1992).

13 In addition to ref. 9, see A.B. Arons, *The Various Language* (Oxford, New York, 1977).

14 L. C. McDermorr et al., *Electric Circuits* from *Physics by Inquiry,* (Physics Education Group, University of Washington, Seattle, WA, 1982-1992). *Physics by Inquiry* consists of a set of laboratory-based instructional modules soon to be available through a commercial publisher.

15 In addition to ref. 9, see E. Mazur, "Improving student understanding in introductory science courses: The peer instruction method," (1992) unpublished.

16 In addition to ref. 1, see L. C. McDermott and M. D. Somers, "Building a research base for curriculum development; An example from mechanics," in *Research in Physics Learning: Theoretical Issues and Empirical Studies,* Proceedings of an International Workshop, Bremen, Germany, edited by R. Duit, F. Goldberg, and H. Niedderer (IPN/Institute for Science Education, Kiel, Germany, 1992), pp. 330-355.

17 D. J. Grauson, Ph.D. Dissertation, Department of Physics, University of Washington, 1990 (unpublished).

18 For results from other research on student understanding of the Atwood's machine, see R. F. Gunstone and R. T. White, "Understanding gravity," *Sci. Educ.* **65**, 291-299 (1981); R. F. Gunstone, "Student understanding in mechanics: a large population survey," *Am. J. Phys.* **55**, 691-696 (1987).

19 L. C. McDermott, P. S. Shaffer and M. D. Somers, "Research as a guide for teaching introductory mechanics: An illustration in the context of the Atwood's machine," to be published in the *American Journal of Physics* (Feb.-Apr. 1994).

20 P. S. Shaffer, Ph.D. Dissertation, Department of Physics, University of Washington, 1993 (unpublished).

21 For other examples of difficulties in relating concepts, representations and physical systems, see Refs. 5, 6, 9, 16, 17, 19 and 20.

22 Results from research indicate that for most student lecturing is an ineffective mode of instruction. In addition to Ref. 1, see S. Tobias in *They're Not Dumb, They're Different,* (Research Corporation, Tucson, AZ, 1990), p. 36; and D. I. Dykstra, C. F. Boyle and I .A. Monarch, "Studying conceptual change in learning physics," *Sci. Educ.* **76**, 615-652 (1992).

23 For other discussions of this instructional strategy, see in addition to Refs. 1, 9, 16 and 19, P. W. Hewson and M. G. Hewson, "The role of conceptual conflict in conceptual change and the design of science instruction," *Instructional Science* **13**, 1-13 (1984).

24 There are several techniques that can be used in large classes to promote active participation by students. In addition to the article by E. Mazur in ref. 15, see A. Van Heuvelen, "Learning to think like a physicist: A review of research-based instructional strategies," *Am. J. Phys.* **59**, 891-897 (1991); P. Group versus individual problem solving," *Am. J. Phys.* **60**, 627-636 (1992); P. Heller and M. Hollabaugh, "Teaching problem solving through cooperative grouping: Part 2: Designing problems and structuring groups," *Am. J. Phys.* **60**, 637-644 (1992).

25 For a more extensive discussion of the development of reasoning skills, see Ref. 8.

WHAT CAN A PHYSICS TEACHER DO WITH A COMPUTER?

Edward F. Redish, Department of Physics
University of Maryland, College Park

Introduction

When Bob Resnick[1] and David Halliday undertook to change the way introductory physics was taught in the late 1950s, the computer was largely restricted to research labs. As an upperclassman in 1960, I was delighted to be allowed to use a Marchant mechanical calculator. It was driven by electricity so you didn't even have to turn a crank! Each multiplication was performed by decimal wheels spinning in repeated additions. To multiply two six-digit numbers took about three seconds.

Today's computers are not much bigger than that Marchant calculator and more than a million times faster. But more than that, the computer can do a lot more than just manipulate numbers. Modern personal multimedia computers seem closer to the television that my students watch these days instead of reading books. This changes a lot. Can these new computers be used effectively in teaching introductory physics?

Teaching and learning physics has always been difficult and using innovative technology, especially at the introductory level, has not been particularly successful. Moreover, research in physics education has shown that at the introductory level, traditional methods do not help the majority of our students achieve the most basic and elementary goals we have for them.[2]

It is in this context that we must assess the application of modern computer technology to physics education. Can the computer help us overcome the difficulties that have become apparent through research in physics education? Or will it simply take us one more step down the path of slow, accumulating deterioration (SAD) that can be seen in the way that many introductory physics classes have changed over the years?

To judge the value of any innovation in physics education, we have to consider three important questions:

(1) What are our detailed goals for our students?
(2) What is the state of our students' knowledge and expectations of learning when they begin?
(3) What can we do to help them change the state of their knowledge?

Only in the context of these answers can we effectively ask the question: Can the computer help?

I first focus on articulating the issues of teaching and learning that might be relevant for including computers, a step often ignored or suppressed. Then, I review some uses of the computer that have either been demonstrated to facilitate learning or seem to have good prospects for doing so.

Physics Education Research

Where are our students when they come to us? Listening

Over the past decade and a half, a number of physicists and educators have begun to treat the problem of student learning as a scientific one. To do this, one must:

1. shift focus in the course from the content to the student;
2. accept the idea that we want to model our students' knowledge, both statically and dynamically;
3. listen carefully to our students and find out what they think and know both before and after instruction.

Adopting this viewpoint requires us to make a significant change in our attitude towards teaching. In physics education research (PER), extensive observations (including long interviews with dozens of individual students) are made to find out what students really think. It has been compellingly demonstrated that our traditional assessments (which are used mostly to certify and hence filter out some groups of students) do not often yield this information (see for example ref. 2 and references therein).

What PER studies have taught us is that most of our students, especially at the introductory level, learn far less in our classes than we had thought. They come to us with strongly held ideas about what they are supposed to do and how the world works, and many of these ideas conflict seriously with what we want them to learn.[3]

From this research, and from research in cognitive science, we learn a number of important and relevant facts.[4] Perhaps the most important is that all students "construct" their ideas and observations — pulling together what they see and hear into a (not necessarily coherent or consistent) mental model.[5] Mental models have some surprising characteristics.[6] They may contain contradictory elements. Different elements of a mental model may become confused because of superficial similarities. Access to elements of the mental model may not be automatic and appropriate. The access links are a part of the model that must also be built.

These results remind us of the fact (well known to good teachers for decades![7]) that we have to worry not only whether students "got" the content but whether they can also get to it when they need it and use it appropriately.

Another fact that has become clear from the result of much detailed research is that standard problem solving does not always help students build the appropriate mental models.[8] Part of the difficulty is that the mapping from a mental model to a problem output is not one-to-one. A sample of the kind of difficulty we can encounter by using only standard problems as assessment is illustrated in Figure 1.

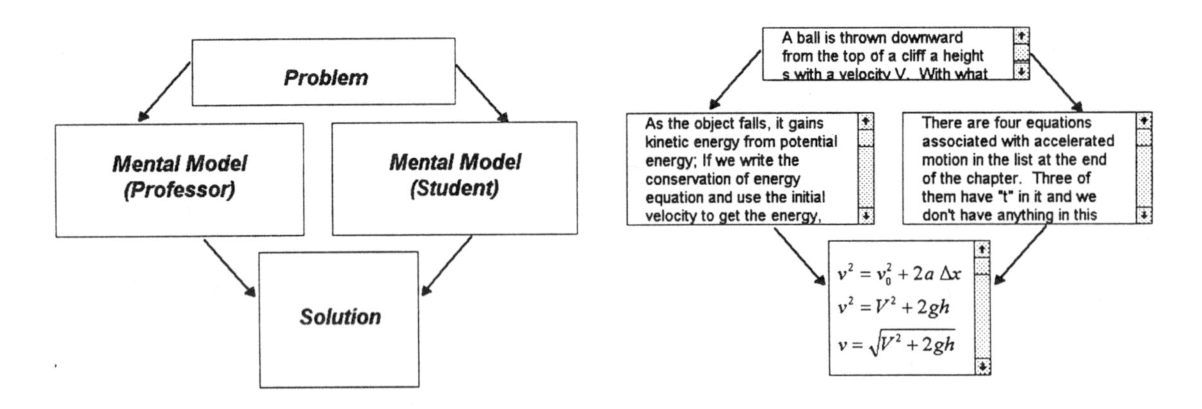

Figure 1: The map from mental model to problem solution is not one-to-one.

Understanding the basic concepts of physics

Students' experience with the world has led them to construct a set of beliefs about how the world works. Students misunderstand basic ideas of physics, in part because of their previous experience and in part because of the way we use language. In addition, "common sense" speech and reasoning are "fuzzy". Similar concepts are not differentiated, resulting in confusions that seem bizarre to a physicist who is accustomed to precise reasoning and operational definitions.

It has been demonstrated by extensive PER experiments that a majority of students entering (and leaving!) a typical calculus-based physics class:

- do not understand the meanings of velocity and acceleration,[9]
- fail to distinguish force from momentum,[10]
- fail to distinguish heat from temperature, and
- retain rigid and inappropriate beliefs about the nature of light, images,[11] and electric currents.[12]

Furthermore, the student's success on typical "end-of-the-chapter" problems has little correlation with developing an understanding of the deeper concepts.[13]

Knowing what science is and how it works

In addition to problems with the basic concepts of science, students in introductory physics courses often have deep misunderstandings of the nature of science. They often have little idea about how science works or how one does it. (Often, a physics major does not learn this last until graduate school!) Not only do many students get the basic concepts of physics wrong; their very idea of how science works and how knowledge is obtained is often wrong.[14, 15, 16]

- Many students fail to distinguish mathematics from physics and hypothesis from experiment.
- Many students do not understand the kind of knowledge we expect them to build nor the kinds of skills we expect them to develop.
- Many students do not understand that science is fundamentally open and non-authoritarian.
- Many students do not understand that science is fundamentally approximate and inexact.
- Many students think that scientific knowledge is absolute and must be taken from wiser authorities (incomprehensible geniuses).
- Many students think that physics is a collection of unrelated pieces that have little to do with the everyday real world.

The activities of a typical introductory class do little to disabuse our students of these deep structural misconceptions.

Re-forming an existing pattern may be hard

Once we've realized what difficulties our students have, why don't we just fix them? Unfortunately, it's not as easy as that — a result which we might well have guessed by the repeated failure of innovative curricula designed without detailed testing and modeling of students' responses.

If you're on the wrong track, it can be very hard to readjust. Getting students to make a "gestalt shift" in the way they see the world can be difficult. A classic example in cognitive psychology is the ambiguous picture of a woman in Figure 2. This picture can be viewed in two ways, as a young woman or as an old woman.[17] If students are presented with this picture in a less ambiguous form (certain features are enhanced to make it clearly the young or old woman) and then shown the ambiguous one, they find it exceedingly difficult to make the transition to the other, despite detailed explanations from members of the other group.

Figure 2: Portrait of a lady

There is extensive documentation in the cognitive literature about the difficulty of transforming a well-established mental model.[18] (See reference 1 and references therein.) By now there is also explicit documentation of how persistent are the mental models that students bring into a physics course and how these mental models make it difficult for students to understand what we are trying to teach them.

What do we want our students to learn? Deciding on our explicit goals

As a physics teacher I am not satisfied to have my students memorize a few equations and algorithms and be able to apply them in limited examples.

- I would like them to understand what physics is about, how it works, and why we believe it.
- I would like them to understand the basic concepts and the different representations used by physicists and to understand how these relate to the real world.
- I want them to see links between the different ideas in physics and to build a strong, accurate, and useful intuition for physical phenomena.

I refer to this collection of ideals as a *robust knowledge structure.*

It is clear to most physics teachers that very few students achieve such a structure by the end of their introductory physics course. Indeed, many graduate students do not possess one. Is this goal ridiculously out of reach? I do not believe so. Educational efforts with non-traditional approaches have demonstrated that a very large fraction of our students can begin to build real understandings of physics and how it is done even at the introductory level.[19] I operate under the hypothesis that most of our students can succeed at developing a robust knowledge structure of physics, even at the introductory level, given appropriate time, materials, and environments.

How can we help them get there? Guidelines for effective instruction from PER

Of course the problem with the sentence at the end of the last section is that we may not be able to give the students the appropriate time, materials, and environments they need to develop this knowledge structure. Furthermore, some of the necessary elements may be beyond our control, such as whether students have the time, motivation, and background to study adequately independent of coterminous jobs taken to support their education or the presence of tempting and diverting social activities.

Given these limitations, it is essential that we do our best to identify those elements which hamper students and about which we may have some control: the cognitive ones. Even students who come to class, who listen to correct and well-delivered lectures, who take notes, who attend and perform laboratories, often do not learn from those activities what we expect them to.

Two of the most important observations are:

1. If students have mental models that differ from the professor's, they may not interpret the presented material in the desired way.
2. If the activity in question can be performed without engaging the student's mental model, the learning that takes place will be disconnected and superficial.

How can we get students to both hear what we are trying to say and change their deeply held ideas? A number of principles have been developed as a result of research in physics education, both as to the kind of activities that tend to be successful and the kind of subtle mistakes and misunderstandings that we have to watch out for. These are documented extensively in the books and papers by Arnold Arons and Lillian McDermott. See their talks in this volume and Arons' book (ref. 3). I will only mention a few that I think are relevant for the topic of this meeting:

1. Go from the concrete to the abstract. (Make the link to the real world. Concept first, name second.)
2. Put whatever is new into a known and understood context.
3. Make students articulate what they have seen, done, and understood in their own words.
4. To change people's ideas, you must first get them to understand the situation, then make a prediction, and finally, to see the conflict between their prediction and their observation.
5. Telling someone something often has little effect in developing their thinking or understanding. You have to get them to do something, but "hands-on" activity does not suffice. It must be "brains-on".
6. "Constructive" activities in which students feel they are in control are much more effective than activities in which the students are being shown results, no matter how eloquently or lucidly the results are presented.

These observations suggest that modern computer technology might help, but they also suggest a division between two kinds of computer activities:

- *"video game" applications* — in which the computer is used to show the students something and to relieve them of carrying out certain tedious activities (which might, however, have provided valuable learning);
- *"constructivist" applications* — in which the computer is used to enable the students' personal explorations by giving them tools (and guidance) to work things through by themselves.

Can the Computer Help? Some Constructivist Applications

Let us now combine:

- our goals for our students,
- the problems they are documented to have,
- the guidelines for effective instruction, and
- the strengths of the computer

in order to identify computer applications which can provide effective elements of an instructional program. We'll consider three types of uses.

First, the computer can capture and display data from the real world quickly and accurately. This helps students make the link between concrete elements in the real world and the abstract representations of physics. This has been demonstrated to be much more effective in producing good learning of concepts than traditional methods.

Second, the computer can perform and display complex simulations. Abstract and inferred concepts may be made real for the student in a way which is difficult without the computer. In some cases the computer has been demonstrated to be effective in helping students build good mental models through letting them display and control abstract visual displays.

Finally, the computer can put modeling tools in the hands of introductory students that let them carry out activities that are much more like real science than are traditional activities in introductory courses. If these modeling tools are combined with data collection tools, students can learn the nature of science as a model of reality and see that they can actually *do* science for themselves.

Collecting Data: New tools for seeing the real world

Microcomputer-based laboratories (MBL)

Numerous researchers have shown that introductory physics students often confuse position and velocity (see references 9 and 10). One result is that they have difficulty interpreting even simple graphs of an object's velocity. The misreading of velocity graphs has been demonstrated to be deeper than simply an unfamiliarity with a particular representation.[20] Many students also use the wrong equations or put the wrong quantity into equations, confusing position, total distance traveled, instantaneous velocity, and average velocity.

In what is by now a classic experiment, Thornton and Sokoloff[21] have developed a series of discovery laboratories using MBL to teach the concept of velocity. They use modular data accumulators to give students a device that easily displays position, velocity, or acceleration of an object on the computer screen. These modular analog to digital converters (ADCs) have a number of advantages. They are easy to assemble (modular), they have quick response time, and they are personally kinesthetic. The students can hold them, move them, or become the object themselves and watch their own motion displayed on the screen.

Even though an introductory student may not understand the details of how an ADC works, these devices are easy to "calibrate psychologically". The student can see what they do and how their output relates to the real world.

Thornton and Sokoloff's labs not only have the students taking data, but they change the mode of interaction between the student and the equipment significantly from that found in the traditional introductory lab. The students work in groups and are required to make predictions — to figure out what a graph will look like before they make specific observations. This is a much more engaging activity than listening to a lecture and appears more effective in getting the students' brains turned on.

In order to evaluate their results, Thornton and Sokoloff developed a series of simple short answer questions[22] — matching velocity graphs to descriptions of motion. The error rates on four of these questions are displayed in Figure 3. For each question, the error rate in a control section (test given after lectures and homework on the subject) is shown in the dark colored bar. The error rate for students in the test group (after MBL) are shown in the lighter colored bar. The improvement is spectacular.

These results are extremely robust. Dozens of classes with and without the Thornton-Sokoloff labs have been tested in many universities and virtually identical results obtained.[23] In my own test in a class of calculus based physics for engineers at the University of Maryland, I tried my best to "teach to the test" without giving the questions beforehand. I focused on the concept of velocity, stressed the importance of learning to read velocity graphs, and gave numerous examples and demonstrations in lecture and for homework. My results were essentially identical with the dark colored bars everyone else has gotten in a traditional lecture setting. Although I was able to get good results by the final exam by having questions like these on every exam, the effort required was substantial. The MBL method is clearly much more effective in getting students to understand the concept of velocity and to interpret its graphical representation than is a traditional lecture-plus-homework approach.

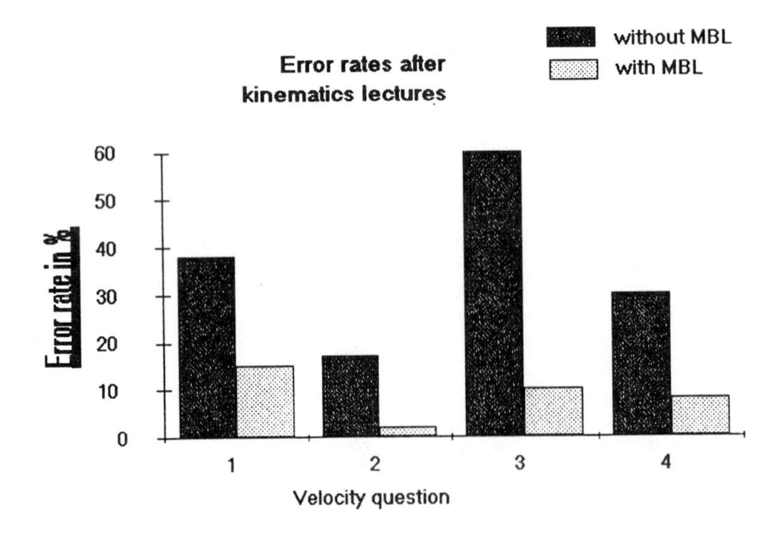

Figure 3: Results of lectures vs. MBL for teaching the concept of velocity.

Video

Computer controlled video is a tool that should be able to help students make the link to the real world. Many students' mental models appear to be strongly influenced by visual imagery. Furthermore, most students today have had extensive experience watching video. The computer can control, capture, and display video images on a monitor and allow students to take data from the screen a frame at a time. Unfortunately, at this writing there is little published PER to document the value of such activities in helping students build models.

Two examples where video might be used effectively are in the study of one dimensional projectile motion and in the cut pendulum.

PER has shown that a large fraction of our students persistently confuse velocity and acceleration for an object thrown straight up when it reaches the highest point in its trajectory. Many will say that the object must have zero acceleration at the highest point since it has stopped for an instant. (See references 2 and 10) Even students who can correctly separate velocity and acceleration for horizontal linear motion often make this error when pressed.[24] Extracting the position by hand from a series of video frames and then constructing the velocity and acceleration — again by hand — should prove an effective and convincing conflict with the student's existing mental model.

Research shows that students in the non-calculus based introductory class have misconceptions about what will happen if the string of a pendulum is cut while the pendulum is swinging.[25] In particular, many think that if the pendulum's string is cut when the pendulum is at the top of its swing, that it will continue on in some curved path rather than falling straight down.

The difficulty in this case is that many students fail to internalize what might be called Newton's zeroth law — that the state of an object at a given instant of time is specified by its position and velocity, and that the change in its velocity at that instant is only affected by the forces acting on it at that instant. From a more general point of view, this can be seen as a problem with local-global thinking. The students apply an argument that refers to the global motion rather than the appropriate local conditions. It is also a problem with their failure to check for consistency. They know Newton's second law but fail to apply it.

In this case, the use of video combined with student predictions should be very effective in strengthening the students' intuitions and understanding of Newton's laws.

At present, the use of video in introductory university physics classes is in its infancy. A number of groups have begun to use it in classes, but to my knowledge, as of this writing (early 1994) there is no published PER study of the effect of video on student understanding.

Simulations: New tools for building mental models

In addition to its value in building and correcting concepts, the computer should be highly effective in helping students to build complex mental models in cases where good visualization is required. Seeing "live" images can help one build correct and appropriate mental models. Particularly important are:

- the effect of seeing a system change
- the effect of user control.

I will discuss two interesting cases: building a mental model of the electric field and building a mental model of the atomic structure of matter.

Case 1: The electric field

A particularly difficult concept for students to understand is the electric field. Despite being given numerous examples in lecture and text, my students firmly resist developing a precise mental model of the electric field.

Many of them insist on continuing to interpret the electric field as "the region of space where the electric influence of a particular charge is felt."

Part of the difficulty is that we throw many different representations at the students at once — in part because we know the concept of electric field is an abstract and difficult one and we are trying to help. But if the students don't understand the concept, they may have difficulty in seeing what the representation is trying to say. In order to describe electric fields we use:

- equations,
- vector field displays,
- lines of force, and
- equipotential surfaces.

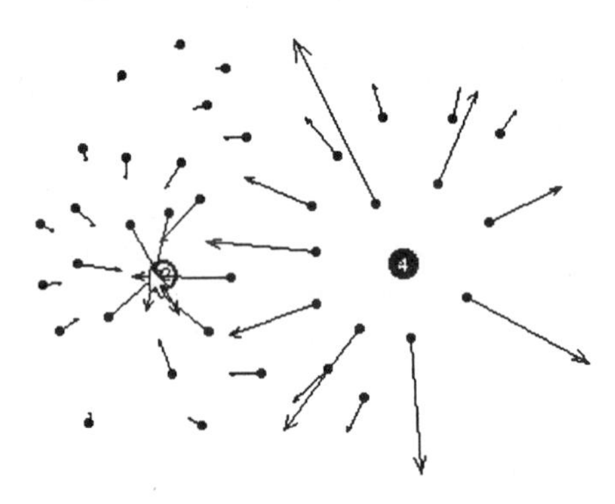

Figure 4: A screen display from EMField

Students have a lot of trouble figuring this out and creating a robust mental model of the electric field.

The newly published computer program, *EMField* by Trowbridge and Sherwood,[26] should prove very useful in helping students build a good mental model of the electric field. The student can pick up charges from a palette and place them on the screen. The student can then press the mouse button and have the cursor become a test charge. As the cursor moves around the screen, it displays an arrow representing the E-field it measures at that point. When the mouse button is released, the arrow is "dropped" — left in place to help remember what was measured at that point.

This program helps students build an E-field for themselves. A set of guided activities using this program should help students overcome a number of traditional misconceptions, including these:

- There is no E-field at points between the field lines.
- Charge flows along field lines.
- The field lines are the paths that charges follow when released

At present, I know of no studies that document the effectiveness of this technique, but the general principles learned from other examples suggest that it will help.

Case 2: The structure of matter

Students in high school (and college students not in the physical sciences) often have strange models of the structure of matter, despite having heard and read many descriptions of the atomic model and having seen many static pictures.[27]

A group at Boston University[28] has developed a powerful molecular simulator that runs on a workstation. With this, students can observe three-dimensional projections of a group of molecules move and interact. The students can control the temperature and angle of the display. The BU group demonstrated that working with this simulation for a few hours had a profound and positive effect on the mental models of matter possesses by high school and non-science college students. Figure 5 shows one frame of their display of water (left) and ice (right).

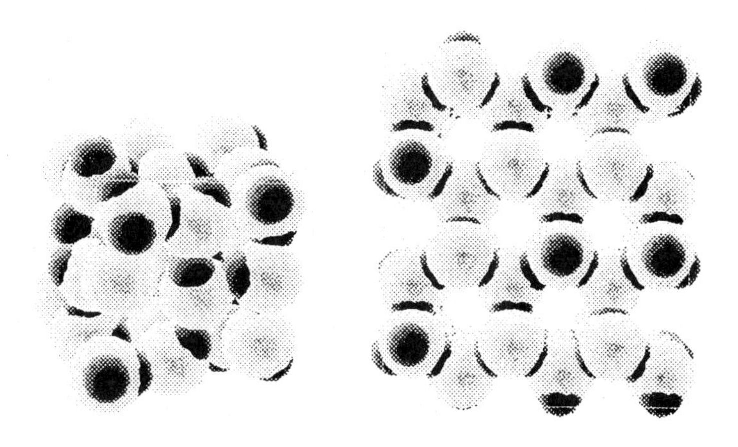

Figure 5: Displays of the molecular structure of water and ice.

Visual tools of the future

The use of good user controlled visual simulations like the two discussed above is just the beginning. Even more powerful visualization tools are now being developed that could have a profound influence on how we can teach physics.

Methods are now being developed to present users with three-dimensional controllable displays through the use of special glasses. These virtual reality representations are already available commercially for game applications.

Imagine how much easier it will be to teach Gauss's law with this device! The student could reach into a basket of charges, place them somewhere in the air in front of them, display the electric field vectors (as in the use of *EMField* described above, but now in true 3D), describe a Gaussian surface using their hand, and rotate the view of the entire representation using the virtual handles below the model.

Figure 6: Gauss's Law using virtual reality display

Modeling: Tools for apprentice scientists

We've discussed the use of the computer to take data from the real world via MBL and video, and the use of the computer to produce visual images via simulations. A third class of constructivist computer applications puts powerful modeling tools directly in the hands of students. Instead of the students seeing themselves as simply taking information directly from faculty and text, they become more active participants in building their own knowledge. They become apprentices rather than students.[29, 30] Two projects that move in this direction are M.U.P.P.E.T. and Workshop Physics.

M.U.P.P.E.T.

M.U.P.P.E.T. stands for the Maryland University Project in Physics and Educational Technology.[31] This project was begun at the University of Maryland in the early 1980s to develop tools to enable introductory physics majors to solve physics problems by programming. We developed a programming environment that provides structured Pascal tools[32] for handling input, output, and branching to help students "get to the physics" quickly without spending a lot of time learning to program.

With the computer as an "enabler" — a tool, not a demonstrator — students can solve problems at the introductory level that

- are self-selected and self-designed,
- are open ended,
- require planning,
- require writing and presentation, and
- are much more like "doing science" than traditional introductory activities.

At the University of Maryland, first and second year physics majors have been able to successfully design and carry out independent research projects using the M.U.P.P.E.T. materials.[33] Figure 7 shows a screen from a project done by a first semester freshman who studied the role of the backboard in basketball.

In the M.U.P.P.E.T. approach, student-apprentices do not simply participate in limited aspects of a real research project but create a complete (albeit scaled down) research project at their own level. This has the advantage that they begin to get experience with many aspects of doing science including planning, evaluating, and communicating. In addition, it begins to help them change their view of what science is about and how it's done.

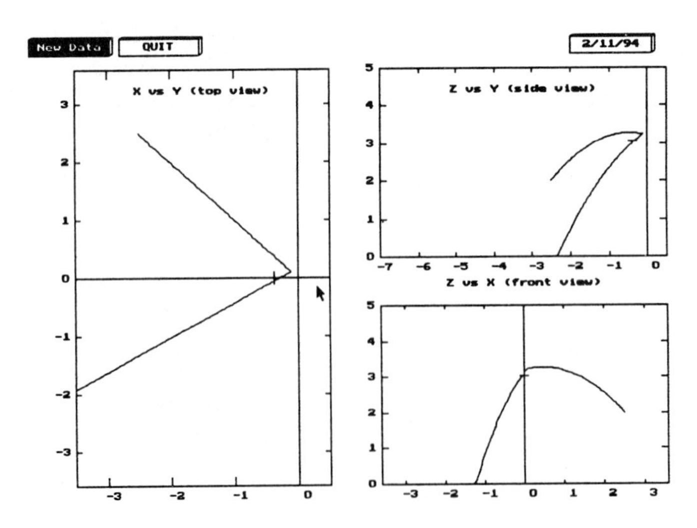

Figure 7: A screen from BALL3D, a student-written M.U.P.P.E.T. program to study the role of the backboard in basketball.

Using the M.U.P.P.E.T. programming environment also lets us introduce more realistic and modern problems at an earlier stage than is traditional. Physics majors at the University of Sydney (Australia) use the M.U.P.P.E.T. environment to extend the kind of problems they can treat in a wide variety of courses throughout the curriculum, including electromagnetism and quantum mechanics.[34]

Workshop Physics

Workshop Physics,[35] developed by Priscilla Laws and her collaborators at Dickinson College is an innovative introductory curriculum that has proven effective in overcoming many of the pedagogical problems that I've discussed above. Although this project is described elsewhere in this volume (in the talk by Priscilla Laws), it is appropriate to mention it here for its effective pedagogical use of computers.

Workshop Physics relies heavily on the use of the computer as a flexible and general purpose tool. Each pair of students has a networked computer for all their work in physics. But students do not run canned physics programs. They learn computer tools that help them investigate issues in physics:

- graphers,
- spreadsheets,
- lab tools (MBL), and
- video tools.

The students learn to use these tools early in the semester. This helps build their confidence and enables them to take greater charge of their own learning.

In Workshop Physics, students are guided by written materials that familiarize them with constructive reasoning by guiding them through the steps one or two at a time. In her talk, Laws cites numerous tests that show the effectiveness of this approach. It would be very interesting to follow the physics majors in this group to see whether they tend to be effective as researchers at an earlier stage in their careers than comparable students who have had a more traditional classroom environment.

Putting it all together: CUPLE

The Comprehensive Unified Physics Learning Environment (CUPLE)[36, 37] is a project to provide an environment for developing, presenting, and distributing educational materials on the computer. In addition to providing the student with data collection tools, simulators, and modeling, the computer also provides the teaching community with the opportunity to share educational innovations in a new way.

The CUPLE environment brings together the ideas of M.U.P.P.E.T., Tools for Scientific Thinking, Workshop Physics, and other innovative educational projects (such as Peter Signell's PhysNet project for modularizing the curriculum[38]).

The CUPLE idea is to put a wide variety of tools and materials into the student's hands in a single, unified environment including:

- modularized text materials,
- reference materials (glossary, data tables, etc.),
- word processors,
- graphers,
- lab tools (MBL),
- video tools,
- programming tools,
- a calculator,
- a symbolic manipulator,

- a spreadsheet,
- and a way to link them all together in a convenient way!

This is an environment in which multiple representations and the links to the real world can become compelling and useful in building understanding. The CUPLE environment allows students to easily display and manipulate:

- photographs and video images,
- numbers,
- graphs,
- equations, and
- simulations.

The CUPLE environment can be used in many pedagogically effective ways, including:

- as lecture demonstrations,
- as a "studio" or workshop environment without lecture,
- as an enhancement of the traditional lab,
- as an exploration tool for permitting apprentice-like activities.

But perhaps the most important aspect of CUPLE is the possibility to change the way that educational materials themselves develop. Although many physics faculty teach innovatively, very few publish their results, methods, and materials. There is little accumulation of innovations, even within a single college or university.

Part of the problem is in the lack of existing mechanisms for sharing and exchange. In the research arena, researchers share results by publishing papers in a research journal. Dozens of such journals exist and more than ten thousand physics research papers are published every year. In the educational arena, the critical products are materials for students to use, not research papers describing educational methods. Essentially all of the materials used in physics teaching are in the form of textbooks.

Part of the problem appears to be the difference in modularity. Physics research papers rarely create an entire field in a single step. The CUPLE system is designed to be open and modular. It lets a teacher easily modify existing material or add additional material of his or her own. The CUPLE engine is a database that contains tools for linking and finding links. It is a structure in which the interactions of the parts are tightly controlled and monitored by the CUPLE engine. Explication and management of the links permit simple substitution of modules by the user. The user can also add new links.

These features open the system for growth and cumulation. Once the first CUPLE course is complete, we hope to establish CUPLE as a publishing environment — one in which teachers using it could develop and submit modules for peer review, testing, and incorporation into the full system. The open structure of the system and the unprecedented modularization of the learning environment should permit many more teachers to add their contributions to an ever-growing and evolving structure, one that can embody a vision of the future that uses computer technology to put new power into the hands of both student and teacher.

CUPLE is currently being used in a discovery laboratory environment without lectures at Rensselaer. See Jack Wilson's talk in this volume for more details. Other notable pre-release applications of CUPLE have been developed including its use in computer-assisted tutorials in recitation sections at Maryland and to enhance lecture demonstrations at Virginia Polytech.[39]

Conclusion

It is clear from these examples that there are many computer applications for teaching introductory physics that can help students develop a good understanding of both the concepts of physics and the structure of science in a way that is more effective than traditional lectures + end of the chapter homework problems + labs. However, there are also serious dangers with using the computer. Computers offer professional physicists new ways of approaching topics they may have considered stale and uninteresting. But presentations that enthrall the expert may bewilder the novice.

There is an important difference between the constructive computer use discussed and throwing a canned program at a class. In all the examples presented above as pedagogically effective, the focus is on the student first and how the student interacts with the content as opposed to our traditional focus on the content alone. This new focus is what should be driving our selection and use of the computer in education.

References

1 *A physics professor at Poly*
Wrote a text in a manner quite jolly.
His book found the key
To the kids' apathy
And continues to rake in the lolly.

2 I. A. Halloun and D. Hestenes, "The initial state of college physics students", *Am. J. Phys.* **53** (1985) 1043-1056; Lillian C. McDermott, "Millikan Lecture 1990: What we teach and what is learned — Closing the gap", *Am. J. Phys.* **59** (1991) 301-315

3 A. Arons, *A Guide to Introductory Physics Teaching* (Wiley, 1990).

4 Edward F. Redish, "The Implications of Cognitive Studies for Teaching Physics", submitted to *Am. J. Phys.*, January, 1994.

5 Many of the principles stated here are stated in similar form in L. C. McDermott's talk in this volume. See ref 1. and McDermott's talk for documentation of these principles and references to the literature.

6 Norman, Donald, "Some observations on mental models", in Gentner and Stevens, *Mental Models* (Lawrence Erlbaum Associates, 1983) 7-14.

7 A. N. Whitehead, *The Aims of Education and Other Essays* (Free Press, NY, 1967).

8 L. C. McDermott, "Guest Comment: How we teach and how students learn — A mismatch?" *Am. J. Phys.* **61** (1993) 295-298.

9 David E. Trowbridge and Lillian C. McDermott, "Investigation of student understanding of the concept of velocity in one dimension", *Am. J. Phys.* **48** (1980) 1020-1028; D. E. Trowbridge and L. C. McDermott, "Investigation of student understanding of the concept of acceleration in one dimension", *Am. J. Phys.* **49** (1981) 242-253.

10 Halloun, I. A. and D. Hestenes, "Common sense concepts about motion", *Am. J. Phys.* **53** (1985) 1056-1065.

11 F. M. Goldberg and L. C. McDermott, "Student difficulties in understanding image formation by a plane mirror", *The Physics Teacher* **24** (1986) 472; F. M. Goldberg and L. C. McDermott, "An investigation of student understanding of the real image formed by a converging lens or concave mirror", *Am. J. Phys.* **55** (1987) 108.

12 L. C. McDermott and P. S. Shaffer, "Research as a guide for curriculum development: An example from introductory electricity. Part I: Investigation of student understanding", *Am. J. Phys.* **60** (1992) 994-1003; erratum, ibid. **61** (1993) 81; P. S. Shaffer and L. C. McDermott, "Research as a guide for curriculum development: An example from introductory electricity. Part II: Design of an instructional strategy", *Am. J. Phys.* **60** (1992) 1003-1013.

13 Eric Mazur, "Qualitative vs. Quantitative Thinking: Are We Teaching the Right Thing?", *Optics and Photonics News* (Feb. 1992)

14 David Hammer, "Defying common sense: Epistemological beliefs in an introductory physics course", Ph.D. thesis, U. of Cal., Berkeley (1991) and to be published in *Cognition and Instruction.*

15 Deanna Kuhn, "Children and adults as intuitive scientists", *Psych. Rev.* 96:4 (1989) 674-689.

16 S. Carey, R. Evans, M. Honda, E. Jay, and C. Unger, "'An experiment is when you try it and see if it works': a study of grade 7 students' understanding of the construction of scientific knowledge", *Int. J. Sci. Ed.* **11** (1989) 514-529.

17 *Hint*: The young woman's chin becomes the old woman's nose. The old woman's mouth is the young woman's necklace. (This picture is given courtesy of the Harvard-Radcliffe Cognitive Science Society, Clay Budin, Pres. It may be used at will and distributed, provided this message remains.)

18 Peter W. Hewson and Marian G. A'Beckett Hewson, "The role of conceptual conflict in conceptual change and the design of science instruction", *Instructional Science* **13** (1984) 1-13.

19 David Hestenes and Malcolm Wells, "A Mechanics Baseline Test", *The Physics Teacher* **30**:3 (1992) 159-166.

20 Dewey Dykstra, "Studying conceptual change: Constructing new understanding", in R. Duit, F. Goldberg, and H. Niedderer (Eds.) *Research in Teaching and Learning: Theoretical Issues and Empirical Studies*, Proc. of Int. Workshop, U. of Bremen, March 4-8, 1991 (IPN Univ. Kiel, 1992) 40-58.

21 R. K. Thornton and D. R. Sokoloff, "Learning motion concepts using real-time microcomputer-based laboratory tools", *Am. J. Phys.* **58** (1990) 858-867.

22 R. Thornton, "Tools for scientific thinking: Learning physical concepts with real-time laboratory measurement tools", in Redish and Risley (Eds.), *Computers in Physics Instruction: Proceedings* (Addison-Wesley, 1990) 177-188.

23 R. Thornton, personal communication. The results for six different sites are shown in ref. 22.

24 Andy diSessa, "Misconceptions reconceived", *AAPT Announcer* 23:4(1993) 94.

25 Michael Ranney and Paul Thagard, "Explanatory coherence and belief revision in naive physics", in *Proc. of the Tenth Annual Conf. of the Cog. Sci. Soc.* (1988).

26 D. Trowbridge and B. Sherwood, *EMField* (Physics Academic Software, Raleigh, NC, 1994).

27 J. Stepans, "Developmental patterns in student's understanding of physics concepts", in *The Psychology of Learning Science* (Erlbaum, 1991) 89-115.

28 Linda Shore, Melissa Erickson, Gita Hakerem, "Student understanding of molecular water", AAPT Announcer, **22**:2 (1992) 80.

29 J. S. Brown, A. Collins, P. Duguid, "Situated Cognition and the Culture of Learning", *Educational Research* (Jan-Feb, 1989) 32-42.

30 Jean Lave and Etienne Wenger, *Situated Learning: Legitimate Peripheral Participation* (Cambridge University Press, 1991).

31 W. M. MacDonald, E. F. Redish, and J. M. Wilson, "The M.U.P.P.E.T. Manifesto", *Computers in Physics* (July-Aug 1988)

32 E. F. Redish, J. M. Wilson, and I. D. Johnston, *The M.U.P.P.E.T. Utilities: Programming Tools for Turbo Pascal with Physics Examples* (Physics Academic Software, Raleigh, NC, to be published).

33 Edward F. Redish and Jack M. Wilson, "Student programming in the introductory physics course: M.U.P.P.E.T.", *Am. J. Phys.* **61** (1993) 222-232.

34 I. D. Johnston, "Standing waves in air columns: Will computers reshape physics courses?", *Am. J. Phys.* **61**:11 (1993) 996-1004.

35 Priscilla Laws, "Calculus-based physics without lectures", *Phys. Today* **44**:12 (1991) 24-31.

36 J. M. Wilson and E. F. Redish, "The Comprehensive Unified Physics Learning Environment: I. Background and System Operation" *Computers in Physics* **6**:2 (March-April 1992) 202-209; J. M. Wilson and E. F. Redish, "The Comprehensive Unified Physics Learning Environment: II. A Basis for Integrated Studies", *Computers in Physics* **6**:3 (May-June 1992) 282-286.

37 E. F. Redish, J. M. Wilson, and C. K. McDaniel, "The CUPLE Project: A Hyper- And Multi-Media Approach To Restructuring Physics Education", in E. Barrett (Ed.), *Sociomedia* (MIT Press, Cambridge, MA) 219-256;

38 Peter Signell, "Computers and the broad spectrum of educational goals", in Redish and Risley (Eds.), *Computers in Physics Instruction: Proceedings* (Addison-Wesley, 1990) 54-69.

39 Dale D. Long, Robert L. Bowden, and S. Raymond Liu, "DemoMedia I: The Magnetic field of a long straight current", *AAPT Announcer* **23**:4 (1993) 88; Robert L. Bowden, Dale D. Long, and S. Raymond Liu, "DemoMedia II: Double slit-interference and diffraction", *AAPT Announcer* **23**:4 (1993) 88.

From Concrete to Abstract: How Video Can Help

Dean Zollman, Department of Physics
Kansas State University

In selecting the title for this talk I made the assumptions that concrete experiences are a good way to start the study of any topic in physics and that our job is to help students move from the concrete to the abstract. The value of that idea has been discussed in some detail in other presentations. Thus, I assume that the most appropriate way to teach physics is the strategy based on the work described by McDermott and Redish. In this presentation I will describe how interactive video can provide one component in the process of helping students move from concrete experiences to the abstract theories of physics.

Interactive Video

By interactive video I mean a type of video in which the teacher and/or the students have control over the visual images which appear on the screen. For a video system to be interactive the learners must be able to have random access to a wide variety of video images or sequences, be able to stop the image on any of the individual frames, play those images at a variety of speeds backward and forward. This student control can be achieved through several devices — a keypad that allows entry of commands, a bar code reader which sends commands based on printed bar codes, or a computer which is connected to the video player via a serial port.

Interactive video can be recorded in a variety of formats. For a long time truly interactive video was possible only with laser-read videodisc. Recent improvements in the quality and types of videotape players now make it possible to use these tape players in a much more interactive environment. While random access is not as convenient as on a videodisc, many of the variable speed playing and stopping options are now available with rather high quality pictures. Finally, digital video systems in which the video is stored on a hard disk or on a CD-ROM provides another way to create interactive video. This digital format also offers the opportunity to go beyond the present concepts of interactive video and start considering image processing techniques for the study of physics.

Modeling and Visualization

Several of the speakers during this conference have already mentioned the importance of modeling to the physicist and the importance of teaching students how we use modeling to understand nature. Students need to understand how one might create a model and why we believe abstract models are valuable in describing nature. Thus, in this paper I will concentrate on modeling as an example of abstractions and used by physicists and demonstrate how interactive video can help students understand the process and use of modeling.

I will divide modeling into three broad categories — mathematical modeling, visual modeling, and visualization. Mathematical modeling is the most standard form which appears in introductory textbooks such as Resnick and Halliday. When physicists or students create a mathematical model, they prepare an equation which describes some event as accurately as possible. Visual models are also used extensively by physicists but are not often introduced explicitly in the introductory physics course. However, we do rely on them frequently in discussing items such as the Bohr model of the atom or an ideal gas model for the behavior of a collection of molecules. Finally, a form of modeling which has become popular with the increased availability of computational power is visualization. With visualization one tries to create a representation of a rather complex event. Visualization techniques, which sometimes can be quite abstract in themselves, can help students move from a concrete event to a more abstract model or representation of that event.

Mathematical Modeling

As physicists we know that mathematical models can be quite effective in describing very complex events. To demonstrate this idea to students we have used a laboratory activity based on a video scene which shows a collision of an automobile with a solid wall (Figure 1). This collision was filmed by General Motors as part of its research on automobile safety and has been included on the videodisc, Physics and Automobile Collisions1.

The goal of this activity is to determine equations which govern the motion of a car as it collides with the wall at a normal city driving speed. Students reach this goal by collecting distance and time data from the video. To set the scale the students use the markers which are separated by 0.5 meters on the automobile. Thus, we have a distance scale built into each frame of the video. From documentation provided with the video we know that the frame rate of the camera which recorded the event was 500 frames per second. With this information we can collect distance and time data from the video sequence.

Figure 1: A picture from the videodisc Physics and Automobile Collisions. The scene from which this picture was taken is used to create a mathematical model of the motion of this car as it strikes a solid wall.

Several methods are available for collecting such data. Here we will consider the method which is least expensive in terms of technology and equipment requirements. (Other methods will be discussed in other section below. They could be applied to this scene.) To collect the data the students place a clear sheet of acetate on the video screen. Using an overhead transparency pen they mark the location of a point of interest on the car. Then, they step the video forward and mark the same point on the car again. Continuing through the entire collision of the automobile with the wall, they collect distance-time data for this collision.

Students may step the video forward by using a key pad. However, a convenient way to control the videodisc player is with a series of bar codes. The bar code shown in Figure 2 plays the videodisc forward by 20 frames (.04 seconds of the automobile collision). With a series of these types of bar codes the students can be given control over the video and can collect the data quite easily without the worry of accidentally pressing a wrong button on the keypad and becoming "lost" on the videodisc.

Figure 2: An example of a bar code for playing video from a videodisc. This bar code will play Frames 14810 - 14830 of Car-Wall Collision

Once the students have collected the data, they may enter them into a computer, graphical calculator or plot them directly on paper. In general, we prefer a spreadsheet or a program such as MathCAD for data analysis. Using rudimentary curve fitting routines available in many different programs the students can see how well a simple mathematical model fits the data which they have collected.

In this example, the simplest approximations work. Before the collision, the car is moving at a constant velocity. A straight line fit of the data gives that velocity. This result is not a surprise because the car was being pulled at a constant velocity before the collision occurred. However, the students can use this curve fitting as a test of their data collecting and analysis ability.

Using the velocity calculated above one can fit with a parabola the data for the time during which the car is in contact with the wall. This parabola is a very good fit to the data. This result is, perhaps, a little surprising to the students. In spite of all of the complex motions and changes which are happening to the car during the time that it is in contact with the wall, a point on the car can be treated as if it undergoes a constant acceleration. This exercise shows the students how a simple mathematical model can be used to describe an extremely complex event. (A model of a constant velocity after the car leaves contact with the wall seems to work for the remaining video that we have.)

Students frequently complain that the standard approach to teaching physics with its frictionless planes have little connection with reality. Here, we have students work with a very real event and one which many students find rather interesting. By completing this experiment before the students are introduced to constant acceleration situations we can show the value of motion under constant acceleration as a good approximation to real-world events.

Visual Modeling

We define a visual model as one in which a sketch, diagram or picture rather than an equation provides the model. Thus, the picture of gas molecules bouncing around as hard spheres would be a visual model of an ideal gas. We use athletic events to help students understand how physicists can use simplified visual models of an object, in this case a person, to understand an event involving that object. Here, we borrow from the research techniques of kinesiologists —people who study the motion of humans, particularly athletes. They frequently use stick figures of those athletes to try to understand why one technique results in a superior athletic performance than another. Using this concept we introduce three models of the human being for consideration. The first model is that we treat the human being as a point mass which is located at the center of mass when the athlete is standing erect. This point is approximately at most people's hips. Two other models involve stick figures of three and five segments as shown in Figure 3. Using these figures students can take data on the different models and see how the motion of the center of mass of the athlete varies with each of the models.

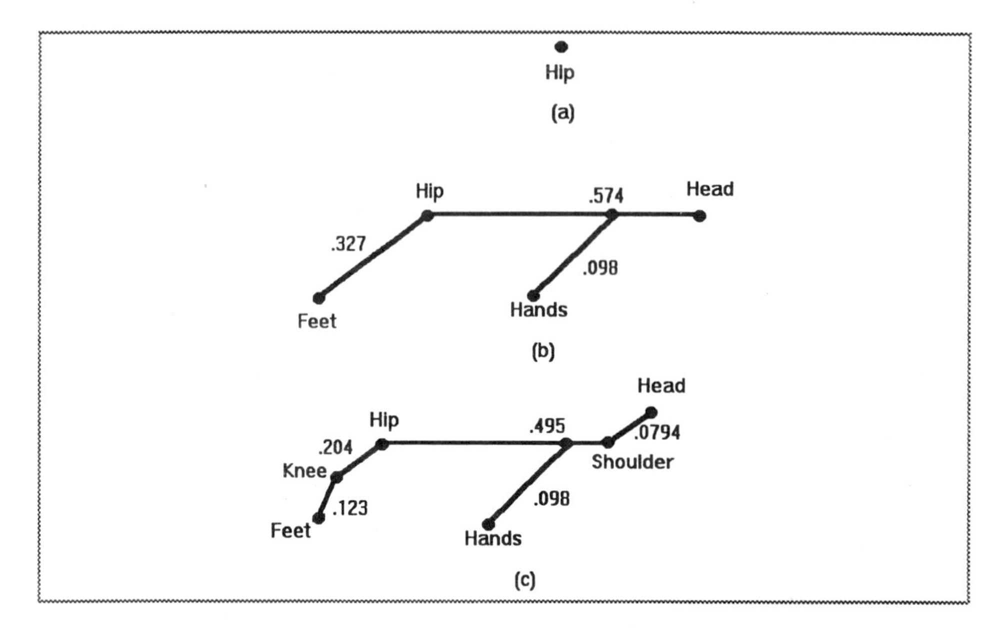

Figure 3: Three models of a high jumper. (a) The high jumper is treated as if she were a point mass with her mass concentrated at her hip. (b) The jumper is treated a three segments. Each segments has the percentage of body mass indicated and has a uniform mass distribution. (c) This five segment model is similar to the three segment one.

Figure 4: A frame from the high jump scene. In this frame the three segment model with bending at the hip is a very good model.

A very interesting event to study in this manner is the high jump. Using a sequence from the Physics of Sports2 we can play the video frame by frame. Students can draw the model on each individual frame of the athlete (Figure 4), and then collect the coordinates for the end points of each segment of the model. Entering these points into a spreadsheet allows the student to compare the motion of the center of mass for each of the models. To most students' surprise they find that the center of mass using a three segment or five segment model will sometimes go under the bar while for the point mass model it must always go over the bar.

For this exercise students can use the same technological system as described for the car collisions data collection. They may place acetates on the screen, draw the models of the athletes for each of the video frames and place the acetates on graph paper to read coordinates. Alternately, we can also use a computer graphics which is placed on top of the video image. Here the students can either draw, using the mouse, the stick figure and thus show the model or click on the location of the athlete which represent the end points of each segment. The computer can draw straight lines connecting them. The computer overlay method, using a video tool such as the CUPLE video tool described by Wilson during this conference, is a way to collect data quickly and easily.

Independently of which of the models or data collection techniques that a student chooses to use, he or she will always finds times when the athlete's configuration does not match that of the model. For example, in the three segment model which we have presented we assume that the athlete does not bend her knees. High jumpers, of course, bend their knees significantly during parts of their jump. However, if we are going to limit ourselves to a three segment model and use one of those segments for the arms and hands, we must limit ourselves either to a bend at the knees or at the hips but not both. When students have chosen a model, they must keep it throughout the entire motion. Thus, by analyzing a real event they quickly see both the value of simplified models and their limitations (Figure 5).

Figure 5: Another frame of the high jump. Here the jumper has bent her knees, so the three-segment model which bends only at the hips does not fit well. Pictures such as this one help students understand the limitations of models.

Visualization

In each of the examples described above, we are creating and collecting digital data from an analog video signal. That digitalization occurs by having the student pick points of interest and collect the coordinates and time for those points. This process is required, in part, because videotape and videodisc provide analog signals. Recent advances in the storage of video as digital information on computer disks offer a new way to use video in teaching. Because video can now be stored digitally, we are able to bring to motion video much of the image processing that has been available for still pictures. These digital video images provide a new way for us to help students to learn about some aspects of physics and to visualize physical events.

As an example of this new visualization technique we have begun the testing of a procedure to help students understand space-time diagrams and graphing by visualizing a collision. Unlike the videodisc, the students collect data from an experiment which they have completed. Thus, the students have more flexibility in determining the scene than with the videodisc because they operate the camera.

The first step in the procedure is for the students to capture onto their hard disk digital video images of a collision. Once they have recorded these images, they can select a section of the video which they consider most interesting. From this selection consecutive pictures of the video can be placed on top of each other to create a still image which contains images of the cart at different times. The students see clear images of the cart but the consecutive images start forming an overall image which looks something like a graph of distance versus time for each of the two carts. By making the size of the image from each of the consecutive frames smaller we lose the recognizable image of the cart but are able to put additional pictures on the screen at one time. Such images are shown in Figure 6 and are, in fact, space-time diagrams of the motion of these two carts. By using this procedure the students are better able to see how a graph or space-time diagram builds up from looking at the motion of an object during consecutive time intervals.

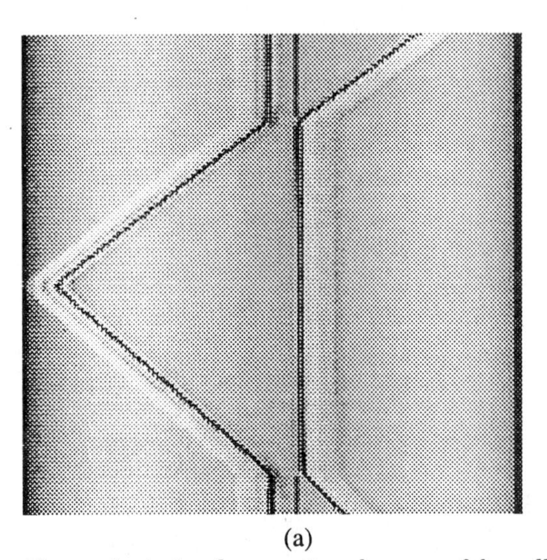

(a)

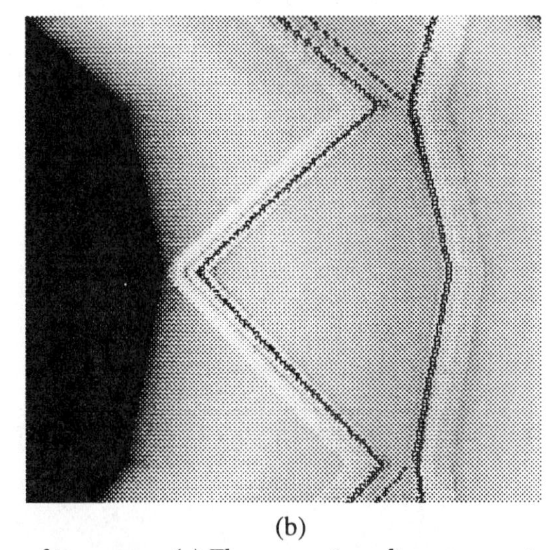

(b)

Figure 6: A visual space-time diagram of the collision of two carts. (a) The space-time diagram created by placing 240 consecutive images on the screen and viewing it from the lab reference frame. (b) The same collision and same number of frames as viewed from a center of mass reference frame.

We can take this a step further and ask the students how the space-time diagram would look if we were to view it from a different reference frame. Figure 6(b) is an example of the space-time diagram for the collision in Figure 6(a) from the center of mass when one cart is much more massive than the other. Using the data from these visual diagrams we can manipulate the video frames without taking any new video to see what the view would be from the reference frame of either cart or from the reference frame from the center of mass. Going beyond this we can return to the video and ask: how would it look if the collision were viewed from each of the different reference frames? Because the video is digital we can in fact manipulate its location on the screen and modify it so that the students can view the image from these different reference frames. We now have a way to allow the students to visualize how this collision would look in each of a collection of reference frames. This visualization of motion, which was never actually recorded but created from digital effects, provides a new way for students to look at information in a variety of different reference frames.

This process of using digital video is now in its first steps of development3. As we explore this new medium further we will find other ways to exploit the advantages of digital video in the teaching of physics. Certainly the ability to capture images onto a hard disk and then give the students control over those images to collect data from their own video is a major step forward over the prepared videotape or videodisc in which the students had to rely on someone else for the images.

Conclusions

Video in a variety of formats - videotape, videodisc, and digital video - offer a way to begin with familiar concrete events. Then, by analyzing these events through modeling and digitalization students can explore real-world events through hands-on experiences and see how physicists use abstract mathematical and visual models to understand nature better. Because the students see that simplified models are effective in understanding complex events, they can be better motivated to work with these simplified models.

We would hope that a further outcome of this type of activity would be for students to begin seeing how physics is applicable in the real world and in other videos that they observe. For example, many special effects in films, television, and music videos are created through an unusual shift in reference frame. By studying reference frames through processes such as capturing digital video onto a hard disk and manipulating the reference frame through the video camera, they can begin to understand these effects and to unravel the puzzle of how special effects are created. By applying models to scenes such as the car collision or the athlete's motion, students may extend the idea to other scenes or events which they see outside the classroom. Thus, video can be a way to create an important bridge between concrete, everyday experiences and the more abstract models which are used by physicists to describe those events.

Acknowledgments

Much of the videodisc work described in this paper was completed in collaboration with Robert Fuller. The digital video work was completed with S. Raj Chaudhury. The study of the use of digital video as a means to teach physics is supported by the National Science Foundation under grant number MDR 915022.

References

1. Dean Zollman, *Physics and Automobile Collisions* videodisc (John Wiley & Sons, New York, 1983).
2. M. Larry Noble and Dean Zollman, *Physics of Sports* videodisc (Videodiscovery, Seattle, 1988).
3. For more details on the use of digital video in physics teaching, see S. Raj Chaudhury and Dean Zollman, "Image Processing Enhances the Value of Digital Video in Physics Teaching", *Computers in Physics,* 8 518-523 (1994).

Learning Physics Concepts In The Introductory Course:

Microcomputer-based Labs and Interactive Lecture Demonstrations

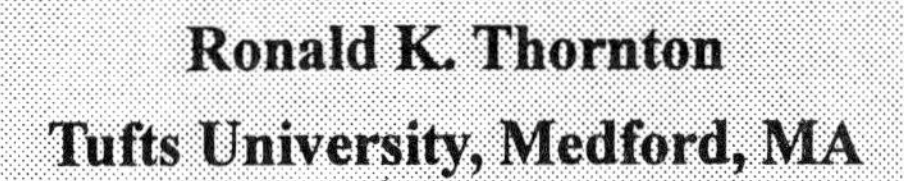

Ronald K. Thornton
Tufts University, Medford, MA

Abstract

This paper will address the questions: What physics concepts do our students understand? How do we know? and What can be done to promote such understanding in introductory courses? Careful and extensive research has shown that a majority of students have difficulty learning essential physical concepts in the best of our traditional courses where students read textbooks, solve textbook problems, and listen to well-prepared lectures. There are an increasing number of physicists addressing the problem of the failure of students to understand physics. Many successes have involved new approaches to learning supported by new technologies with an emphasis on active learning by students. This paper will focus on computer-based student learning tools and a number of curricula that support active learning by students in the areas of kinematics and dynamics that we have developed. By using Microcomputer-Based Laboratory (MBL) sensors and software, students can simultaneously measure and graph such physical quantities as position, velocity, acceleration, force, temperature, light intensity, sound pressure, ionizing radiation, current and potential difference. Such MBL tools and carefully designed curricula have been used successfully to teach physics concepts to a wide variety of students in the laboratories of universities and colleges and in the classroom and lecture hall through computer-supported *Tools for Scientific Thinking Interactive Lecture Demonstrations (ILDs).* Research data show substantial and persistent learning of basic physical concepts, not often learned in lectures. The paper will discuss problems with introductory courses, introduce MBL, give examples of curricula and techniques for mechanics instruction, and examine student learning in traditional introductory courses and in those using new methods.

Introduction

Are most students in physics courses acquiring a sound conceptual grasp of basic physics principles? For many years physicists teaching introductory courses have believed that they are, but those doing research in physics education have been convinced that they are not. There is evidence that traditional methods of teaching science are unable to bring a majority of students, even those intending to become scientists, to understand the physical world. Recently, extensive studies of students' basic conceptual knowledge before and after introductory physics courses have convinced some in the larger community of physics teachers that there is less basic understanding than they had believed. The results of these studies show that students in selective universities fail to agree with physicists when they answer the simplest conceptual questions. These same students are able to solve many traditional problems involving the solution of algebraic equations or even those requiring the methods of the calculus.

Traditional science instruction in the United States, refined by decades of work, has been shown to be largely ineffective in altering student understandings of the physical world. Even at the university level, students who take physics courses, whether they be science majors or not, enter and leave the courses with fundamental misunderstandings of the world about them essentially intact. Their learning of facts about science remains within the classroom and has no effect on their thinking about the larger physical world. The ineffectiveness of these courses is independent of the apparent skill of the teacher, and student performance does not seem to depend on whether students have taken physics courses in secondary school.

There are a number of barriers to an effective science education for non-scientists and potential scientists. Introductory courses physics often discourage potential majors and not only because they are difficult. Students often perceive science as difficult, boring, and overly concerned with detail. This is due in part to stereotypes and in part to the courses actually offered. Science is exciting to scientists because they are engaged in discovery and in creatively building and testing models to explain the world around them. Yet scientists rarely *preach what they practice*. Science courses rarely reflect the practice of science. In most courses, students do not "do" science. Instead they only hear lectures about already validated theories. Not only do they not have an opportunity to form their own ideas, they rarely get a chance to work in any substantial way at applying the ideas of others to the world around them. The worst courses consist of the presentation of collections of unrelated science facts and vocabulary with no attempt to develop critical thinking or problem solving skills.

Student Conceptual Understanding Before and After Traditional Instruction

Consider, as an example, traditional instruction in kinematics and dynamics. These topics consume a large portion of the time spent on instruction in the first semester of most introductory courses. Physics professors often assume that their students understand kinematics when they come into the course or certainly from the textbook and lecture presentation. Over many years we have evaluated student understandings of simple motion (kinematics) concepts using a series of multiple choice questions on velocity and acceleration which are part of the *Motion Conceptual Evaluation* developed as part of the *Tools for Scientific Thinking* project. Some of these kinematics questions are included in the *Force and Motion Conceptual Evaluation (FMCE)*[1,29] which has been developed to probe student understanding of dynamics. The choices available to the student on these carefully constructed multiple choice questions were derived from student answers on open-ended questions and from student responses in interviews. We have now administered the motion questions to many thousands of college and university students in the United States and other countries. The results, some of which appear in previous publications,[4,7,18,19] indicate that these students, mostly future engineers and scientists in calculus-based courses at major universities and colleges, do not understand simple kinematics concepts *after experiencing traditional introductory physics instruction (that is, after attending lectures and traditional laboratories, and working standard problems)*.

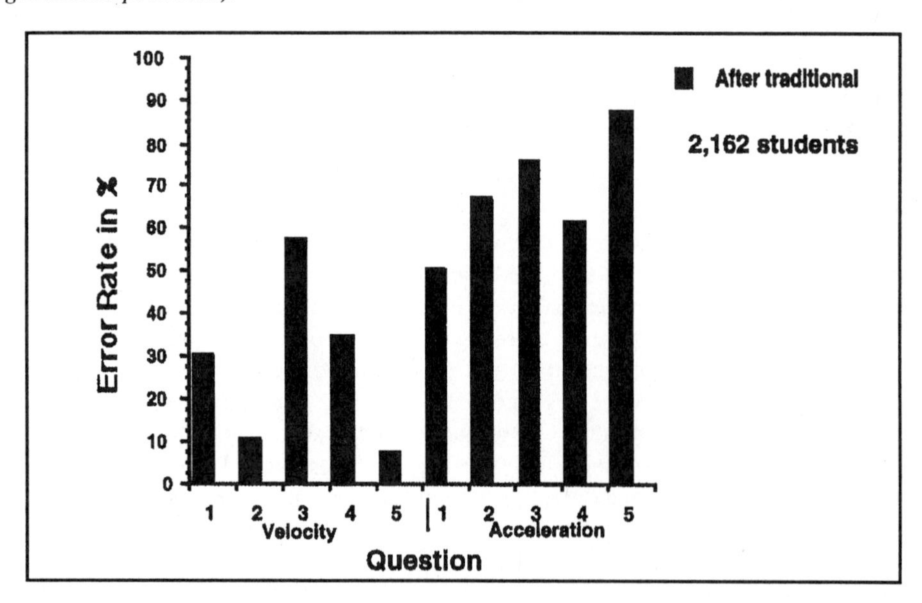

Figure 1. Composite data for over 2000 students after traditional instruction in introductory physics courses (mostly calculus-based courses) at selective universities. The figure shows error rates on simple questions intended to measure student understanding of kinematics from the Tufts Force and Motion Conceptual Evaluation . The results for questions on Velocity and ***Acceleration*** *are labeled. Most professors predict that no more than 10% of their students would miss these simple questions. The actual questions asked are discussed in the text.*

Figure 1 shows typical results for more than 2000 physics students after traditional instruction in introductory physics courses at selective universities. These are straightforward questions for which physics professors predict error rates of 5-10%. The actual error rates are 40% or above on velocity concepts, and 70-95% on acceleration concepts. For example, one of the velocity questions, on which the error rate was almost 60%, asks students to choose from among eight velocity-time graphs the one which describes motion of an object "toward the origin at a steady (constant) velocity." (velocity question 3 in Figure 1.) One of the acceleration questions, on which the error rate was nearly 80%, asks for the acceleration-time graph for an object moving "toward the origin at a constant velocity." (acceleration question 3 in Figure 1). Our research has shown that the problem is not student understanding of graphs but of motion concepts. In Figure 1 almost all students answer correctly velocity question 5 "Which velocity graph shows the car increasing its *speed* at a steady (constant) rate?" The result indicates that the students in this sample understand graphs.

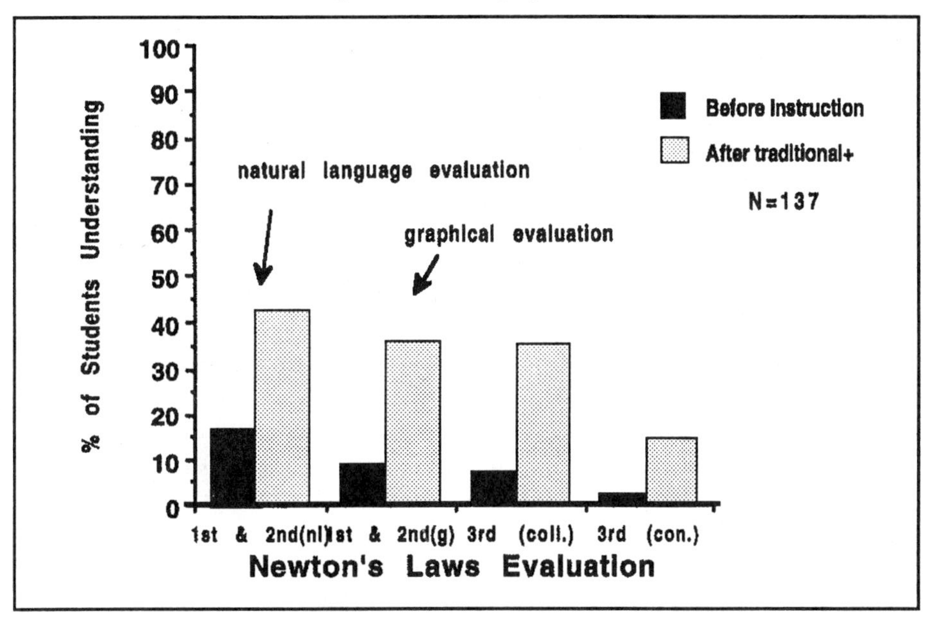

Figure 2. Student understanding of dynamics before and after traditional instruction. Percent of University of Oregon non-calculus introductory physics students who understood dynamics concepts related to Newton's 1st and 2nd laws before and after good traditional instruction which included lectures, problems, quizzes and exams. The measure of "understanding" used was the average of correct answers to particular groups of questions on the Force and Motion Conceptual Evaluation. The average labeled "natural language evaluation" consisted of a group of Force Sled questions (questions are shown in Figure 3) and the average labeled "graphical evaluation" from the Force Graph questions (questions are shown Figure 4). The same 240 students in the 1989 and 1990 were evaluated before and after instruction.

Although a Newtonian framework is essential to understanding non-relativistic (and later relativistic) motion, it is common for 80% of students to answer most questions from a non-Newtonian point of view after an introductory physics course. Such students may believe, for example, that a net force is required to keep an object in motion at a constant velocity, that there is a residual force on an object that has been pushed and released that keeps it moving, and that force and acceleration must increase as velocity increases. In contrast, students and physicists who believe the world behaves in a Newtonian manner (for every day speeds) use a conceptual framework based on Newton's laws of motion. They understand that a body moving at constant velocity requires no net force to keep it moving and so no residual forces are required. They also understand that a constant acceleration produces a uniformly increasing velocity. Learning to substitute values into the equations of motion seldom results in Newtonian conceptual understanding. Our research has shown that it is common for standard instruction to only change the conceptual point of view of 5 to 15% of the students in the area of dynamics.

A sled on ice moves in the ways described in questions 1-7 below. *Friction is so small that it can be ignored.* A person wearing spiked shoes standing on the ice can apply a force to the sled and push it along the ice. Choose the one force (A through G) which would **keep the sled moving** as described in each statement below.

You may use a choice more than once or not at all but choose only one answer for each blank. If you think that none is correct, answer choice J.

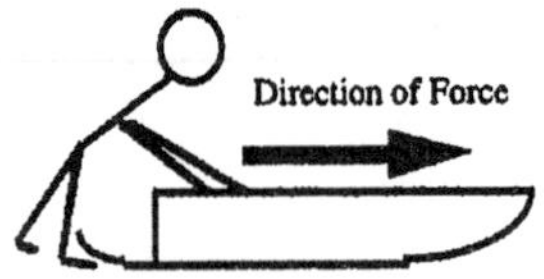

Direction of Force

A. The force is toward the **right** and is **increasing** in strength (magnitude).

B. The force is toward the **right** and is of **constant** strength (magnitude).

C. The force is toward the **right** and is **decreasing** in strength (magnitude).

D. No applied force is needed

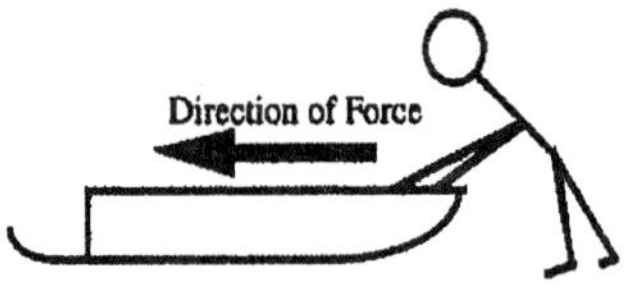

Direction of Force

E. The force is toward the **left** and is **decreasing** in strength (magnitude).

F. The force is toward the **left** and is of **constant** strength (magnitude).

G. The force is toward the **left** and is **increasing** in strength (magnitude).

______1. Which force would keep the sled moving toward the right and speeding up at a steady rate (constant acceleration)?

______2. Which force would keep the sled moving toward the right at a steady (constant) velocity?

______3. The sled is moving toward the right. Which force would slow it down at a steady rate (constant acceleration)?

______4. Which force would keep the sled moving toward the left and speeding up at a steady rate (constant acceleration)?

______5. The sled was started from rest and pushed until it reached a steady (constant) velocity toward the right. Which force would keep the sled moving at this velocity?

______6. The sled is slowing down at a steady rate and has an acceleration to the right. Which force would account for this motion?

______7. The sled is moving toward the left. Which force would slow it down at a steady rate (constant acceleration)?

Figure 3. Questions from the Force and Motion Conceptual Evaluation that probe student understanding of Newton's First and Second Laws using ordinary language. Student results for an average of questions 1-4 and 7 are labeled ***1st & 2nd (nl)*** *(where* ***nl*** *denotes natural language) in figures in this paper.*

Figure 2 shows the result of student responses on a composite set of questions the *Force and Motion Conceptual Evaluation. (FMCE)*[1,29] intended to measure dynamics concepts. The students were evaluated before and after they experienced good traditional introductory physics instruction. This paper will present student average results from four sets of questions from the *FMCE* as a measure of student understanding of force and motion concepts described by Newton's Three Laws. The "Force Sled" questions (questions 1-7) shown in Figure 3 and the "Force Graph" questions (questions 14-21) shown in Figure 4 evaluate understanding of the Newton's First and Second laws in two rather different ways. The Force Sled questions use simple English, do not explicitly use coordinate systems, and display the forces involved. The student results shown for these questions (labeled **1st & 2nd (nl)** where **nl** denotes natural language) is an average of Questions 1-4 and 7. The Force Graph questions use force-time graphs and explicitly require knowledge of coordinate systems. The student results shown for these questions (labeled **1st & 2nd (g)** where **g** denotes graphical) is an average of Questions 14 and 16-21. A discussion of why particular questions were chosen for the averages can be found in references 1 and 18.

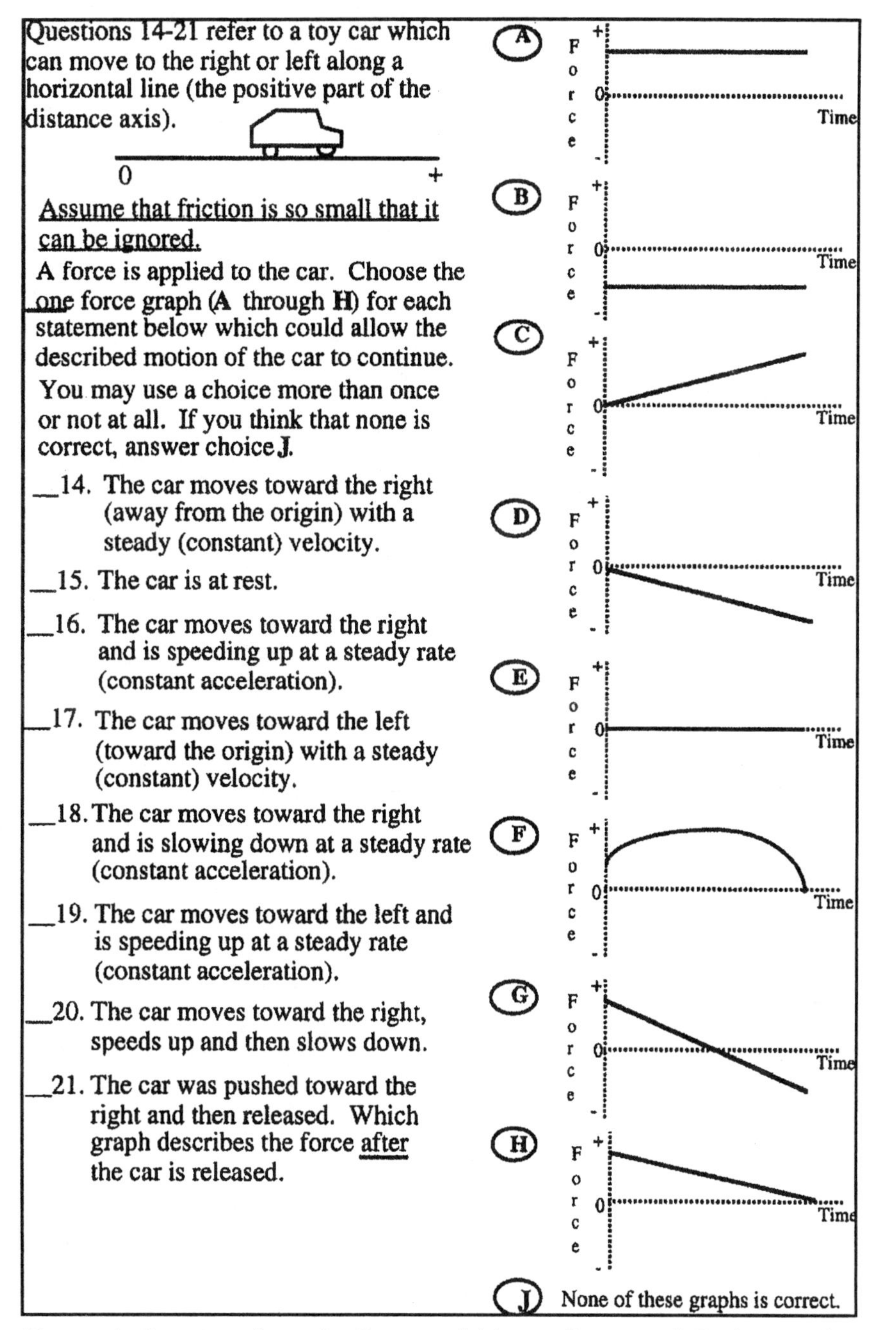

Questions 14-21 refer to a toy car which can move to the right or left along a horizontal line (the positive part of the distance axis).

0 +

Assume that friction is so small that it can be ignored.

A force is applied to the car. Choose the one force graph (A through H) for each statement below which could allow the described motion of the car to continue. You may use a choice more than once or not at all. If you think that none is correct, answer choice J.

__14. The car moves toward the right (away from the origin) with a steady (constant) velocity.

__15. The car is at rest.

__16. The car moves toward the right and is speeding up at a steady rate (constant acceleration).

__17. The car moves toward the left (toward the origin) with a steady (constant) velocity.

__18. The car moves toward the right and is slowing down at a steady rate (constant acceleration).

__19. The car moves toward the left and is speeding up at a steady rate (constant acceleration).

__20. The car moves toward the right, speeds up and then slows down.

__21. The car was pushed toward the right and then released. Which graph describes the force after the car is released.

*Figure 4. Questions from the Force and Motion Conceptual Evaluation that probe student understanding of Newton's First and Second Laws using a graphical format. Student results for an average of questions 14 and 16-21 are labeled **1st & 2nd (g)** (where g denotes graphical) in figures in this paper.*

The averages for two additional sets of questions on Newton's Third Law will be used in Section VII. Questions 30 through 34 shown in Figure 5 probe student understanding of the forces in collisions. The student results for these questions (labeled **3rd(coll)** where **coll** denotes collision) are an average of questions 30-32 and 34. Questions 35 through 38 shown in Figure 6 probe 3rd Law concepts when two objects are in contact for an extended period of time. The student results for these questions (labeled **3rd(con)** where **con** denotes contact) are an average of questions 36 and 38. Understanding of Third Law concepts by students is even less common than the first two laws after traditional instruction.

Questions 30-34 refer to collisions between a car and trucks. For each description of a collision (30-34) below, choose the one answer from the possibilities A though J that best describes the size (magnitude) of the forces between the car and the truck.

A. The truck exerts a larger force on the car than the car exerts on the truck.
B. The car exerts a larger force on the truck than the truck exerts on the car.
C. Neither exerts a force on the other; the car gets smashed simply because it is in the way of the truck.
D. The truck exerts a force on the car but the car doesn't exert a force on the truck.
E. The truck exerts the same amount of force on the car as the car exerts on the truck.
F. Not enough information is given to pick one of the answers above.
J. None of the answers above describes the situation correctly.

In questions 30 through 32 the truck is ***much heavier*** *than the car.*

_____30. They are both moving at the same speed when they collide. Which choice describes the forces?
_____31. The car is moving much faster than the heavier truck when they collide. Which choice describes the forces?
_____32. The heavier truck is standing still when the car hits it. Which choice describes the forces?

In questions 33 and 34 the truck is a small pickup and is the ***same weight*** *as the car.*

_____33. Both the truck and the car are moving at the same speed when they collide. Which choice describes the forces?
_____34. The truck is standing still when the car hits it. Which choice describes the forces?

Figure 5. Questions from the Force and Motion Conceptual Evaluation that probe student understanding of Third Law forces for objects which collide. Student results for an average of questions 30-32 and 34 are labeled ***3rd (coll)*** *(where* **coll** *denotes collision) in figures in this paper.*

As with the kinematics questions, most physics professors thought initially that these questions were much too simple for their students and expected that few students would answer in a non-Newtonian way after traditional physics instruction. After seeing typical student responses to these questions where only 5 to 15% of students seem to change their views of dynamics after traditional instruction, some professors suggested that perhaps the questions are not significant (or valid and reliable) measures of student knowledge. Our research does not support this point of view, and discussions of the validity of the *FMCE* can be found in references 1, 18, and 19.

The failure of beginning physics courses to convince students that the Newtonian view of motion makes more sense of the world than their fragmentary, childhood views, has broader implications than the fact that the students do not understand force and motion. It calls into question, particularly for students not intending to become scientists, the validity of the scientific process. If science does not make "sense" for students, then there is no good reason to accept conclusions arrived at through the process of science. Of course some will "accept" science merely on the perceived authority of scientists without any expectation of understanding it. Such a result is clearly not desirable.

There is more widespread agreement on the ineffectiveness of traditional instruction than there is on the solutions to the problems of traditional instruction. For some time, substantial agreements among researchers in physics education on the ways that traditional instruction is not working have been masked by real and apparent disagreement over particular ways of defining physics learning and over disagreements about the appropriate pedagogical response. Such disagreement has too often meant that much work in research and pedagogy goes on as a series of separate efforts, so that projects with the potential to have widespread impact on physics teaching and learning remain isolated. What is needed to change the state of physics education is agreement on a set of underlying principles about the teaching and learning of physics that will support the integration of the work of many different groups into a coherent educational response based on careful research and with the potential to influence the larger physics and science community.

Questions 35-38 refer to a large truck which breaks down out on the road and receives a push back to town by a small compact car.

Pick one of the choices **A** through **J** below which correctly describes the size (magnitude) of the forces between the car and the truck for each of the descriptions (35-38).

A. The force of the car pushing against the truck is equal to that of the truck pushing back against the car.
B. The force of the car pushing against the truck is less than that of the truck pushing back against the car.
C. The force of the car pushing against the truck is greater than that of the truck pushing back against the car.
D. The car's engine is running so it applies a force as it pushes against the truck, but the truck's engine isn't running so it can't push back with a force against the car.
E. Neither the car nor the truck exert any force on each other. The truck is pushed forward simply because it is in the way of the car.
J. None of these descriptions is correct.

_____35. The car is pushing on the truck, but not hard enough to make the truck move.
_____36. The car, still pushing the truck, is **speeding up** to get to cruising speed.
_____37. The car, still pushing the truck, is at cruising speed and continues to travel at the **same speed**.
_____38. The car, still pushing the truck, is at cruising speed when the truck puts on its brakes and causes the car to **slow down**.

Figure 6. Questions from the Force and Motion Conceptual Evaluation that probe student understanding of Third Law forces for objects which experience extended contact. Student results for an average of questions 36 and 38 are labeled ***3rd (con)*** *(where* ***con*** *denotes contact) in figures in this paper.*

At a meeting entitled "The New Mechanics" at Tufts University in August of 1992, a consensus began to emerge concerning the inadequacies of traditional instruction. The physicists who attended the meeting and formed the New Mechanics Advisory Group are all researchers in physics education. They were brought together by Priscilla Laws, David Sokoloff and the author to work toward a number of goals. One purpose of the group's work was to establish general agreement on the effect of traditional instruction on student learning and to identify those methods of teaching physics that have been shown through research to enhance student learning.

During meetings of the New Mechanics Advisory Group, researchers came to agreement on some generalizations about student learning in physics that were originally drafted by Lillian McDermott. Each generalization is supported by research from different sources using different techniques. These sources include, for example, the Physics education group at the University of Washington, [9-13,16] which elicits detailed accounts of student understanding through interviewing; the analysis done by the Center for Science and Mathematics Teaching at Tufts University on responses from thousands of students in introductory physics courses at many different institutions to short-answer questions which are part of the *Force and Motion Conceptual Evaluation* discussed above; and the work done by David Hestenes at Arizona State University in developing benchmark conceptual exams. [14,15] A list of these points of agreement about student learning in physics follows.

- Facility in solving standard quantitative problems is not an adequate criterion for functional understanding.
- A coherent conceptual framework is not typically an outcome of traditional instruction. Rote use of formulas is common.
- Certain conceptual difficulties are not overcome by traditional instruction.
- Growth in reasoning ability does not usually result from traditional instruction.
- Connections among concepts, formal representations (algebraic, diagrammatic, graphical), and the real world are often lacking after traditional instruction.
- Teaching by telling is an ineffective mode of instruction for most students.

What teaching methods work in introductory physics courses?

Each of these generalizations about student learning has strong implications for the changing of physics teaching. It will be difficult for scientists who look at the evidence and who accept these results to find justifications for continuing to teach in a traditional manner. But what is a teacher to do? Even physics education researchers have disagreements about the "best" way to proceed. While most believe that students must be intellectually engaged and actively involved in their learning and that traditional instruction is failing to provide a context in which a majority of students can learn, there is more debate about which methods of teaching and what learning contexts will help students learn most effectively. Can educational technology improve physics learning? Under what conditions does collaborative learning work well? What role should experimentation play in student learning?

While the New Mechanics Meeting did articulate agreements about current student learning, the time was too short to agree upon generalizations for suggested methods of physics teaching; and because of limited resources, participants were only from the United States. To see how to start changing our teaching, it is useful to look at agreements about productive methods of physics teaching reached at an earlier international meeting of physicists and physics education researchers on a more specific topic. A NATO Advanced Study Workshop entitled *STUDENT DEVELOPMENT OF PHYSICS CONCEPTS: THE ROLE OF EDUCATIONAL TECHNOLOGY* was organized by Robert Tinker and the author and held at the University of Pavia in Italy during October of 1989. The participants included researchers from nine different countries and some members of the New Mechanics Advisory Group. The participants were researchers in physics and science education and included major developers of curriculum and pedagogical tools for the teaching of students and teachers. Many of the participants were interested in incorporating recent findings in physics education research and cognitive psychology into new instructional models made possible by the use of interactive technologies.

The NATO workshop was concerned with student conceptual learning and the pedagogical uses of interactive educational technologies in physics teaching and learning. One major focus was on the uses of technologies that allowed students to construct physics concepts successfully from their own experiences of the physical world. Some of the interactive educational technologies demonstrated and discussed at the conference were real-time, computer-based, data-logging tools (often called microcomputer-based laboratory or MBL tools); the use of robotics for teaching science concepts; interactive video disk/CD ROM systems; student-directed software pedagogical tools; telecommunication as a means for students to share scientific discoveries; and constructivist intelligent tutor systems.

After examining the evidence, participants were in substantial agreement that students of all ages learn science better by actively participating in the investigation and the interpretation of physical phenomena; that listening to someone talk about scientific facts and results was not an effective means of developing concepts; and that well-designed pedagogical tools (generally computer-based) that allow students to gather, analyze, visualize, model and communicate data can aid students who are actively working to understand physics. In particular, there was evidence from a number of countries (Italy, Germany, UK, USA, USSR) that real-time Microcomputer-Based Laboratory tools in appropriate learning environments resulted in successful student learning of physics concepts. It was also agreed that to best develop their understanding, students need the freedom and ability to pursue interesting scientific investigations; the opportunity to interact with their fellow students; and the means to communicate their findings. (Unfortunately most introductory courses have none of these features.)

The international NATO Workshop resulted in substantial agreement on ways physics teaching can be changed to improve student learning. These conclusions have subsequently stood the test of time and research. The New Mechanics meeting at Tufts, in addition to building agreement about student physics learning, began the work of refining curricular and instructional strategies that will help introductory physics students acquire a conceptual understanding in one specific area of the curriculum—Newton's Laws of Mechanics. The results of this meeting, including a revision of the dynamics sequence described by Arnold Arons in Chapter 2 of his book A Guide to Introductory Physics Teaching,[8] are being incorporated into a new curricular project, *RealTime Physics: Active Leaning Laboratories in Mechanics* [21] and into major revisions of the *Workshop Physics* [24] and *Tools for Scientific Thinking* [20] curricula. The revision of the dynamics sequence is discussed by Priscilla Laws in these proceedings[26] Extensive research on learning, which is part of these projects, is helping to establish productive instructional techniques and sequences that work for almost all introductory physics students. The next sections of this paper will introduce some of the specific tools and instruction techniques that have been successful. We will focus on MBL tools and on the laboratory active-learning curricula, and the interactive lecture demonstrations that make use of them.

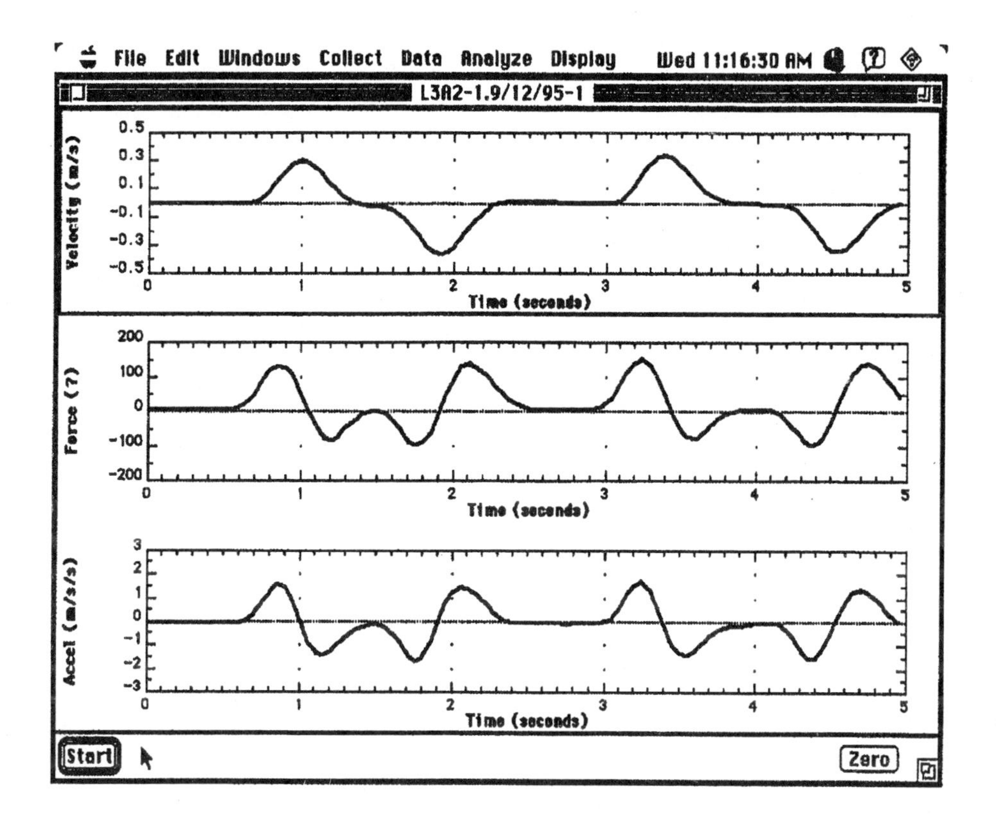

Figure 7. Force probe and motion detector used with MacMotion to measure the force, acceleration, and velocity of a mass being acceleration along a vertical line. The result shows that force is proportional to acceleration not velocity. This is an activity from the RealTime Physics Mechanics curriculum described in the next section.

Tools for Scientific Thinking Student Learning Tools: Microcomputer-Based Laboratory Software and Hardware.

Using the *Tools for Scientific Thinking (TST)* Microcomputer-Based Laboratory (MBL) sensors and software,[6] students can simultaneously measure and graph such physical quantities as position, velocity, acceleration, force, temperature, light intensity, sound pressure, ionizing radiation, current and potential difference. Such MBL tools and carefully designed curricula have been used successfully to teach physics concepts to a wide variety of students in the laboratories of universities and colleges and in the classroom and lecture hall through the *TST* microcomputer-supported *Interactive Lecture Demonstrations (ILDs)* which will be discussed below. Here we will only discuss the motion and force probes and associated software.

We have previously reported on the development and use of microcomputer-based laboratory (MBL) tools and curriculum to teach motion (kinematics) concepts.[4] These materials were developed as part of the *Tools for Scientific Thinking* project[5] at the Center for Science and Mathematics Teaching (CSMT) at Tufts University. Students use an ultrasonic motion detector (interfaced with a Macintosh or MS-DOS computer) and user-friendly software, and follow a specially designed curriculum.[6] Using these tools students are able to observe the motion of any object—including their own bodies—and to display graphs of position, velocity and/or acceleration. Such materials create an active learning environment in the laboratory which has been demonstrated to be highly successful in teaching these concepts in introductory courses at the college and high school levels.[4,7]

The force probe,[6,22] developed at CSMT, when used with the Universal Laboratory Interface (ULI)[6] and appropriate software allows students to apply forces to objects and display their magnitudes on a Macintosh or MS-DOS computer. The probe consists of a Hall effect transducer and a magnet. The magnet is mounted on a flexible diaphragm, which holds it a small distance away from the transducer. When a force is applied to the hook mounted on the opposite side of the diaphragm, the magnet is moved toward or away from the transducer producing a voltage proportional to the force. The force probe can be calibrated by applying a known force to the hook with a spring scale or hanging weight. When the force probe is used in conjunction with the motion detector and *MacMotion* software (the MS-DOS version is called *Motion*)[6] force-time graphs can be displayed along with position-, velocity- and/or acceleration-time. Figure 7 shows the velocity, force, and acceleration-time graphs of a moving mass. Graphs of any of these quantities against any other are also possible, e.g., phase plots of position vs. velocity. The software allows a wide variety of display options, including the ability to compare graphs from two different measurements. It also contains complete graphical fit and statistics packages, and allows data tables to be exported to a spreadsheet if more extensive analysis is desired. Any function of the measured variables may be displayed in real time using the *New Column* feature. The usefulness of this feature is demonstrated in Figure 8 which shows the kinetic, potential, and mechanical energies of a cart rolling up and down an inclined ramp.

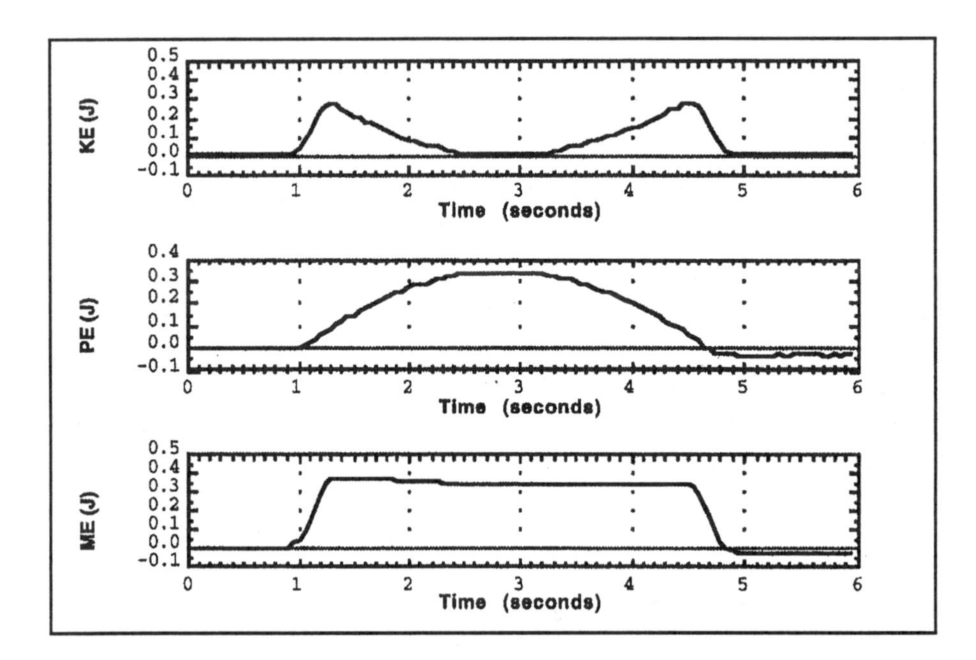

Figure 8. Motion detector used with MacMotion to measure the position and velocity of a low friction Pasco Cart which is given a quick push up an inclined ramp. The New Column feature of the software is used to calculate (in real time) the kinetic energy (KE) from the velocity, the potential energy (PE) from the position, and to add them together to produce the total mechanical energy (ME) as the cart moves up and down the ramp. It is clear that ME is essentially conserved after the initial push and before the catch. This is an activity from the TST Interactive Lecture Demonstration Sequence on Energy. Interactive Lecture Demos are described in Section VI.

The force probe also may be used alone with sampling rates of up to 6000 data points per second. The software has a triggered mode, in which data collection begins only after a specified force is applied to the force probe. Two force probes may be used simultaneously to examine the forces during short-duration collisions between two objects. This mode is used to teach Newton's Third Law. Example data can be seen in paper by Priscilla Laws in these proceedings.[26]

The following characteristics of the MBL tools appear to enhance the ability of students to learn in the contexts we have investigated. (1) The tools allow student-directed exploration but free students from most of the time-consuming drudgery associated with data collection and display. (2) The data are plotted in graphical form *in real time*, so that students get immediate feedback and see the data in an understandable form that can be discussed. (3) Because data are quickly taken and displayed, students can easily examine the consequences of a large number of changes in experimental conditions during a short period of time. The students spend a large portion of their laboratory time observing physical phenomena and interpreting, discussing, and analyzing data with their peers. (4) The hardware and software tools are, in general, independent of the experiments. The variety of probes use the same interface box and the same software format. Students are able to focus on the investigation of many different physical phenomena without spending a large amount of time learning to use complicated tools. (5) The tools dictate neither the phenomena to be investigated, the steps of the investigation, nor the level or sophistication of the curriculum. Thus a wide range of students from elementary school to university level are able to use this same set of tools to investigate the physical world.

The Tools for Scientific Thinking Motion and Force and the Real Time Physics Mechanics Laboratory Curricula

The discovery-based curricula developed at CSMT for college and secondary schools allow students to take an active role in their learning and encourage them to construct physical knowledge for themselves from actual observations. They make substantial use of the results of educational research.[8-19] They use a guided discovery approach and are intended for student groups of two to four. They support the peer learning that is possible when data are immediately presented in an understandable form. They also use predictions to engage students and to provide a vehicle for discussion. In addition, they pay attention to student alternative understandings that have been documented in the research literature.

The *Tools for Scientific Thinking (TST) Motion and Force Curriculum* was developed at CSMT in 1988-90, and revised in 1992.[20] It is divided into five labs. The first two labs, *Introduction to Motion* and *Introduction to Motion—Changing Motion*, dealing with kinematics concepts, have been described previously.[4] In these labs, students explore the concepts of kinematics with constant and uniformly accelerated velocity by using the motion detector to examine the motion of their own bodies and of a dynamics cart or toy car. The next two labs, *Passive Forces* and *Force and Motion* will be summarized below. The fifth lab, Periodic *Motion: Simple Harmonic Motion,* will not be described here.

Student descriptions of physical phenomena which physicists describe using Newton's laws of motion are extremely persistent and usually far from Newtonian.[13-19] The curriculum is designed to offer students experiences that change their views of dynamics. The availability of a microcomputer-based force probe, coupled with the motion detector, provides unprecedented ability to simultaneously measure and display the force applied to an object and its motion (position, velocity and/or acceleration) in real time. The *TST Passive Forces* laboratory includes activities designed to explore the means of measuring forces, calibration of the force probe, and the concepts of tension, normal, and frictional forces.

The *Force and Motion* lab leads students to discover Newton's First and Second laws. Students use the force probe and motion detector to discover the forces needed to produce different accelerations. As in all of our curricular pieces, we explore more familiar motions with friction, as well as the nearly frictionless cases. We use a cart which has been modified with an adjustable friction pad on the bottom,[23] so that motion can be examined with different amounts of friction.

The *Tools for Scientific Thinking* curriculum has been designed to be incorporated into introductory physics courses found at most colleges and universities by replacing a subset of the traditional labs. In such a setting, laboratory sections are often taught by teaching assistants with varying pedagogical skills, and lecturers often pay little attention to the laboratory. In place of classroom discussions—which under the best of circumstances would be used to consolidate the concepts learned in laboratory—each of these laboratories is accompanied by a homework assignment which the students complete after the laboratory session. They are asked to draw and interpret a number of graphs similar to and different from the ones they have produced in the laboratory. This curriculum has also been used successfully at many high schools, where discussions of the lab observations are much more easily incorporated into teaching.

The *TST Force and Motion* curriculum was designed to improve the conceptual understanding of mechanics, as quickly as possible, by substituting a small number of active learning laboratories, supported by MBL tools, for traditional laboratories. The *RealTime Physics (RTP) Mechanics* curriculum, released in 1994,[21] has the more ambitious goal of completely replacing the entire introductory laboratory with a sequenced, coherent set of active learning laboratories. The origins of *RTP Mechanics* lie in *TST Motion and Force* and *Workshop Physics (WP).*[24] In *WP* separate laboratories and lectures have been completely replaced by hands-on learning and mathematical synthesis in a "workshop" environment. Like *TST*, the *RTP* curricula are designed to infiltrate traditional introductory physics courses by restructuring the laboratory portion of the course. Like *WP*, the *RTP* curricula are designed to develop both the conceptual understanding and quantitative skills of students by the use of guided-discovery learning in a laboratory environment.

RTP Mechanics includes 12 laboratories covering kinematics, dynamics (Newton's three laws of motion), gravitational forces, passive forces, impulse and momentum, energy and projectile motion. Each lab has a core designed for weekly two-hour laboratory sessions, with extensions to provide more in-depth coverage if more time is available. A further description of the *RTP* labs can be found in these proceedings[26] and in reference 1. Several years of classroom experience with *TST* and *WP*, careful study of the difficulties of teaching mechanics and the availability of quality MBL tools have led us to adopt a modified sequence for teaching mechanics in *RTP* called the *New Mechanics Sequence.*[26] The major changes from the traditional approach to mechanics include 1) teaching all of one-dimensional kinematics and dynamics before considering any two-dimensional situations, 2) teaching Newton's Second Law before the First Law, 3) teaching momentum before energy and 4) teaching Newton's Third Law last through interaction forces between objects, e.g., in collisions and by pushing one object with another. More details on the origin, rationale and features of the *New Mechanics Sequence* are given elsewhere.[26,27]

Tools for Scientific Thinking Microcomputer-Based Interactive Lecture Demonstrations (ILDs).

Despite considerable evidence that traditional approaches are ineffective in teaching physics concepts, most physics students in this country continue to be taught in lectures, sometimes in smaller classes but often in large lectures with more than 100 students. Also, while effective MBL laboratory curricula like those discussed above (*Tools for Scientific Thinking (TST)* [20] and *RealTime Physics (RTP)*[21] have been developed, many high school and college physics programs have only a few computers and are unable to support hands-on laboratory work for large numbers of students. Over the past six years David Sokoloff and the author have worked at creating successful active learning environments (like those associated with our laboratory curricula) in large (or small) lecture classes primarily at the University of Oregon and at Tufts University. The result of these investigations has been the development of a teaching and learning strategy called *Tools for Scientific Thinking Microcomputer-Based Interactive Lecture Demonstrations (ILDs).* [2,3,30]

After several years of research, we formalized in 1991 a procedure for *ILDs* which engages students in the learning process and, therefore, converts the usually passive lecture environment to a more active one. The steps of the procedure are:

1. The instructor describes the demonstration and does it for the class without MBL measurements.
2. The students record their individual predictions on a Prediction Sheet, which will be collected, and which can be identified by each student's name written at the top. (The students are assured that these predictions will not be graded, although some course credit is usually awarded for attendance at these *ILD* sessions.)
3. The students engage in small group discussions with their one or two nearest neighbors.
4. The students record their final predictions on the Prediction Sheet.
5. The instructor carries out the demonstration with MBL measurements displayed on a suitable display (multiple monitors, LCD, or computer projector).
6. A few students describe the results and discuss them in the context of the demonstration. Students fill out a Results Sheet, identical to the Prediction Sheet, which they may take with them for further study.
7. The instructor discusses analogous physical situation(s) with different "surface" features. (That is, different physical situation(s) based on the same concept(s).)

The instructor must use her/his judgment in controlling the amount of time devoted to steps (2) and (3). The small group discussions in a large lecture class are initially quite animated and "on task." After awhile, the discussions may begin to stray into extraneous matters, at which point it is time to move on to steps (4) and (5). The instructor must also have a definite "agenda" for steps (6) and (7), and must often use guidance to move the discussion towards the important points raised by the individual *ILDs*.
The *ILDs* are a series of short, simple experiments. We have used two basic guidelines in designing experiments for *ILD* sequences: 1) the order and content of the sequences are based on the results of research in physics learning, and on our experience with hands-on guided discovery laboratories and 2) the *ILDs* must be presented in a manner such that students understand the experiments and "trust" the apparatus and measurement devices used. Many traditional exciting and flashy demonstrations are too complex to be effective learning experiences for students in the introductory class. Our experiences in developing laboratory curricula and evaluating the learning results[1,3,7,18,19] have been invaluable in selecting simple but fundamental lecture demonstrations. For example, in kinematics and dynamics we start with the most basic demonstrations to convince the students that the motion detector measures motion and the force probe measures force in an understandable way, and to begin to solidify student understanding, before moving on to more complex and concept-rich demonstrations.

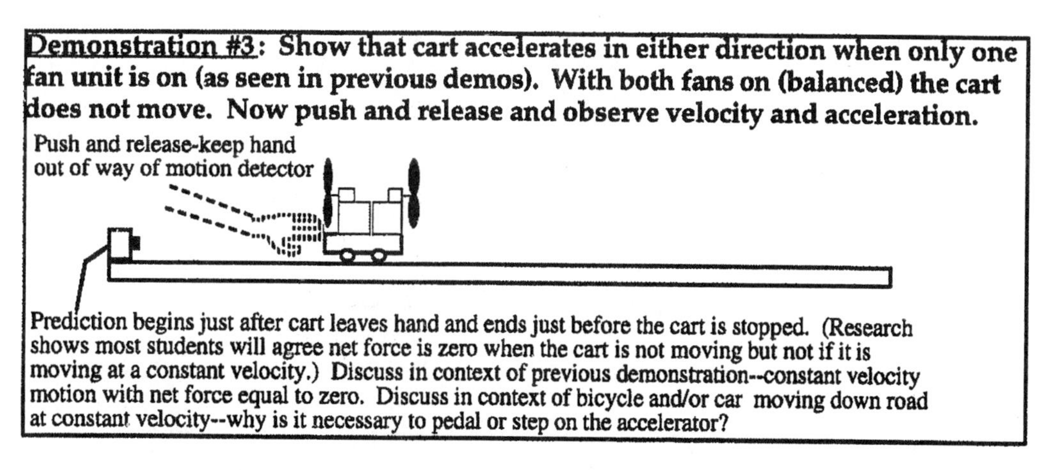

Figure 9. Teacher summary for third demonstration in the Newton's First and Second Laws TST Interactive Lecture Demonstration sequence.

We have developed four forty-minute mechanics sequences[30] of *ILDs* to enhance the learning of kinematics and dynamics including Newton's Three Laws: 1) *Human Motion:* introductory, constant velocity kinematics using the motion detector to explore walking motions, 2) *Motion with Carts:* kinematics of uniformly accelerated motion using the motion detector to observe the motions of a low-friction cart[23] pushed along by a fan unit,[28] 3) *Newton's First and Second Laws:* dynamics using the force probe and motion detector to measure forces applied to a low and high friction cart and the resulting velocity and acceleration, and 4) *Newton's Third Law:* using two force probes to allow students to examine the interaction forces between two objects during collisions and when one object is pushing or pulling another.[3] One experiment in the dynamics sequence is shown in Figure 9.

What is a physics professor to do? What learning results can be expected?

This section will discuss the effect on student learning of instituting various pieces of the computer-supported lab curricula and/or the *Interactive Lecture Demonstrations* into the introductory physics course. The same averages of student responses to questions from the *Force and Motion Conceptual Evaluation (FMCE)* described in Section II (text and Figures 2-6) will be used to evaluate student knowledge about physical processes described by Newton's laws. Careful evaluation of student learning gains illustrate the possibility of large learning gains in introductory courses through the use of new methods developed through educational research.

Because the *Tools for Scientific Thinking* guided-inquiry conceptual labs described above are based on research on what students know; make use of real-time data collection and display (MBL); and encourage students to work collaboratively, even a few hours of laboratory can lead to significant results. By introducing such labs into otherwise conventional introductory courses, it is possible to change student beliefs about the physical world to those held by physicists. We saw in Figure 1 that traditional instruction was not teaching kinematics concepts well. After good traditional instruction, only 30% of over 1200 students in calculus-based physics courses at five different universities were able to answer simple questions on fundamental acceleration concepts. When, for the first time, two *Tools for Scientific Thinking* active-learning kinematics labs were offered at these universities, more than 75% of students understood these concepts. At universities where there is more experience with the labs, such as the University of Oregon or Tufts University more than 90% of students (even in non-calculus introductory courses) understand these kinematics concepts.

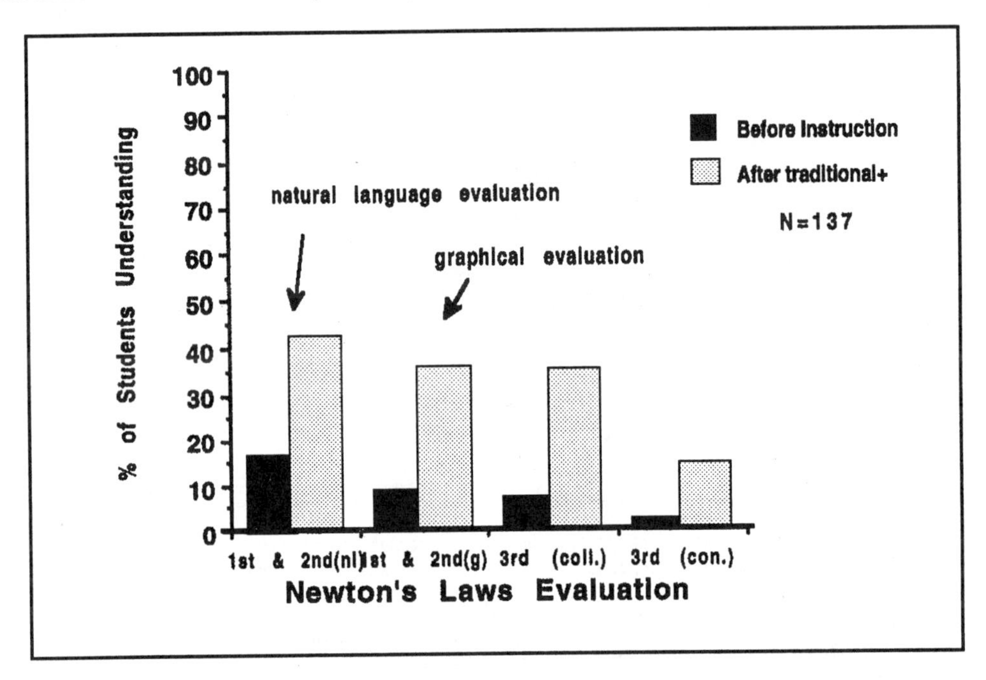

Figure 10. Student understanding of Newton's Laws before and after slightly enhanced traditional instruction. Percent of Tufts University non-calculus introductory physics students who understood Newton's Laws before and after enhanced traditional instruction in the fall of 1994. Instruction was traditional except for two TST MBL kinematics labs. Students enhanced knowledge of kinematics improved their learning of dynamics taught using traditional methods. The measure of "understanding" used was the average of correct answers to particular groups of questions on the Force and Motion Conceptual Evaluation. which are explained in Section II.

What is even more gratifying is that increasing student knowledge of kinematics using new methods, results in more students learning dynamics concepts even when traditional instruction is used. Early in this paper, we documented a 7% improvement in student understanding of dynamics after all traditional instruction. Figure 10 shows that 20 to 25% of Tufts students learned dynamics (for a total of 35 to 40% of students including those who knew it coming in) from traditional methods after experiencing only two *TST* kinematics labs. Much better results can be had if students experience all four of the *TST Motion and Force* labs or the *RTP Mechanics* labs. About 80% of the students understand Newton's Laws using the questions we have chosen as benchmarks. Figure 11 shows the results of students in the non-calculus introductory course experiencing the *RTP* labs at Oregon. Students in calculus-based physics classes at Oregon or Tufts do about 5% better.

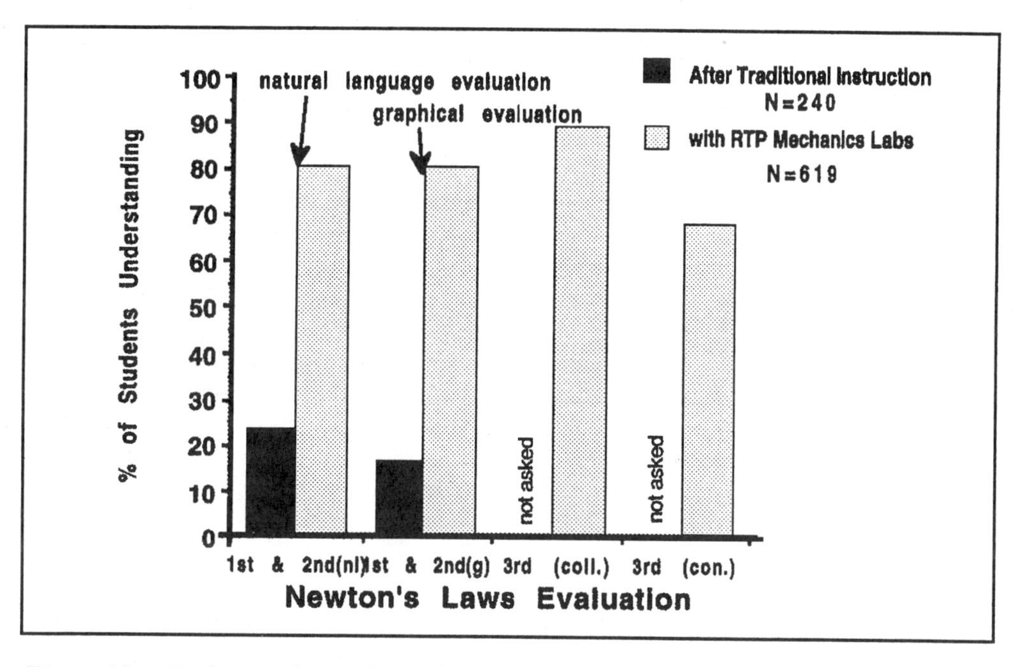

Figure 11. - Student understanding of Newton's Laws after RealTime Physics Mechanics Labs. Percent of University of Oregon non-calculus introductory physics students who understood Newton's Laws after doing RealTime Physics Mechanics laboratories in place of traditional laboratories during 1992-4. Striped bars show the results after good traditional instruction as described in Figure 2 above. The measure of "understanding" used was the average of correct answers to particular groups of questions on the Force and Motion Conceptual Evaluation. which are explained in Section II.

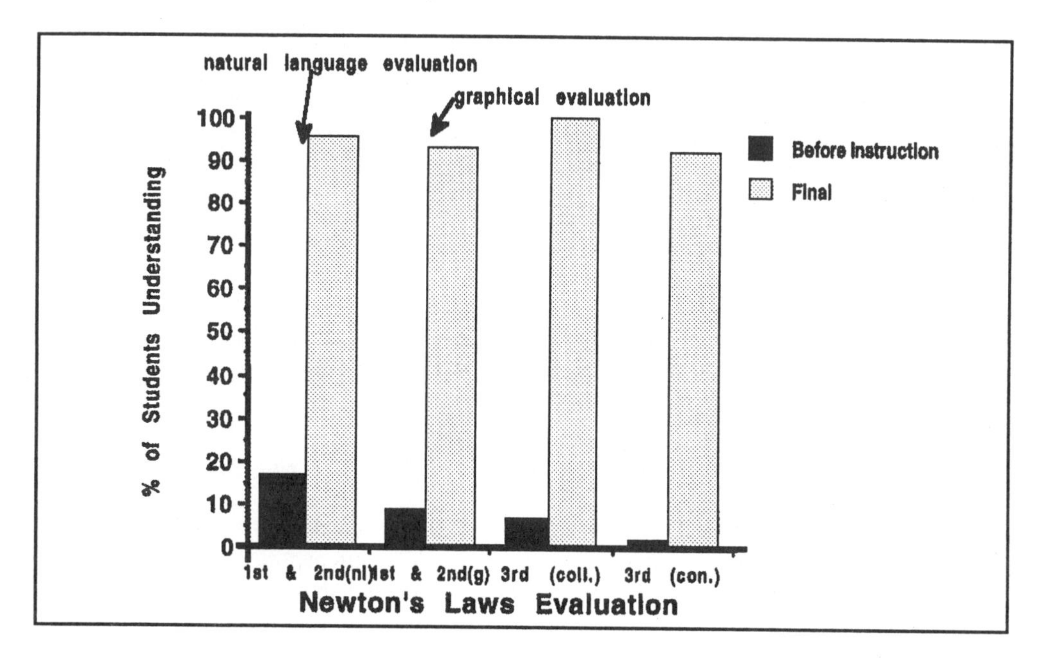

Figure 12. Student understanding of Newton's Laws before and after all instruction. Percent of Tufts University non-calculus introductory physics students who understood Newton's Laws before and after all enhanced traditional instruction in the fall of 1994. Instruction was traditional except for two TST MBL <u>kinematics</u> labs and three 40-minute sequences of Interactive Lecture Demonstrations. The measure of "understanding" used was the average of correct answers to particular groups of questions on the Force and Motion Conceptual Evaluation. which are explained in Section II.

We have had similar success in changing the large lecture environment into an interactive environment where students can learn force and motion concepts using the *Tools for Scientific Thinking computer*-supported *Interactive Lecture Demonstrations.* (*ILDs*) At Oregon in 1991, students in the non-calculus class, who experienced only two forty-minute ILD sequences on kinematics and dynamics, averaged 70% on the 1st and 2nd Law questions compared to approximately 20% from only traditional instruction in the previous two years. Tufts students in the non-calculus class in 1994 experienced three 40 minute ILD sequences (one each on kinematics, 1st and 2nd law, and the 3rd law). They also did the two *TST* kinematics labs. Approximately 90% of these students understand Newton's Laws by the measures we have been using. Figure 12 shows these results. These data demonstrate dramatic changes in student conceptual understandings of Newton's laws when these laboratory curricula or Interactive Lecture Demonstrations are used. There is good evidence that this conceptual understanding is retained. Such limited implementation of new methods is not enough, but begins to address the problem of changing instruction in traditional environments. Similar positive results are achieved in the more comprehensive *Workshop Physics* program[24] at Dickinson College which has replaced lectures with a combination of student-oriented activities using similar active learning-techniques and the educational technology described above.

Conclusions

In summary, there is considerable evidence collected by researchers in physics teaching and learning that traditional instructional methods, largely lecture and problem solving, are not generally effective methods for promoting student conceptual learning in physics. There is also widespread, but not total, acceptance by researchers of evidence that interactive-learning methods, some of which are mentioned above, work well in many different environments. There is enough agreement among careful researchers that the physics community would do well to begin changing traditional teaching methods dramatically for the wide variety of students that take physics and begin examining in a scientific way the learning results of these changes.

Acknowledgments

Parts of this paper were adapted from "Why Don't Students Learn Physics?,"[25] and "Using Large-Scale Classroom Research to Study Student Conceptual Learning in Mechanics."[19] from the proceedings of the NATO Advanced Research Workshop on Microcomputer-Based Laboratories, Amsterdam, November, 1992. I am grateful for support from the National Science Foundation and other agencies listed in reference 4. I would particularly like to thank David Sokoloff who participated fully in producing the research results quoted here and who is an author of the curricula described and Priscilla Laws for her work on RealTime Physics and Workshop Physics.

References

1. R. K. Thornton and D. R. Sokoloff, "Assessing and improving student learning of Newton's Laws part I: the Force and Motion Conceptual Evaluation and active learning laboratory curricula for the First and Second Laws," submitted to the Am. J. Phys. 1996.
2. D. R. Sokoloff and R. K. Thornton, "Assessing and improving student learning of Newton's Laws part II: Microcomputer-based Interactive Lecture Demonstrations for the First and Second Laws," submitted to the Am. J. Phys. 1996.
3. R. K. Thornton and D. R. Sokoloff, "Assessing and improving student learning of Newton's Laws part III: active learning of the Third Law in lecture and laboratory," in preparation.
4. R. K. Thornton and D. R. Sokoloff, "Learning motion concepts using real-time, microcomputer-based laboratory tools," Am. J. Phys. **58**, 858-867 (1990).
5. This work was supported in part by the National Science Foundation under grant number USE-9150589, *"Student Oriented Science,"* grant number USE-9153725, *"The Workshop Physics Laboratory Featuring Tools for Scientific Thinking."* and grant number TPE-8751481, "*Tools for Scientific Thinking: MBL for Teaching Science Teachers,"* and by the Fund for Improvement of Post-secondary Education (FIPSE) of the U.S. Department of Education under grant number G008642149, *"Tools for Scientific Thinking,"* and number P116B90692, "*Interactive Physics*".
6. The MBL probes, interface, software and curricula discussed here are available from Vernier Software, 8565 Beaverton-Hillsdale Highway, Portland, OR 97225-2429.

7. R. K. Thornton, "Changing the physics teaching laboratory: using technology and new approaches to learning to create an experiential environment for learning physics concepts," *Proceedings of the Europhysics Study Conference, The Role of Experiment in Physics Education*, Seta Oblak, Nada Razpet, ed. (Ljubljana, Slovenia, 1993).
8. A. B. Arons, *A Guide to Introductory Physics Teaching*, (Wiley, New York, 1990).
9. L.C. McDermott, "Research on conceptual understanding in mechanics," Phys. Today **37**, 24-32 (July, 1984)
10. L.C. McDermott, M.L. Rosenquist, and E.H. van Zee, "Student difficulties in connecting graphs and physics: Examples from kinematics," Am. J. Phys. **55**, 503-513 (1987).
11. D.E. Trowbridge and L.C. McDermott, "Investigation of student understanding of the concept of velocity in one dimension," Am. J. Phys. **48**, 1020-1028 (1980) and "Investigation of student understanding of the concept of acceleration in one dimension," Am. J. Phys. **49**, 242-253 (1981).
12. L. C. McDermott, "Guest comment: How we teach and how students learn—a mismatch?," Am. J. Phys. **61**, 295-298 (1993).
13. L. C. McDermott, "Millikan lecture 1990: What we teach and what is learned—closing the gap," Am. J. Phys **59**, 301-315 (1991).
14. J. A. Halloun and D. Hestenes, "The initial knowledge state of college physics students," Am. J. Phys. **53**, 1043-1056 (1985).
15. D. Hestenes, M. Wells and G. Schwackhammer, "Force Concept Inventory," The Physics Teacher **30**:3, 141-158 (1992).
16. L. C. McDermott, P. Shaffer and M. D. Somers, "Research as a guide for teaching introductory mechanics: an illustration in the context of the Atwood's machine , " Am. J. Phys. **62**, 46-55 (1994).
17. M. Wells, D. Hestenes and G. Schwackhammer, "A modeling method for high school physics instruction," Am. J. Phys **63**, 606-619 (1995).
18. R. K. Thornton, "Conceptual dynamics: changing student views of force and motion," chapter in *Thinking Physics for Teaching* , C. Tarsitani, C. Bernardini, and M. Vincentini, ed., (London, Plenum Publishing, 1995) in press.
19. R. K. Thornton, "Using large-scale classroom research to study student conceptual learning in mechanics and to develop new approaches to learning," chapter in *Proceedings of the NATO Advanced Research Workshop—Microcomputer-Based Laboratories, Amsterdam, November 9-13, 1992, NATO ASI Series* (Berlin-Heidelberg-New York, Springer Verlag, 1996) in press.
20. Ronald K. Thornton and David R. Sokoloff, *Tools for Scientific Thinking—Motion and Force Laboratory Curriculum and Teachers' Guide, Second edition*, (Portland, Vernier Software, 1992).
21. David R. Sokoloff, Priscilla W. Laws and Ronald K. Thornton, *RealTime Physics*: *Mechanics V. 1.40* (Portland, Oregon, Vernier Software, 1994).
22. The Student Force Probe, which is also available from Vernier Software, may also be used with the ULI, *TST* software and *TST* and *RTP* curricula.
23. PASCO Scientific, P.O. Box 619011, 10101 Foothills Blvd., Roseville, CA 95678-9011.
24. P. W. Laws, "Calculus-based physics without lectures," Physics Today **44**:12, 24-31 (December, 1991).
25. Ronald K. Thornton, "Why don't students learn physics?," *Physics News in 1992*, American Institute of Physics , 48-50 (1992).
26. P. W. Laws, "A new order for mechanics," *Proceedings of the Conference on the Introductory Physics Course, Rennselear Polytechnic Institute, Troy New York, May 20-23, 1993* (Wiley, New York, 1995).
27. A paper on the New Mechanics Sequence is in preparation.
28. See R. Morse, "Constant acceleration experiments with a fan-driven dynamics cart," The Physics Teacher **31**, 336-338 (1993).
29. Copies of the latest version of the *Force and Motion Conceptual Evaluation*, are available from the Center for Science and Mathematics Teaching, Tufts University, Sci-Tech Building, 4 Colby St., Medford, MA 02155.
30. Copies of the complete Microcomputer-Based *Interactive Lecture Demonstration* package including teacher instructions and student Prediction and Results sheets are available from the Center for Science and Mathematics Teaching, Tufts University, Sci-Tech Building, 4 Colby St., Medford, MA 02155.

Reforming and Constructing University Physics in China

Yun Ying
Southeast University, Nanjing, PRC

Introduction

In China, the number of institutions of higher learning is above one thousand. In the past, they have always been divided into three categories: so-called "comprehensive university", institute of technology, and normal college. But this division is becoming less and less distinct as times change. Now, some universities have the disciplines of engineering and medicine in addition to science. In contrast, colleges of engineering may have disciplines of liberal arts or economics.

There are 36 key universities directly affiliated with the State Education Committee of China, some are Qinghua University, Peking University, Fudan University, Southeast University where I have my teaching post. These universities are first-class in China and they cover the fields of science, technology, agriculture, and medicine, respectively. In order to promote the reforming and constructing of courses, such as mathematics, physics, radio engineering, etc., the committee is composed of about 15 professors from the key universities.

Take the course of "Fundamental Physics" as an example. There are two such committees: one is for students majoring in physics and the other is for the students in the engineering department. The former is two-year course and the committee has several subordinate groups which are in charge of the teaching of mechanics, modern physics, etc.. The class hours for the course is about 300. The course for the engineering students is only for a one-year period and the class hours are about 200 in general: 140 hours for lecture and 60 for lab. The committees always lay down a "Basic Requirement" respectively so that physics teachers can have a blue-print when they are teaching. Although this is not a rigid rule, the responsible institutions often take it as a criteria of measurement. This method has advantages and disadvantages. It can refrain from divergement in physics teaching, but it also can restrict teacher's creativity and imagination.

Let me give some examples in the reforming of the Introductory Physics course.

Written and Interposed Video-tapes Unified University Physics

A. Five Parts of the New Mode 'University Physics'

In 1987, under the concern of Jiangsu Education Commission, the research project "Application of Modernized Teaching Method in 'University Physics'" was carried out by 16 physics teachers from 7 universities (Southeast University etc.) and the staffs of the Audio Visual Education Centers. We developed a new teaching mode which unified the modern technology with the traditional one, we compiled a new mode of textbooks for "University Physics" which includes:

(1) Written teaching materials — textbooks

The principles for selecting the contents of the written teaching materials are to broaden the knowledge of students after High School physics, to help the students really learn something new, and compress suitably classical physics in order that we can pay more attention to modern physics and contemporary physics and the relationship between physics and developing technology.

The difference between this and the conventional one is that the contents of the Interposed Videos have been written into the new book as part of the teaching contents. In the beginning of each chapter there is a catalogue and the major contents of the combined Interposed Videos and the last part of each chapter is the question for discussion. The beautiful and practical pictures in the interposed video-tapes were also used as illustrations in the book. Two pictures of "Precession" are shown in the Fig. 1

(2) Interposed Video-tapes

The characteristics of the Interposed Videos are: (a) closely combined with the class instructions, forming a series set; and (b) each video-tape is very short, about 3-6 minutes, but the teaching aim of the tape is very clear. The whole book is fitted with 70 relevant short and clear interposed video-tapes totaling 320 minutes. The contents include the basic physics concepts, important laws and focal points such as the subtract three dimensional images, microscopic mechanisms and applications.

(3) Computer Software
(4) Slides and Transparency
(5) The Supplementary Teachers' Manual and Students' Study Guide.

We compiled the teachers' manual and students' study guide to make it convenient to teach and to make the tapes more effective and efficient.

In our 6 years of teaching practices, this new mode has proven to be a successful one. It received an award from The State Education Committee of China in 1990.

B. Organical Combination between the books and videos

One of the important characteristics of this new teaching mode is that the short, vivid videotapes are used as a very important teaching aid in the classroom. It is merged organically with class teaching, not a substitution. It makes better use of the superiority of the video and supplements the ordinary textbook, in which the individual factor of the students is always omitted. It organically combines the modern education media with the traditional one and optimizes teaching construction, therefore the teaching effect is enhanced.

(1) The contents of the Interposed Videos is written into the textbook. In the beginning of each chapter there is a catalogue and the major contents of the videos, e.g.

Chapter 4 Mechanical Vibration

4-1	Rotating Vector Diagram for SHM and phase	How to describe a simple harmonic oscillation and phase by using the rotating vector diagram is intro duced.
4-2	Damped Vibration	Three kinds of damped vibration are introduced: damped vibration and over damped vibrations.
4-3	Forced vibration and resonance	Appearances of forced oscillation and resonance are introduced. The relation between the ampli tude of a resonance and the damping coefficient, and instances of resonance are analyzed.
4-4	Beats	Beats phenomenon is shown by computers.
4-5	Superposition of Vertical SHMs	Two vertical vibrations of SHMs are described by computers, so is Lissajou's Figures

(2) The Interposed Video is part of the class instruction, and it plays an active role in class teaching. With these succinct tapes, the teacher can clearly explain a physical phenomenon, a physical law, a special principle, or some other application in technology.

(3) The role of the Interposed Videos in class is very important to the teachers, they must pay attention to using it and unifie it with the instruction, explaining and helping students to understand. To show the video from the beginning to the end at one time is not a good teaching format, however to stress a certain point the teacher should stop the tape at appropriate times and give an explanation being made on the video. By using this kind of teaching method, we have achieved the following effective results:

(a) Showing vivid physical phenomena, deepening students' understanding of concepts.
(b) Deepening the image information, and making the connection between image thinking and abstract thinking,
(c) Remedying some effects of the demonstration experiments, and
(d) Stimulating students individual interests.

C. Feedback and Effects

This new mode of "University Physics" is now used in 23 universities in 14 cities.

Our 7 universities are:

Southeast University
Air Force College of Meteorology
Nanjing Institute of Communication
Nanjing Institute of Aviation
Nanjing Institute of Chemistry
River and Sea University
Yanzhou Institute of Technology

The other 16 universities:

Wuhen University of Mapping Technology
Northeast Institute of Technology
Mining University of China
Institute of Air Force Service
Changzhou Normal Institute of Technology
Zhengzhou Institute of Light Technology
Nanjing Institute of Engineering
Nanchang Institute of Ariation
Nanjing University of Forest Industry
Jiangnan University
Suzhou Institute of Silk Technology
Meteorology Institute of China
South China University of Science and Technology
Zhejiang Institute of Silk Technology
Daqing Institute of Oil Industry
Changsha Institute of Railway

Feedback information from students: In our six years using this teaching method, it has been proven to be a successful reform in teaching, course content and in improving teaching quality. According to the feedback information received from students from 1987, 1988, and 1991 classes, most students say that this new teaching method is vivid, deepens impressions, attracts interests, strengthens understanding, broadens knowledge, and broadens their ways of thinking.

This work is also approved by the State Education Committee of China. We have finished these teaching materials in the first round in 1990, and reformed it in 1992 and 1993. It will be published by the Higher Education Press in Beijing in 1994. We have been getting support from many experts and universities in China and abroad.

"New concept Physics" by Kaihua Zhao

"New concept Mechanics" by Kaihua Zhao, Weiying Luo

A. Some View Points

(1) To examine the conventional teaching material with modern a viewpoint
(2) Should the wording of the basic concept of conventional physics be revised?
(3) Has the related positions of each conventional physics law changed?

Original: The Newton three law is the kernel, Mass and Force — basic concept.
Now: (a) The three conservation laws of Momentum, Energy and Angular Momentum is the kernel.
(b) To link up Modern Physics with the viewpoint, and improve some of the wording. Introduce the concept of Mass from the Law of Momentum Conservation. The exchange of momentum among bodies — Force.

B. Window and Interface

(1) Relating he modern and frontier fields.
(2) To appropriately open a 'window', encourage the students to look outside and broaden their horizon. Such as the conservation of angular momentum — the Galaxy is flat; the escape velocity — Black Hole.
(3) Interface, some part of Modern Physics is the extension of the connotation of General Physics — say, Chaos, Nonlinear effects, Fourier analysis.

C. Foster the students' ability of intuition

By intuition reach a quantitative of semiquantitative conclusion.
For example: introduce the quantity estimation and dimensional analysis earlier.
Reform teaching methods: foster the students' ability for independent study and creation.

Reforming of Laboratory Work

A. Physical Experiments for Science Department

We take the Modern Physics Experiment in Peking University as an example.

(1) Modern Physics experiments are an important fundamental course for each specialty in the field of Physics.
(2) Make comprehensive renovation, offer nearly 50 new teaching experiments, provide chief source for national modern physics experiment courses outline.
(3) According to the characteristics of Physics courses in our country and consulting the experience of the famous universities of other countries, we have completely renovated the original courses. Up to now, our courses have offered 31 fundamental and 18 selective experiments in the fields of atomic physics, nuclear physics, optics, vacuum technology, X-ray and electron diffraction, magnetic resonance, microwave, low temperature physics, semiconductor physics and the theory of relativity.
(4) When deciding courses, we not only choose the experiments that are famous in the history of modern physics to help students to understand the importance of physical experiments, but also pay attention to the

achievements of scientific research, which forms 1/4 of our experiment courses. For instance, after the declaration of high temperature super conductivity in 1987, we offered the experiment of 'Hight Temperature Superconductor Test' in 1989.

(5) Based on current conditions in China, we have built up teaching laboratories to advanced world level, and developed many excellent experimental instruments.
(6) Many of them were made by ourselves. In order to save money, we make full use of the equipment that was eliminated in scientific research but is still usable in teaching. And we have successfully offered many Nobel Prize winning modern experiments with Chinese-made equipment.
(7) We have published <<Modern Physics Experiment>>, the first of its kind in China.

Based on our many years' teaching practice, we published <<Modern Physics Experiment>> in 1986, edited by Sicheng Wu, Zhusuan Wang.

B. Engineering College Physical Experiments

(1) The State Education Committee has organized professors to work out the "Basic Requirement" for teaching. As a prerequisite of satisfying this request, each school offers various kinds of experiments to meet their specific needs. For colleges of engineering, we arranged 20 experiments in a period of 60 hours.
(2) The general trend of reforming is to actively combine each experiment, and gradually make the individual ones constitute a general system.
(3) Another trend is to increase the synthetical experiments and those that provide the students opportunities for designing, and open the laboratories to students.

CAI in Quantum Physics

Quantum Physics is one important basis of Modern Physics. Since this course studies the microcosmos, no direct experimental observation can be made. There are no demonstrative experiments for this course, therefore most students feel it is hard to understand.

To solve this problem, Mr. Luo Jin of Chongqing Teachers College uses computers in their teaching. They have developed <<Quantum Physics CAI System>>, which achieved great effects in the problem solving and renovation of teaching methods.

A. The design guiding ideology of the system:

(1) Closely connect the teaching practice, lay emphasis on focal points, be scientific, interesting, effective and applicable.
(2) Use structured and blocked design, make full use of simulating and dynamic graphing.
(3) Based on the concept of planned step-by-step progress, to make the system flexible, adaptable, and easy to handle.

B. The teaching contents of software

In order to be systematic, integrated and focused on the difficult points, the contents of software includes the basic experiments of quantum theory, fundamental concepts and theory, and other related fields. For example: Rutherford scattering, Bohr Atomic model, Diffraction of electron, Electron cloud of hydrogen atom, Spectral quantum characteristics of atom etc.

C. Periodic achievements of practicing effects

This system has been used in the atomic physics course for the students of grade 87, 88, 89, and 90 in the Chongqing Teachers College. Most of the students and teachers think that the system has the advantage of being cognizant; well arranged contents, excellent screen effects and easy to use, helps students understand the physical concepts and related laws and also increases their interests.

Conclusion

I would like to say that Qinghua University and Zhe-jang University have done a lot of work on demonstrations, and students are very interested in it.

Thirteen universities have graduate students in the field of Physics Education where they are trained to be teachers in high school and university. We do hope they will be excellent physics teachers.

Curriculum Issues: The German Perspective

Günther Kurz, Fachhochschule für Technik Esslingen, Germany

Introduction

When John Rigden[1] [1986] discussed the impact and influence of physics on the cultural life in the United States some eight years ago at the Communicating Physics Conference at the Comprehensive University of Duisburg he also addressed the role of physics in European/German culture: *"Let me mention the Michelin Tourist Guide to Germany. The authors of this guide provide the visitor to Germany pages of information about the artists of Germany, the writers and literacy events of Germany, the musicians and music of Germany, and the legends and folklore of Germany. But they give no hint that, in both the past as well as the present, scientists and science are also a vital part of German culture".* Nothing has changed and a mistaken quotation of a Latin proverb is still regarded to be a sign of lack of education whereas illiteracy in physics and technology is excusable.

In Germany physics seems to be under pressure on the secondary and the tertiary level. But the major problems in education lie beyond the field of physics. I intend to address the problems in the German educational system with respect to the tension between the traditional system and its changes, the demands of the job market, the developments in industry and the feedback on the curriculum, the influence of the changes due to the reunification of Germany, the future of education within the European Community with the implications of the Maastricht Treaty and the impact of the greater Europe due to the political changes in Central Eastern Europe. The EC's memorandum on Higher Education was released in September 1991, it states: *"The success of the international market depends on having people who have the capability to operate across national and cultural boundaries. A European dimension in higher education is perceived as a practical economic necessity apart from its desirability on cultural and political grounds".* All these developments and changes have an impact on the educational system in Germany and Europe and therefore on the curriculum.

The German Educational System — An Overview

The German educational system is a public one with only minor exceptions. Education is under the auspices of the Länder (States) and the Federal Government has only guideline responsibilities. On the secondary level the traditional Gymnasium used to emphasize either ancient languages, modern languages or mathematics & natural sciences and awarded after 13 years of general schooling the Abitur certificate as the general admission ticket to higher education. The universities were devoted to the unity of research and teaching according to Humboldt's ideals. The traditional task was to educate an elite i.e. the main aim was to reproduce their own personnel; it was not until 1971 when the Framework Act of Higher Education stated that a course of studies should prepare for a professional career. In the same year the Fachhochschulen were established in the system of higher education. They offer mainly courses of studies in engineering, business administration and social work. Admission to a Fachhochschule requires a certificate of Fachhochschulreife which is awarded after 12 years of general schooling upon passing a final examination at a Fachoberschule which is a special type of vocational school. Applicants holding a certificate of Fachhochschulreife usually have completed an apprenticeship in a related training field. The courses of studies at a Fachhochschule are geared to practical applications in professional life rather than to basic research.

In the fields of physics and engineering respective courses are offered by Universities, Fachhochschulen and Comprehensive Universities. The curricula of the Universities are set up for five-year studies those of the Fachhochschulen for four-year studies (including one or two semesters of practical work in industry) and the Comprehensive Universities offer both types of studies. The degree awarded is in each case a "Diplom" i.e. *Diplom-Physiker* and *Diplom-Ingenieur,* respectively. In engineering two thirds of the graduates receive their degree from a Fachhochschule.

Up to the fifties only a small percentage of an age group was eligible to the traditional elite university education. The (West) German educational system changed drastically during the fifties and sixties. Both the secondary and the tertiary system were expanded and today about one third of an age group enters the system of higher education. At the same time tuition and fees have been waived. The system of higher education changed accordingly from an elite university to a mass university and up to now the resulting overflow problem is a highly controversial political issue. In the seventies the secondary system was further diversified by introducing a college-type course system in grades 12 and 13 of the Gymnasium. Basic and intensified courses are offered to be combined in quite different patterns to meet the individual needs and desires of a student. As the Abitur certificate is still the general admission certificate to higher education it is quite common that students in the field of physics and engineering are underprepared in mathematics and physics as they did not take the respective courses in the upper level of the Gymnasium.

Demands of Industry on Graduates in Physics and Engineering

Danielmeyher[2,3] (1992 a, b) addressed the demands of industry from the point of view of a multinational company. I will try to summarize briefly his ideas: Whereas the structure of our economic system has changed dramatically the system of higher education did not (besides of a change from an elite university to a mass university). Economy became a global issue whereas the universities still follow Humboldt's ideals and the courses of studies are still structured in specific traditional fields (physics, engineering, computer science) which do not match the needs of industry. University research is mainly oriented towards epistemology and pure mental comprehension. Industry requires the fast and effective transfer of results of basic research to innovative products in a world wide competitive market and thus the capability to operate across national and cultural boundaries. In an electronics company like Siemens, R & D integrate physics, chemistry and materials science for basic technologies and mathematics and information technology are added in digitizing technologies. This integrating task is not stressed in a field of studies in any specialized academic field. Danielmeyer claims that after about two years in a creative team in industry the former field of studies is virtually of no interest any more and any unneeded specialization should be strictly avoided in higher education. As the curricula in physics, electrical engineering and information technology reveal profound similarities, interdependencies should be used to free mental potential in the area of responsibility in complex problem settings rather than in a highly specialized academic discipline. An entry step might be textbooks revealing the width and flair of related and interdependent disciplines, asking for integration and synergy rather than for differentiation. Essential topics of interest for the present and future of mankind are health issues, labor security, traffic problems and environmental issues. In all those fields the research tasks of higher education and industrial needs do not match. Danielmeyer also stresses the problem of the high age of German graduates in the European comparison. To stay abreast as an industrial nation it is a necessity to have shorter duration of studies, an earlier age of entering professional life for the graduates, and the readiness to take risks and the willingness for continuing education and life-long learning.

Physics in the Curricula of Secondary Education

In 1992 the German Physical Society (DPG) dedicated a one-day symposium to the problem "What are our students learning in physics at the Gynmasium?" [DPG[4], 1992]. Physics courses are offered in the syllabuses from grade 6 through the final year of the Gymnasium (which is grade 12 in the new Länder and grade 13 in the old Länder). The upper-level system of basic and intensified courses yields a large spread of lessons per week taught in a student's school career ranging from 4 to 23 as indicated by the figures of Table 1. This holds for the old Länder as well as for the new Länder. It is estimated that more than one half of the students holding an Abitur certificate had chosen to take no physics courses in grades 12 and 13. They rather decided to take chemistry or biology courses instead. This selection behavior yields distinct differences in the knowledge of mathematics and physics in any entry cohort in physics and engineering. The difficulties are revealed by an increasing number of remedial courses mainly in mathematics offered by the system of higher education [Kurz[5], 1985]. Problems of the secondary system are passed on to the system of higher education. Any reform has to address secondary and higher education.

Bundesland (State)	min	max
Baden-Württemberg *	7	19
Bayern *	6	22
Berlin * / **	8	22
Brandenburg **	6	23
Bremen *	5	21
Hamburg *	8	18
Hessen *	8	18
Mecklenburg-Vorpommern **	7	18
Niedersachsen *	6	21
Nordrhein-Westfalen *	8	26
Rheinland-Pfalz *	7	21
Saarland *	9	23
Sachsen-Anhalt **	9	19
Sachsen **	7	21
Schleswig-Holstein *	8	22
Thüringen **	4	20

*Table 1: Physics in secondary education - grades 6 through 12 ** /13* . Stated are the minimum and maximum numbers of lessons taught per week in a student's school career. * old Länder, ** new Länder [Mierické[6],1993]*

In the middle level of the secondary schools a new course "introduction to technological subjects (Techniklehre)" was recently introduced to the syllabus. This was mainly at the expenses of the formally allocated hours in physics. The development is similar to the cut of hours in mathematics when the new topic "introduction to information technology" was added. A further reduction of physics hours in the curriculum is to be feared in the old Länder if the number of years of general schooling of 12 years up to the Abitur is realized. To summarize this fear I quote von Rhöneck: *"Avoid physics becoming the Latin of the 20th century"* [DPG[4], 1992].

The Main Problems of the System of Higher Education in Germany

The remarks which follow are restricted to the situation in the old Länder due to the lack of data available from the former German Democratic Republic. I will address the following general topics and I will try to give illustrations from the physics courses:

- Overcrowded system of higher education
- High age of new entrants
- Inhomogenuities in basic knowledge and manipulative skills in mathematics and physics of an entry cohort
- Duration of studies
- Drop-out rate
- High age of graduates

Overcrowded system of higher education

The German system of higher education is comprised of (Technical) Universities, Comprehensive Universities, Fachhochschulen and Colleges of Art and Music. In the winter semester of 92/93 a total of 1 830 000 students have been enrolled — 1 289 000 (or 92.3 %) in the old Länder and 141 000 (or 7.7 %) in the new Länder. The system, however, is set up to accommodate 900 000 students as determined by the number of faculty and staff and by space available. Since 1975 in the old Länder the number of students nearly doubled whereas the number of faculty and staff rose only by about 15 % [BMBW, 1993].

Age of new entrants

As mentioned before the general admission to higher education is the Abitur certificate awarded after 13 years of general schooling (in the old Länder), i.e. at age 19 or 20. In male dominated fields like physics and engineering the compulsory army service or an equivalent social work requires another year of life time. An additional apprenticeship in a study-related field or practical work as a prerequisite for admission results finally an average age of the entrants of M = 22.0 y. The corresponding age distribution is given in Table 2. An entry age of 22 years comes close to the graduation age of students in other European countries.

Age	<19	19	20	21	22	23	24	25	26	27	28	29	30	>30
(%)	0.5	12.9	21.9	19.8	14.2	10.1	6.6	4.0	2.6	1.7	1.2	1.0	0.8	2.7

Table 2: Age distribution (percentage) of new entrants (winter semester 1990/91 in the old Länder). Total number of new entrants is N = 213 500 [BMBW[7], 1993].

Underprepared entry students

Investigations of the entry knowledge of students in mathematics and physics reveal pronounced deficiencies in basic knowledge and manipulative skills. This holds for beginning students in physics [Krause & Reiners-Logothetidou[8], 1979] as well as for those in engineering [Kurz[9], 1988] and medicine [Kern[10], 1993]. The main reason is the selection behavior in the upper level of the Gymnasium. Therefore many institutions of higher education offer remedial courses prior to the first study semester as well as additional mathematics courses to support the physics lectures [Kurz[5], 1985].

Duration of studies

The curricula of the courses of studies at universities are designed for five years, those at Fachhochschulen for four years. Fachhochschulen come close to this aim but the excessive duration of studies at universities is a major political issue. There is much freedom when to take an examination. Examinations are mostly comprehensive i.e. after a two-semester physics course this means just one written examination. According to the traditional German system each student is regarded to be a full-time student. But the present student generation is different: students might possess the certificate of a completed apprenticeship, might have work experience, their life style is governed by individual self-determination as far as housing, family, and extra- curricular activities are concerned. In the US such students would definitely choose a part-time course of studies. With no tuition and fees to be paid in Germany this part-time status is made relatively easy; a student's ID guarantees public health care, reduced fees for public transportation, etc.. At present the fear of receiving a degree and not finding employment might be another reason to stay under the protective umbrella of the system of higher education.

Table 3 shows the number of students and their respective year of enrollment in the field of physics and astronomy. It clearly indicates that the students do not finish in the assigned period of 10 semesters. Schwoerer[12] (1992) lists the duration of studies in the study year of 91/92: the median was 12.8 semester (= 6.4 years). They varied in the old Länder from M(min) = 10.8 semesters at the Comprehensive University of Dortmund to M(max) = 15.1 semesters at the Technical University of Berlin. At all universities in the new Länder the degree was awarded after 10.0 semesters in agreement with the designed five-year curriculum. The federal statistics for the old Länder in 1990 states for graduates in physics and astronomy at universities a mean of M = 6.8 y and in mechanical engineering M = 6.8 y at universities and M = 4.6 y at Fachhochschulen [BMBW[1], 1993].

Unfortunately there are no official statistical data available but it is estimated that up to 40 % of the students drop out from the physics track before the intermediate examination within two years. This is indicated in table 3 by the drop of student numbers up to the third year of studies.

year of enrollment	1st	2nd	3rd	4th	5th	6th	7th	8th	9th	> 9th
students	7 852	5 802	4 996	4 252	3 643	3 359	2 740	1 926	1 062	2 039

Table 3: Number of university students enrolled in the field of physics & astronomy (N(total) = 37 671 in winter semester 90/91) [BMBW[1], 1993]

Age upon graduation

The high age of new entrants and the duration of studies yield an average age of the graduates which is probably the highest in Europe. The average age upon graduation with a "Diploma" degree or an equivalent was for the old states 28.1 years in the study year of 1990 (Universities: 28.7 y, Comprehensive Universities: 28.9 y and Fachhochschulen 27.1 y) [BMBW[1], 1993]). This high age in the European comparison is the main reason for the lawmakers to demand a curriculum reform to assure curricula which can be studied and completed in the prospective time of four and five years respectively.

Physics in the Curricula of Higher Education

Courses of study in physics for physics majors

(five-year university programs)

The public university system leads to quite comparable curricula across all physics departments in Germany. The study and assessment regulations of an individual university follow the "Guidelines for degree courses in physics" published by the joint coordinating commission for regulations concerning study and assessment of the conference of physics departments. The tasks, responsibilities, and perspectives of a physicist, the courses of study, and comments on the present situation and the estimated developments are described in detail in "Diploma-Physiker - his/her professional qualification and education" [Danielmeyer & Schwoerer[12]].

	Experimental Physics	Theoretical Physics	
(1)	Mechanics	Introduction 1	(1)
(2)	Electricity and Magnetism	Introduction 2	(2)
(3)	Optics - Heat and Thermodynamics	Classical Mechanics	(3)
(4)	Atomic Physics and Nuclear Physics	Quantum Mechanics 1	(4)
(5)	Molecular Physics - Solid State Physics 1	Electrodynamics	(5)
(6)	Solid State Physics 2	Thermodynamics and Statistics	(6)
(7)	Nuclear Physics and Elementary Particles	Quantum Mechanics 2	(7)
		Solid State Physics	(8)

Table 4: The sequence of physics courses for physics majors. The example is taken from the syllabus of the University of Bayreuth [Danielmeyher & Schwoerer[12]]

Traditionally in the first year besides of mathematics and chemistry the physics course program starts with experimental physics. In the second year (after possible introductory courses) the corresponding series in theoretical physics courses start in parallel. It is an inherent feature of the German system to divide physics rather strictly in experimental and theoretical physics; this is reflected in the division of the faculty into the institutes of theoretical and experimental physics. The foundation studies (first and second year) are completed with an intermediate examination *(Vordiplom)*, the third and forth year are dedicated to intensified studies and after the final examination *(Hauptdiplom)* in the fifth year a diploma thesis has to be completed. About one half of the Diplom-Physiker go on for a doctorate which takes about another four years of thesis work.

An interesting new model for the course studies was launched in winter semester 91/92 at the University of Munich [Fischer & Luchner[13]]. The traditional experimentally-oriented introductory course (mechanics, electricity and magnetism, heat and thermodynamics) is now taught by a team of two professors, one experimentally and the other theoretically oriented. About two thirds of the lecture are oriented to experimental phenomena, the remaining third is devoted to theoretical modeling and the presentation of the mathematically sound descriptions of the underlying physics. Both professors are present during all lectures. The recitation sections are also team-work designed. To ease the mathematics problems in the first course of theoretical physics (classical mechanics) additional help sessions are offered where students are tutored by graduate students.

These changes are directed towards a curriculum reform with respect to a more carefully structured set of topics and their related time-sequencing to ease the learning and comprehension by the students. Introducing small group work contributes to the individualization of the learning process and increases the interactive mode of learning.

Scenario of the Introductory Physics Course for Non-physics Majors

Courses of study in physical sciences, engineering, and medicine used to include traditionally an introductory physics course in the 1st year. Depending on the size of the entry cohort the audience might vary from 50 students up to a crowd of 700 in the large physics auditorium. It is a German tradition that this course used to be presented as "experimental physics" and the traditional contents of mechanics, electricity and magnetism, heat and thermodynamics and some aspects of modern physics are presented in a series of well designed and prepared experiments. The lecture might be accompanied by recitations (usually optional with no influence on the grade) and by experiments to be performed in groups of two in laboratory exercises. It is inherent to the academic freedom to recommend always a number of textbooks for students use rather than to follow the track of just one textbook which is standard in the US. There are traditionally no midterm examinations and there is only one final examination which is held in the examination period after the 1st year.

Under the political pressure of reducing the duration of studies some schools reduced the credit hours serviced by the physics department for the introductory physics courses. The reasoning is that some contents of the traditional physics syllabus are taught in different context as well in the courses of the respective school. Developments of cutting physics will have in the long term implications even on the number of faculty and staff members in physics when service is no longer demanded by other departments.

The school of engineering of the (Technical) University of Stuttgart has recently shortened the introduction to physics to a one semester course of four contact hours; to be followed by performing five experiments in the laboratory. Not enough to foster the acquisition of a structure of physical phenomena and the underlying physics concepts. The assessment is done by a short quiz which seems to be only slightly more than a statement of successful attendance of the physics course.

Students of medicine also have to take an introductory physics course. At the University of Tübingen for instance this is a two semester course with a total of six hours of lectures and two optional hours of recitations. In the 2nd semester the students have to attend laboratory exercises in addition. But this physics introductory course is also endangered in its service to medical schools. The medical courses are governed by the guidelines for the licensing of physicians. At present the amendments to these guidelines are discussed by an expert commission of the Federal Agency of Health. The tendency of this draft seems to be to reduce time spent on general basic, field specific topics and to increase the time spent on medical applications. Thus in the future the "relevant" mechanics and optics for MD's might well be integrated into the respective orthopaedics and ophthalmology courses taught by medical school experts. This development and especially the process of transformation of the amendments into new curricula of the medical schools will have to be watched carefully.

Educational Issues and German Reunification

As mentioned before, the traditional Gymnasium awards the Abitur certificate as general admission to higher education after a total of 13 years of general schooling. This still holds for the old Länder. In the new Länder the Abitur certificate is awarded after 12 years of general schooling as a heritage of the former German Democratic Republic. At present both educational systems coexist but according to the unification treaty this must be standardized in all Länder by 1996 at the latest. The discussion is highly controversial but it is likely that the future will be the 12-year solution. As mentioned before this might have strong implications on the number of weekly hours allocated to physics courses in the Gymnasium in the old Länder.

Astronomy was a separate teaching topic in the Gymnasien in the former German Democratic Republic and its survival in the new curricula is not yet finally decided.

The universities in the former German Democratic Republic were teaching institutions. Basic research was conducted at the Academies of Science. This will change as well as the system of higher education itself. Admission to higher education was (mainly politically) restricted to the prospective needs of the socialist society. Individualized tutoring was an essential part of physics teaching. Therefore the five-year studies could be completed precisely in time with nearly no drop-out. The staff/student ratio was small and tutors in recitations and laboratory exercises often held a Ph.D. in physics, which meant an excellent investment for the next generation. This situation is being changed dramatically by the politically and financially based reduction of faculty and staff. This will undoubtedly have implications on the duration of studies in the new Länder.

As an illustration of the situation after the change I would like to give an example of the developments at the Technical University of Dresden an institution dedicated formerly to technical courses of studies [Willemer[14], 1993]. Among the recent changes new schools have been established i.e. the Schools of Law, Economics, Social Sciences and Education. A Medical Academy is intended to be incorporated as a School of Medicine. At present students pour into those new schools rather than into the traditional school of engineering assuming the new fields are prestigious and career-oriented; a comparison of the number of new entrants in the winter semesters of 91/92 and 92/93 respectively has resulted a drop from 1000 to 200 in mechanical engineering and in a drop from 750 to 200 in electrical engineering. The numbers of weekly hours allocated for the introductory physics course have been cut, especially for recitations, i.e. phases of active work on the part of the students.

Quite essentially for the system of higher education in the new Länder the concept of Fachhochschulen was adopted from the old Länder to indicate the need for shorter studies geared to professional applications.

Implications of the Growing Together of Europe on Higher Education

According to the demographic development in Europe the total population will decrease but the fraction of elderly people will increase. The job market will require an increasing number of academically educated and well-trained people. There is a contradiction between the dividing elements of cultural heritages and the unifying elements of economic necessities. One has to compare the different countries' higher educational systems, which have naturally evolved against different social, cultural, and industrial backgrounds. Higher education will have the task of generating an European identity and to preserve and to pass on the heritage of the different branches of European traditions.

The Treaty of Maastricht signed February 7, 1992 includes conceptions for the future of European higher education and research. The "Task Force Human Resources, General and Vocational Education, Youth" prepared the *"EC Commission's memorandum on higher education in the European Community"* which was published November 5,1991. The memorandum gives some basic information about the educational systems and their interdependencies, it states the growing need in the EC for higher education, it addresses new challenges, requirements and politics for higher education and it lists all EC programs in higher education and cooperation between universities and industries.

The following aspects will have an impact on higher education and thus on the curricula and they need special consideration:

- entry qualifications (and mutual recognition) and participation in higher education
- cooperation and partnership with industry
- continuing education and life-long learning
- open learning systems and distance learning
- European dimension in higher education
- mutual recognition of professional titles in the EC

The European dimension could be characterized by headings as

- issue of languages
- mobility of students
- cooperation and joint programs of institutions of higher learning
- European culture in the curricula
- teacher's education and training
- international role of higher education
- information and communication

To foster the integration process the European Community has already launched programs to enhance mobility and exchange. Some of the programs and their aims are described briefly here.

ERASMUS: European Community Action Scheme for the Mobility of University Students.

Promotion of academic recognition throughout the European Community in order to allow students to circulate freely between the twelve Member States. Grants to allow a year of study abroad.

TEMPUS: Trans-European Mobility Scheme for University Studies.

A program for quality improvement in social, economic and cultural development in Eastern European Countries. Cooperation programs between one University in an EC country, one University in a supported country and another University or industrial company in an EC country. Expands the ERASMUS program to the post-socialist countries.

COMETT: Community Program on Cooperation between Universities and Industry regarding Training in the field of Technology.

LINGUA: An additional language program to support the mobility programs by reducing language problems for students participating in exchange programs.

ECTS: European Community Course Credit Transfer System.

This program addresses specifically aspects of comparison and adjustment of curricula in quite different European educational systems. The pilot phase of this program runs from the academic year 89/90 to 94/95 and covers the subject areas of business administration, chemistry, history, mechanical engineering and medicine. The aim is to give full credit — a new term in the German educational systems - for all academic work successfully carried out at any of the ECTS partner institutions and transfer of those credits from one participating institution to another. The FHT Esslingen is one of 30 institutions of higher learning in all 12 EC countries (plus four EFTA countries) participating in the subject area of mechanical engineering. The FHT Esslingen can award a total of ten student mobility grants every year. Engineering courses of studies run in the participating institutions from three-year to five-year programs. The courses of studies in all participating institutions had to be carefully assigned to credit units totaling 60 per academic year, the contents of the courses had to be analyzed to be equivalent for the respective headings, the assessment (requirements, duration of examinations, procedure upon failing, etc.) and the grades and their conversion (i.e. letters to numbers and vice versa) had to be agreed upon to assure that a student can spend one year abroad earning credits for transfer to his/her home institution. One essential is that the duration of studies is not to be prolonged as a result of abroad study.

Professional national and international organizations and international associations of college teachers interested in specific fields have taken it upon themselves to promote higher education in Europe. These organizations are for instance in physics the German Physical Society (Deutsche Physikalische Gesellschaft — DPG) and the European Physical Society (EPS) and in engineering the European Society for Engineering Education (Societé Européenne pour la Formation des Ingénieurs — SEFI); the International Society for Engineering Education (Internationale Gesellschaft für Ingenieurpädagogik — IGIP) and the European Federation of National Engineering Associations (Federation Européenne des Assocations Nationales d'Ingénieurs — FEANI).

A curricular task on setting European wide standards was launched by the European Society for Engineering Education (SEFI). The Mathematics Working Group of SEFI developed in a series of six seminars throughout Europe a common European core curriculum for the mathematics part of all three to five year engineering programs in Europe [SEFI[15], 1992]. To ease the dissemination of the core curriculum in Europe it has been translated into Czech, French and German.

A Physics Working Group is intended to be founded at the forthcoming annual meeting of the SEFI in June. As the post-socialist countries will be cooperate fully from the beginning it will be an exciting task to collect information and to judge the role of physics in the different educational systems of the greater Europe.

The fall of the socialist countries in Europe has changed the world and this also will have an impact on educational issues. The transition from a centrally planned economy and "society" to market economy and "civil society" poses economic and ecological problems. The issues in the transition process of higher education may be described by lack of resources, major financial constraints, lack of motivation of academic staff and the lack of information. The European Society for Engineering Education has devoted a seminar on European Cooperation East-West [SEFI[16], 1993]. The European Community has already taken action and supports mobility and exchange programs with the post-socialist countries and as a result the presence of colleagues at international conferences has already increased markedly. The change of the economic system and the build up of an infrastructure have high preference but environmental issues, energy consumption and security of nuclear plants will undoubtedly change the curricula which require physics as a base and a desirable mobility in universities and industry.

The political changes in Europe will have further implications on the educational systems and the role of physics in the respective societies. Let's face those challenges on the threshold of the next century.

References

1. Rigden J.S. (1986): Conveying the truth we live by, In: Born, G., Euler, M. & Sexl, R.U. (eds): Communicating Physics - Proceedings; p. 1-21; Duisburg: Comprehensive University of Duisburg
2. Danielmeyer, H.G. (1992 a): Die Struktur der Wirtschaft ändert sich — Kommt die Ausbildung nach?, Spektrum der Wissenschaft 12/1992, pp. 119
3. Danielmeyer, H.G. (1992 b): Plädoyer für eine neue Qualität der Lehre an Hochschulen. Physikalische Blätter 48(10), p. 823-825
4. DPG (1992): 12. Tag der Deutschen Physikalischen Gesellschaft: "Was lernen unserer Schüler im Physikunterricht?", Physikalische Blätter 48 (3), pp. 169-182. Mit Beiträgen von K. Luchner ("Zum Stellenwert der Schulphysik"), D. Nachtigall ("Was lernen unsere Schüler im Physikunterricht?"), and Chr. von Röhneck ("Schwierigkeiten beim Verstehen von Physik")
5. Kurz,G. (1985): Remedial courses in mathematics — scopes and problems — a survey for the Federal Republic of Germany. In: International Journal of Mathematical Education in Science and Technology 16 (2), pp. 211-225

6. Miericke, J. (1993): Anzahl der Physikstunden am Gymnasium in den Ländern der Bundesrepublik Deutschland, In: Das Gymnasium in Bayern — Zeitschrift des Bayerischen Philologenverbandes, Heft 5, 1993, S. 24.
7. BMBW (1993): Basic and structural data 1992/93. Bonn: Bundesministerium für Bildung und Wissenschaft.
8. Krause, F. & Reiners-Logothetidou (1979): Der bundesweite Studieneingangstest Physik 1978, Physikalische Blätter 35 (11), S. 495 ff
9. Kurz, G. (1988): Das Eingangswissen von Studienanfängern in Mathematik und Physik — Wiederholte Querschnittuntersuchungen an der FHT Esslingen. Empirische Pädagogik 2 (1), p. 5-32
10. Kern, W. (1993): Sind wir bei der Ausbildung von Nichtphysikern entbehrlich? Physikalische Blätter 49 (2), S.130- 131
11. Schwoerer, M. (1992): Statistiken zum Physikstudium in Deutschland. Physikalische Blätter 48 (9), p. 741-743
12. Danielmeyer, H.G. & Schwoerer, M.: Diplom-Physiker/Diplom-Physikerin. Blätter zur Berufskunde, Band 3; Herausgegeben von der Bundesanstalt für Arbeit, Nürnberg, im Einvernehmen mit der Deutschen Physikalischen Gesellschaft e.V. Nuremberg: Federal Employment Agency
13. Fischer, G. & Luchner, K. (Ludwig-Maximilians-Universität München, Sektion Physik): private communication.
14. Willemer, W. (Technische Universität Dresden, Institut für Angewandte Physik und Didaktik): Private communication.
15. SEFI (1992): SEFI Document No. 92.1 (March 1992): "A Core Curriculum in Mathematics for the European Engineer". Drafted under the editorship of Michael D.J. Barry, Royal Naval Engineering College, Manadon, Plymouth, England & Nigel C. Steele, Coventry Polytechnic, England. Bruxelles: SEFI
16. SEFI (1993): SEFI Seminar on European Cooperation East-West, held at the Technical University of Warsaw, 04/05 April 1990, European Journal of Engineering Education 18 (1), special issue, 120 pages.

Moving the Mountain:
How to Get the Physics Community to Change

Sheila Tobias

Introduction

Among my colleagues in the social sciences there is a riddle that goes something like this: How many experts does it take to change a light bulb? And the answer, provided by the psychologists (with a twinkle) is that no amount of experts can be expected to change a light bulb unless the light bulb itself *wants* to change. This, to me, is our dilemma.

In trying to reform undergraduate instruction in physics, we can appeal to the physicists' empiricism — after all, there is now a great deal of evidence that even the best physics students do not routinely "cross over into the Newtonian world," that is, do not take away from their first-year physics course any profound understanding of your subject, even after they have learned to solve standard numerical problems.

Or, we can appeal to the physicists' moral sense — for it is a fact that too many students, particularly women and historically underrepresented minorities, do not survive college-level physics unscathed and unalienated.

Or, we can rattle the physics community's complacency, as I have tried to do, by enrolling intellectually superior students (my "second tier") in introductory physics and chemistry courses to find out *as they experience these courses* what's wrong.

Or, and this is to be the essence of my message to you this morning, we can appeal to the *self-interest* of the physics community with the argument that to educate *more* and *more different* kinds of students is to grow a future power elite for science.

It seems to me, then, that the question is not (only): What kinds of changes to make, so much as how to persuade the physics community that educating only the elite in physics is no longer adaptive to a future where the nation's understanding and commitment to science will matter quite as much as the quality and training of its practitioners. From this it follows that only when the physics community seriously addresses the issue of "target audience," namely *who* you are educating in physics (and *who* you might want to), will your colleagues, competently and on their own, develop serious and wholesale strategies for change.

The Tier Analysis

If we ask "who we are educating in science," there is little question in my mind that, with some exceptions, only one undergraduate population is and has traditionally been ably served by the college physics programs, namely students who are, by their attitude, their motivation, and their ability to teach themselves, essentially younger versions of yourselves.

But such a tradition excludes other categories of students, students who ought to feel at least some measure of welcome and success if the value of physics to the wider community is to be properly understood and served. These include future teachers (both of science and nonscience subjects), future lawyers, future politicians, future journalists, future bankers, future economists, future businessmen and women, even future Congressional staffers.

Such are the careerists who will one day stand between an inventor and his or her venture capital, a corporate manager and his or her willingness to fund a company research group, a citizen and his or her ability to understand (and tolerate) "risk." Yet, such students are not normally courted by scientists in the market place of college. Students, it is assumed, who *could* do science, will do science; students who don't choose to continue in science, probably shouldn't. Is it any wonder, as it is often said in Washington, that apart from the defense and health establishments, "There are no votes in science."

How do your colleagues think about who they educate? And how can you get them to think differently about whom they attract or weed out of physics? It seems to me that many of your colleagues, looking out on a sea of faces in the introductory course, identify only two types of students. (See Figure 1) One, a group they think of as "them," the majority, not especially committed to science and not particularly gifted or interested in playing the game, and the other, an elite minority, designated "us" on my diagram who, in the way they approach the subject and in their single-mindedness, make their professors think of them as uniquely "talented" in science.

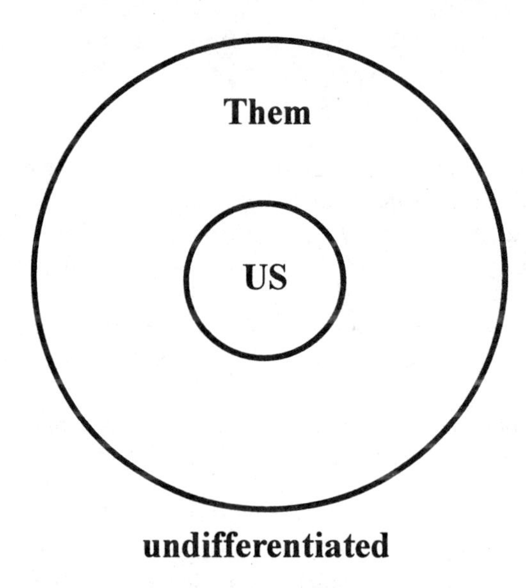

undifferentiated

Figure 1

My view is this: Just as the "us" group includes some highly diverse individuals, so the "them" group is not as homogeneous as one might infer from the convergence of their average grades. The way I would have the faculty think about their students, instead, especially in the first-year course, is depicted, rather, in Figure 2.

Here, in the innermost circle of the disk, remains the "us" of Figure 1 (in my naming scheme "Tier 1") — a group of students so determined to do science that they are essentially "teacher proof," "curriculum proof," and "classroom culture proof." And yet most teaching, most grading schemes, and most instructional strategies are designed to benefit this group. Beyond the inner circle hover my "second tier," students who, as my research indicates, are often disappointed in their introductory physics courses, not because they are too "hard" (whatever "hard" may mean to you), but because they are in many ways not "hard" enough.

My description of "the second tier" experiment has been published elsewhere. And so I will only quickly review the essence of these students' response to first-year science teaching: First, they comment on the absence of an intellectual overview, what John Rigden calls the missing "story line" and Robert Romer the "connecting argument." Second, they complain about (their term) a "tyranny of technique" — too many scales and not enough music — which results in too little opportunity, as they see it, for them to "creatively interact with the material." There is, again, in their words, too much "mimicking the instructor's approach." As one of my experimenters, a graduate student in philosophy, put it, her IQ was sufficient for physics (indeed, she did very well), but her OQ — her obedience quotient — was too low. Third, the courses do not as a rule "play to their verbal strengths." Said Eric, now a third-year Stanford graduate student in literature of his physics exams, "Only once we were asked to comment on or explain something instead of merely having to complete a calculation." If I were a physicist I would worry very much about all of this, and seek to reinject creativity and independent thinking into introductory physics, since, as you know better than I, physics does not in the long run benefit as much from obedience as from contrariness, not as much from mimicry as from new, better yet, wholly new ideas.

1. 1st Tier

2. 2nd Tier

3. Utilitarians

4. Underprepared

5. Unlikelies

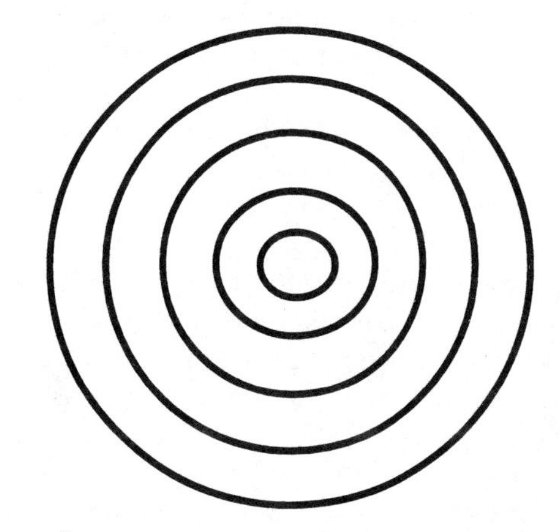

Figure 2

But it is not just the "second tier" who are lost to science through the gateway courses. There are as well in that sea of faces a tier of "utilitarians," a tier of "underprepareds," at least a minority of students hostile to science, and some number of "unlikelies," students who no amount of accommodating or cajoling will bring back for another year.

The "utilitarians" should give you pause (they do me) because these are the students who have figured out how to do the bare minimum to "succeed" in introductory physics, but are not intellectually engaged and do not appear to want to be. They find their way into introductory physics largely because it meets some other requirement for their major or for graduation. Therefore, they will resist curricular and pedagogical change if it in any way affects familiar "rules." Still, there is no reason to dismiss them out of hand. First, because in a different environment they might be shown to be both interested and recruitable to physics. Second, because some of them may be first-tier students in disguise. As I have been told by serious students of science, the way the first-year courses are structured and the way examinations have the effect of shaping as well as driving student behavior, first-tier students are very often forced to become "utilitarians" just to survive. If I were a physicist, I would worry very much about that.

Beyond the "utilitarians," still more distant from the center, are the students who are underprepared, particularly, though not always exclusively, in mathematics. Many of your faculty are aware of the problem of underpreparedness for science and would agree, I am sure, that until and unless those students' mathematics skills are improved, physics is inevitably beyond their grasp. But how many of your faculty believe as fervently as I do, that mathematics disabilities are entirely reversible — and at the college level; and, even more, that such students might well have an unmeasured (and hitherto unmeasurable) long-run talent for physics. The key phrase in that sentence is "long run." For, as Harvard chemist Dudley Herschbach has observed, science needs long-distance runners quite as much as sprinters, but the introductory courses as currently structured, favor the sprinters.

Programs that Work

Two years ago, I left the second-tier analysis to begin a second study for Research Corporation, this time of exemplary undergraduate programs and courses in physical science, programs and courses that "work." For a program to "work," I decided early on in my research, it had to meet three conditions of my own devising: high recruitment, high retention, high morale.

It was not hard to locate the eleven exemplars featured in *Revitalizing Undergraduate Science*; everyone pointed me in the same directions. But not many people had analyzed in any systematic detail *why* or *how* they worked, what they were doing that was so effective. I found out why the lack of previous analysis. Contrary to what I now like to call the "dominant paradigm" of science education reform (where a single-minded search is undertaken for a curricular or pedagogical "solution" to a problem), what I found, among the eleven institutions I chose to describe, was closer to "quality management" than to science. To summarize:

1) Where programs work, quality of instruction is considered to be a departmental responsibility. Faculty feel accountable not just for the courses they themselves teach, but for all the courses and the ancillary support services offered majors and nonmajors alike.

2) Where programs work, effective feedback mechanisms have been put in place so that students, junior and senior staff members, outsiders and insiders, can make their feelings and opinions known. As a result, problems surface and are targeted long before end-of-semester course evaluations and are treated immediately and effectively by the department chair or by the department, under the chair's leadership, as a whole.

3) Where programs work there is overall commitment to the undergraduate teaching enterprise. Everyone "buys in" to the program as a whole, not just the local "educational reformer," or the one who has managed to get outside funding for some "innovation." Indeed, many of the programs I studied were not especially innovative. The faculty saw itself as doing just a little better each year than the last.

4) In programs that work the department makes available small rewards (I call these "little r's") to stimulate exemplary teaching and course improvement. Difficult or important courses are provided with extra funds for super T.A.'s, instrumentation, summer assistance, undergraduate research. No one is waiting for the grand reward system to change. Efforts are made constantly and imaginatively to improve things even under current constraints.

5) As a result, among these programs that work, improvement is the norm, not the exception. The department has some system for setting goals for itself and for evaluating (locally and by means of its own techniques) how it is doing and where it should put further effort.

If, then, one of my most important findings is that the department is the locus of improvement, then the corollary must be that the department have the means to control its own enterprise. This entails control over enrollment, room size, laboratory resources, academic credit, teaching assignments, advising and other academic support services, and faculty rewards, beginning with a recognition that some of the most crucial elements in making change may be (until the department fights to retrieve these powers) outside departmental control. Where does the department garner the *will* for the struggle to implement change? Most of all from consensus-building, from goal-setting, from internal evaluation, from continuous and serious discussion of teaching overall.

Where and how does this discussion begin? And, more importantly, how does it end? Are we to continue to allow only the "creative loners" in the department to pursue innovation and change? Are we to continue to allow central administration to control most instructional resources that is, to set the boundaries and constraints to change? And are we to continue to allow the outside funders to decide (apart from research) what is worth changing in physics instruction and what is not? From a social scientist's perspective, the issue of *change* is inseparable from the issue of *control.* And once a program for change moves from the ideal and the outside-funded to the realizable here at home, the issue of *control* looms larger still.

Brief History of Change

If we date the first postwar round of physics education reform as a response to Sputnik in 1957, then we can date the beginning of the second round in about 1980, with a revival of interest in some metacognitive issues associated with teaching and learning introductory physics at college. My reading of that history suggests that we are now in Phase III of the second round.

Phase I was characterized by the recognition, called "constructivism," that students bring with them to introductory physics not just a *tabula rasa*, but a physical world view (resting on pre-Newtonian ideas) that, unless brought to consciousness, will stubbornly persist even after a first-year course in physics. In that period, investigators like Arnold Arons, Lillian McDermott, Fred Reif, David Hestenes, Alan van Heuvelen, Richard Hake, Fred Goldberg, Jill Larkin and others, identified barriers to learning physics that they carefully documented and explored ways to overturn. Further they found that, unless directly instructed, students will not think *physically* about numerical problems, but will develop algorithmic, plug-and-chug methods for dealing with (that is, avoiding) conceptually interesting and difficult issues. These investigators' writings on "conceptions and misconceptions," "novice to expert problem-solving" and "ordering new knowledge" all led to serious attention on the part of the physics community to both the curriculum and pedagogy of introductory physics courses.

In Phase II, a period I designate as the "applications period," beginning in about 1988, a number of innovators began to seek fundamental change in the introductory course. In response (and for those who were not yet cognizant of the ferment in the physics teaching community, very much in the lead), the American Institute of Physics convened a task force, the IUPP, to consider various models for change in the first-year course, changes that would meet the now better-understood needs of students and of the physics community. The IUPP's and others' efforts seemed to converge on a reduction of coverage — in Philip Morrison's well-worn phrase, "cover less and uncover more" —which corresponded to similar pursuits among mathematicians seeking a "lean and lively calculus," and among chemists. During Phase II, many of those initiatives began to be tested and disseminated.

The question I want to pose is this one: Will there be a third phase of still wider applications? Or, to say it differently, once these innovations leave the realm of the "converted," will they be able to penetrate the wider community?

One way to think about this question — a social scientists' way — is in terms of pockets of resistance to change. Such resistance can be anticipated among at least three populations (not to mention textbook publishers and their authors).

> First, faculty, for whom something new, something they don't know how to do, and something that may take more time and effort, is quite threatening;
>
> Second, students, many of them utilitarians, for whom any change in the "rules of the game" means harder and a different kind of study for the grades that, previously, they could count on by simple plug-and-chug; and
>
> Third, deans and other administrators who will be reluctant to add staff, resources, lab space, instrumentation, and modify the teaching/learning environment in any way that costs money.

Conclusion

I believe, at this stage of the improvement process, we need to pay at least as much attention to *pockets of resistance* as to new curriculum, course content and pedagogy. In doing reform, it is not enough, as it might be in science itself, simply to present a finding, or a model, however "true" or more elegant these might be. Not even an incontrovertible piece of evidence that the old is failing to achieve the community's goals will, by itself, stimulate anything new.

For change to take place, science faculty must prepare themselves to argue for what they believe to be better undergraduate teaching of physics, to wrest back control from those who would put barriers in their way, and, for the sake of their students, to win. Still, as I write in *Revitalizing Undergraduate Science: Why Some Things Work and Most Don't*, scientists may not have either the skills or the stomach for the battles that lie ahead. But until and unless those battles are joined, much of what has been accomplished so far in the rethinking of undergraduate physics teaching may be undone.

The most dangerous words in finance and investment, I am told, are "This time it's different." After several previous cycles of failed or incomplete science education reform, we can't simply assume that this time it will be. It will take hard work and much leadership from the physics community to make sure that this time it is.

References:

1. Arnold Arons, *A Guide to Introductory Physics Teaching*, New York: John Wiley, 1990.
2. Fred Goldberg and Lillian McDermott "An Investigation of Student Understanding of the Real Image Formed by a Converging Lenses or Concave Mirror," *Am. J. Phys.* 55, 1987, pp. 108-119.
3. Richard R. Hake, "Promoting Student Crossover to the Newtonian World," *Am. J. Phys.*, 55, 1987, pp. 878-884.
4. I. Halloun and David Hestenes, "The Initial Knowledge State of College Physics Students," and "Common Sense Concepts about Motion," *Am. J. Phys.*, 53, 1985, pp. 1043-1055 and 1056-1065. See also an update in *The Physics Teacher*, March 1992, pp. 141-166.
5. Jill H. Larkin, Lillian McDermott, Dorothea Simon, and Herbert A. Simon, "Expert and Novice Performance in Solving Physics Problems, *Science*, 208, 1980, pp. 1135-1342.
6. Priscilla Laws, "Workshop Physics: Learning Introductory Physics by Doing It, " Change, July/August, 1991, pp. 20-27. "Calculus-Based Physics Without Lectures," *Physics Today*, 24, (12), 1991, pp. 24-31.
7. Lillian McDermott, "Research in Conceptual Understanding in Mechanics," *Physics Today*, 37 (7), 1984, pp. 24-32. Milikan Lecture, 1990: "What we Teach and What is Learned—Closing the Gap," *Am. J. Phys.*, 59, 1991, p. 301.
8. Fred Reif, "Teaching Problem Solving — A Scientific Approach," *Physics Teacher*, 19, 1981, pp. 310-316.
9. Fred Reif and J.I. Heller, "Knowledge Structure and Problem Solving in Physics," *Educational Psychology*, 17, 1982, pp. 102-127.
10. John Rigden and Sheila Tobias, "Tune in, turn off, drop out," *New York Academy of Sciences*, Winter 1991.
11. Robert Romer, "Reading the equations and confronting the phenomena — The delights and dilemmas of physics teaching," *Am. J. Phys*, **61** (2) February, 1993.
12. Sheila Tobias and Carl T. Tomizuka, *Breaking the Science Barrier*, New York: The College Board, 1992.
13. Sheila Tobias, *Revitalizing Undergraduate Science: Why Some Things Work and Most Don't*, Tucson, Ariz.: Research Corporation, 1992, see especially Chapter 7.
14. Sheila Tobias, *Succeed with Math: Every Student's Guide to Conquering Math Anxiety*, New York: The College Board, 1987.
15. Sheila Tobias, *They're not Dumb, They're Different: Stalking the Second Tier*, Tucson, Ariz.: Research Corporation, 1990.
16. Alan Van Heuvelen, "Learning to Think Like a Physicist: A Review of Research-Based Instructional Strategies," and "Overview Case Study Physics," *Am. J. Phys.*, 59, October 1991, pp. 891-907.

IUPP: An Example of Curriculum Reform

D. F. Holcomb, Department of Physics
Cornell University

One of the most stable of all human enterprises is the content and pedagogical style of the introductory university physics course. As with all stable systems, it firmly returns to its equilibrium position when perturbed. Or, to quote from Bob Hilborn[1] of Amherst College:

> "Most introductory physics courses are like coral reefs: They have grown more by accretion than by design. Admired for their beauty and intricate architecture, they are nevertheless primarily the skeletal remains of the activities of individually insignificant organisms. Only a minute part is made up of currently active organisms...."

There are, however, visible and significant sources of instability. One of these lies in the dawning realization that students do not carry away from either classroom or lab what instructors imagine. A second lies in the realization that choice of material and emphases have not responded to the major changes in the worldview of our discipline which have taken place over the last 50 years. A third lies in the changing composition of the college student population. I will suggest ways in which IUPP and similar projects can contribute to harnessing these sources of instability for the benefit of students as they prepare for various career paths.

The Introductory University Physics Project, supported by the Directorate of Education and Human Resources of the National Science Foundation, seeks to encourage the growth of introductory physics course curricula which incorporate three key features. (1) The course should display a theme or themes which convey(s) to students a coherent structure of the course, (2) Contemporary physics should be a prominent and integrated component in the subject matter, (3) The total number of topics treated should be reduced in comparison with the typical introductory course.

Four different course models, based on non-standard syllabi, have served as the basis of instruction at nine different institutions[2] in academic years 1991-92 and 1992-93. A progress report on the Project has recently appeared[3] in *Physics Today*. My comments will assume that listeners to this talk will have that report at their disposal.

It is important to note that a number of other efforts to modify the introductory university physics course guided by goals similar to those of IUPP are underway around the country. I list some institutions which are sites for those efforts with which our Project members are familiar. I'm sure there are many others which we don't know about.

Bucknell University
California Polytechnic State University, San Luis Obispo (Randy Knight)
Colgate University
Ohio State University at Marion (Gordon Aubrecht)
Miami University (Ohio)
SUNY at Buffalo (Jonathan Reichert)
Warren Wilson College (Donald Collins).

The *Impossible Dream* of IUPP and similar efforts is to assemble a one-year calculus-based physics course, out of which comes a diverse student population with

(a) An awareness of and some mastery of the models and methods of physics
(b) Honed transferable skills and habits of mind
(c) Memories of a stimulating and agreeable exposure to new worlds.

These goals are certainly not new — the implication of my title is that "change" is in order — that the canonical model of such courses is typically not meeting the goals.

Q: Why *do* we need change?
A: In our hearts, we know that students do not carry away what we fondly imagine that they do, that non-physics majors do not typically find physics their favorite subject of study, that the demography of the student population is changing. Instinct tells us that different approaches to the introductory course are going to be needed.

I preface the rest of my remarks with a disclaimer: There is almost never a really new idea in physics pedagogy. For example, the "Less may be more" goal of IUPP is certainly not new. It figured strongly in the deliberations[4] of the Carleton Conference of 1956. From that conference came a spare list of Seven Basic Principles and Concepts which were considered essential components of the one-year introductory course. They were

Conservation of Momentum
Conservation of Mass and Energy
Conservation of charge
Waves
Fields
The Molecular Structure of Matter
The Structure of the Atom

Unfortunately, 35 years later, the "less may be more" flag is frequently raised but seldom followed.

The Textbook

IUPP's efforts have been primarily focused on subject matter and the textual and laboratory materials which have traditionally defined the content of the introductory course. Its goal has been to develop alternate curriculum models and to develop a constituency of support for promising candidates among those models. We do not have aspirations to dictate a new standard syllabus or to write and produce the new "standard textbook" of introductory university physics. Rather, we wish to promote the existence of attractive alternatives to the current array of quite similar texts. In the long run, only a first-class textbook can provide permanence to a successful story line and array of topics. The IUPP argument has been that as long as the textbook does set the syllabus, its contents should reflect the content and modes of thought of physics in the closing years of the 20th century, not the opening years.

Let me make a small aside. In considering the textbook issue, it is important not to jump too quickly to the assumption that the dimensions of the current array of fat and heavy textbooks necessarily mean that the author(s) have fallen into the trap of encyclopedicity. The following table compares a popular text by Robert Bruce Lindsay, published in 1939, with two editions of the Halliday and Resnick text whose contributions to physics instruction we celebrate at this Conference. I have chosen sections which cover an almost identical array of topics from Newtonian mechanics. I find that to the compact exposition of Lindsay have been added many more examples, elaborate graphics, and a far wider array of questions and problems.

A Comparison of Physics Textbooks			
	1939	**1962**	**1988**
Text	R. B. Lindsay	Halliday Resnick	H & R Funds III
Pages	520	1122	1149 (x1.5*)
pp. Mech'cs	132	295	306 (x 1.5)
pp. Q's&P's	15	44	97 (x 1.5)
Mass	0.7 kg	1.8 kg	2.4 kg

This factor of 1.5 is included to take account of the fact that the larger pages of the 1988 edition permit placing roughly 1.5 times as much material on a given page.

The encyclopedic factor has probably NOT been increasing with time in a significant way — the increasing size of the books stems primarily from the addition of elements designed to improve the pedagogical quality of the material and to make the books more appealing to the modern student.

How Much And What Kind Of Contemporary Physics?

What does it mean to integrate more contemporary physics, and what are some examples from IUPP course models?

It is important to note that by "contemporary physics" we do not necessarily mean physics at the research frontier. The phrase can, indeed, include study of the largely microscopic, quantum physics-oriented physics of the 20th century but it also covers such areas as (1) applied physics of current interest, (2) the enhanced role of conservation laws in contemporary physics compared to 100 years ago, or (3) topics made more accessible by the computer.

The time available in this session does not permit an extended review of how the various IUPP models have dealt with the "contemporary physics" issue. I'll give a few examples.

Mechanics

- The "Particles Approach" model generated at the U. S. Air Force Academy uses spreadsheet-based numerical calculations to broaden the reach of Newtonian mechanics. An example is given in Reference 1.

- Several models place a heavy emphasis on the conservations laws, with a correspondly lighter stress on F = ma problems.

- The carefully limited introduction of momentum and energy relations for v near c permit using particle physics examples in collision and explosion problems. The "Particles Approach" model uses the relativistic expressions to support a section on nuclear physics, fission and fusion, treated from a phenomenological point of view.

Thermal Physics

- The "Global Warming" segment of "Physics in Context" gives a nice story line which is based on a topic of current interest.

- The introduction of quantum physics before thermal physics in the "Six Ideas" model permits natural use of simple quantization ideas in treating the Einstein solid in a quantitative and understandable way.

Quantum Physics

All four models which have been tested include a serious attempt to include some quantum Mechanics. We expect our study of the evaluation materials which have been collected to tell us how successful these attempts have been.

Finally, Back To The IUPP Goals

How are we doing in pursuing our three goals? — (1) Coherent story line, (2) More contemporary physics, (3) Less may be more. One confirmed skeptic has colorfully suggested that ambitious projects such as IUPP may be usefully described in the complex plane, with the input effort and dollars spent lying along the real axis and the results firmly placed on the imaginary axis. Those of us working on IUPP are aware of the disappointing lack of permanent effect from similar projects in the past. But the accompanying figure may suggest our optimism. The F = ma vector symbolizes a traditional syllabus — firmly placed on the real axis but not very far reaching. The IUPP courses may indeed stray well up into the complex region of the plane, with Q > 0. But we think there is a good chance that the real component, IUPP cos Q, will be satisfyingly large.

We've tried to collect extensive evaluation data through machinery which has little vested interest in the success of any particular course model or in the accomplishments of the faculty at any particular test site. We expect to spend most of the next year analyzing, synthesizing, reporting. We think we're going to have some interesting results to report at the end of the Project in the summer of 1994. Obviously, we hope that these results will provide useful ingredients for the next wave of university physics textbooks and the next wave of modifications in learning settings. Stay tuned.

References

1. R. C. Hilborn, *Am. J. Phys* 56, 14 (1988).
2. Amherst College, California State at Fullerton, Georgia Tech, Minnesota, Smith College, Southwest Missouri State University, Tulane, U. S. Military Academy, and Virginia Tech,
3. "The Introductory University Physics Project," by John S. Rigden, Donald F. Holcomb and Rosanne DiStefano, *Physics Today* 46, 32, April 1993.
4. "Improving the quality and effectiveness of introductory physics courses," *Am. J. Phys.* 25, 417 (1957).

UNDERSTANDING OR MEMORIZATION: ARE WE TEACHING THE RIGHT THING?

Eric Mazur, Department of Physics
Harvard University

Introduction

When people I meet ask me what I do for a living and I tell them I am a *physicist*, I frequently hear horror stories about high school or college level physics — almost to the point of making me feel embarrassed about being a physicist! This general sense of frustration with introductory physics (mechanics, electricity and magnetism) is widespread among non-physics majors who are required to take physics courses. Even physics majors are frequently dissatisfied with their introductory courses and a large fraction of students initially interested in physics end up majoring in a different field.

Frustration with introductory physics courses has been commented on since the days of Maxwell and has recently been publicized by Sheila Tobias.[1] Tobias asked a number of graduate students in the humanities and social sciences to audit physics courses and describe their complaints. One may be tempted to brush off complaints by non-physics majors as coming from students who are *ipso facto* not interested in physics. Most of these students, however, are *not* complaining about other required courses outside their major field.

The way physics is taught in the 1990's is likely not much different from the way it was taught — to a much smaller and more specialized audience — in 1890. The basic approach of introductory physics textbooks has not changed in over one hundred years, yet the audience has. Physics has become a building block for many other fields including chemistry and the engineering and life sciences. As a result, the enrollment in physics courses has grown enormously, with the majority of students not majoring in physics. This shift in constituency, from physics majors with an interest in the subject to non-physics majors required to take physics — "captives" as Richard Crane calls them,[2] has caused a significant change in student attitude towards the subject and made the teaching of introductory physics a considerable challenge. While traditional methods of instruction have produced many successful scientists and engineers, far too many students are unmotivated by the conventional approach. What, then, is wrong with the traditional approach to introductory physics?

For the past nine years I have been teaching an introductory physics course for engineering and science majors at Harvard University. Until a number of years ago I taught a fairly traditional course in an equally traditional lecture-type of presentation, enlivened by classroom demonstrations. I was generally satisfied with my teaching during these years — my students did well on what I considered difficult problems and the response I received from them was very positive.[3] As far as I knew there were not many problems in *my* class.

A number of years ago, however, I came across a series of articles[4] by David Hestenes of Arizona State University, which, to put it bluntly, "opened my eyes". In these articles, Hestenes shows that students enter their first physics course possessing strong beliefs and intuitions about common physical phenomena. These notions are derived from personal experiences, and color students" interpretations of material presented in the introductory course. Hestenes" research shows that instruction does very little to change these "commonsense" beliefs.

For example, after a couple of months of physics instruction, all students will be able to recite Newton's third law — "action is reaction" — and most of them can apply this law in numerical problems. A little probing beneath the surface, however, quickly shows that many of these students lack fundamental understanding of the law. Hestenes provides many examples in which the students are asked to compare the forces of different objects on one another. When asked, for instance, to compare the forces in a collision between a heavy truck and a light car, a large fraction of the class firmly believes the heavy truck exerts a larger force on the light car than vice versa. When reading this, my first reaction was "Not *my* students...!" Intrigued, I nonetheless decided to test my own students" conceptual understanding, as well as that of the physics majors at Harvard.

The first warning came when I gave the test to my class and a student asked "Professor Mazur, how should I answer these questions? According to what you taught us, or by the way I *think* about these things?" Despite this warning, the results of the test came as a shock: the students fared hardly better on the Hestenes test than on their midterm examination on rotational dynamics. Yet, the Hestenes test is *simple* — yes, probably too simple to be considered seriously for a test by some colleagues — while the material covered by the examination (rotational dynamics, moments of inertia) was, so I thought, of far greater difficulty.

I spent many, many hours discussing the results of this test with my students one-on-one. My previous feeling of satisfaction with my teaching accomplishments turned more and more into sadness and frustration. How could these undoubtedly bright students, capable of solving complicated problems, fail on these ostensibly "simple" questions?

To understand these seemingly contradictory facts, I decided to pair, on the students" remaining examinations, "simple", qualitative questions with more "difficult", quantitative problems on the same physical concept. Much to my surprise some 40% of the students did *better* on the quantitative problems than on the conceptual ones — on the subject of dc-circuits half a dozen even managed to receive *full marks* on a complex quantitative problem involving a two-loop circuit while getting *zero* points on a related "simple" conceptual question! Slowly, the underlying problem revealed itself: many students concentrate on learning "recipes", or "problem solving strategies" as they are called in textbooks, without considering the underlying concepts. Plug and chug! Many pieces of the puzzle suddenly fell into place. The continuing requests by students to do more and more problems and less and less lecturing — isn't this what one would expect if students are tested and graded on their problem solving skills? The unexplained blunders I had seen from apparently "bright" students — problem-solving strategies work on some, but surely not on all problems. Students" frustration with physics — how boring physics must be when it is reduced to a set of mechanical recipes that do not even work all the time! And yes, Newton's third law is second nature to me — it's *obviously* right, but how do I convince my students? Certainly not by just reciting the law and then blindly using it in problems... After all, it took mankind thousands of years to formulate the third law.

Before I had been oblivious to this problem. By the traditional measures — quantitative problem skills and student feedback — I had been fooled into believing that I was succeeding in teaching and that the students were succeeding in learning introductory physics. Now the picture looked quite different. While several leading physicists have written on the students" lack of fundamental understanding,[5], I believe many are still unaware of the magnitude of the problem — as I was until just a few years ago.

An important problem with the conventional teaching method is that it favors problem solving over conceptual understanding. As a result, many students memorize "problem solving strategies"; for these students introductory physics becomes nothing more than problem solving by rote and little understanding of the fundamental principles is gained. This practice of memorizing algorithms and equations without understanding the concepts behind the manipulations is intellectually unrewarding and results in poor student performance and frustration with the material. And what good is it to teach just the mechanical manipulation of equations without achieving understanding?

Another problem lies in the presentation of the material. Frequently, it comes straight out of textbooks and/or lecture notes, giving the students little incentive to attend class. The fact that the traditional presentation is nearly always in the form of a monologue in front of an entirely passive audience compounds the problem. Only exceptional lecturers are capable of holding students" attention for an entire lecture period. It is even more difficult to provide adequate opportunity for the students to *critically think* through the arguments being developed and in introductory classes few students have the motivation and the discipline to do this on their own after class. Consequently, the lectures only reinforce the students" feeling that the most important step in mastering the material is solving problems. One ends up in a rapidly escalating loop whereby the students will request more and more example problems (so they can learn better how to solve them), which in turn further reinforces their feeling that the key to success is problem solving.

In the past three years I explored new approaches to teaching introductory physics. In particular, I was looking for ways to refocus students" attention on the underlying concepts without sacrificing the students" ability to solve problems. During this period of time, I developed a method of teaching, called *Peer Instruction*[6] which I"ll describe in the remainder of this paper. It has become clear that *Peer Instruction* is very effective in teaching the conceptual underpinnings in introductory physics and leads to better student performance on traditional problems. This has been verified not only at Harvard University, but also in a number of other schools, ranging from state schools, to liberal arts colleges, to military academies. Most interestingly, I have found it makes teaching easier and more rewarding.

Peer Instruction: Getting Students To Think In Class

The basic goal of the method is to exploit student interaction in class and to focus the students" attention on underlying concepts. Instead of presenting the material sequentially (as it is in the textbook and in the notes), lectures consist of a number of short presentations of the key points of the material, each followed by a so-called *ConcepTest* — a short multiple-choice conceptual question on the subject being discussed. The students are first given some time to formulate an individual answer, and then asked to discuss their answers with each other in the classroom. This process *a*) forces the students to critically think through the arguments being developed and *b*) provides them (as well as the teacher) with a way to assess their understanding of the concept.

The students" answers to the *ConcepTests* also provide a continuous assessment of the students" understanding of the material *during* the class. If the students" performance on the *ConcepTest* is satisfactory, the lecture can proceed to the next topic. Else, the teacher should slow down, lecture in more detail on the same subject, and re-assess the students with another *ConcepTest* on that subject. This prevents a gulf from developing between the teacher's expectations and the students" understanding — a gulf, which once formed, only increases with time until the entire class is "lost".

1.	Lecture on concept 1 (including demos, etc.)	7-10 minutes
	ConcepTest 1: do students understand concept 1?	5 minutes
	If no: back to square 1!	varies
	If yes: continue	
2.	Lecture on concept 2 (including demos, etc.)	7-10 minutes
	ConcepTest 2: do students understand concept 2?	5 minutes
	If no: back to square 2!	varies
	If yes: continue	
3.	Lecture on concept 3 (including demos, etc.) *etc. etc.*	

Table 1. General outline for a lecture introducing three (or more) new concepts. An outline for each ConcepTest *is shown in Table 1.*

Table 1 shows an outline for a lecture introducing three or more new concepts. Notice how the students" understanding of each concept is verified with at least one *ConcepTest* after approximately 10 minutes of lecturing. Each of these has the general format outlined in Table 2. The total duration of a single lecturing–*ConcepTest* cycle is approximately 15 minutes. Central to the entire method, therefore, is a set of simple qualitative questions for the *ConcepTests*, each dealing with a *single* fundamental concept. These questions help focus the students" attention on understanding *before* problem-solving. In the traditional approach to introductory physics, on the other hand, understanding is assumed to follow from mechanically working problems.

1.	A simple conceptual question is posed	1 minute
2.	Silence: the students are given time to think	1 minute
3.	Students record their answers (optional)	
4.	Chaos: the students are asked to "convince" their neighbors	1 minute
5.	Students record their answers (optional)	
6.	Feedback to instructor: Tally of answers	
7.	Explanation of answer to question	2+ minutes

Table 2. Format of a single ConceptTest.

Let's consider for a moment a specific example — a lecture on fluids. Suppose the concept we want to get across is that of Archimedes" principle. We first lecture for about 7-10 minutes on the subject of Archimedes" principle — emphasizing the concepts and the ideas behind the proof and avoiding (or even omitting) any equations and derivations. This short lecture period could include a demonstration (the Cartesian diver, for instance). Then, before going on to the next topic (Pascal's principle, perhaps), we project the following multiple-choice question[7] (step 1):

> Consider a bathtub brimful of water. Next to it is a second, identical bathtub, also brimful of water, but with a battleship floating in it. Which one weighs more?
>
> 1. A bathtub brimful of water
> 2. A bathtub brimful of water with a battleship floating in it
> 3. Both weigh the same

It is important to read the question with the students and to make sure there are no questions as to the precise meaning of the question itself (strange as this may sound!). Next, we tell the students that they have one minute to come up with an answer — more time would allow the students to fall back onto equations and manipulate equations rather than *think*. It is now silent in the class-room as the students concentrate on the question (step 2). After about a minute we ask the students to record their answer (step 3; see also Appendix 2), and then to try to convince their neighbors of their answers. The silence turns into chaos as *everyone*, intrigued by the question, tries to argue with surrounding neighbors (step 4). After giving the students a minute to argue, we ask them to record a revised answer (step 5). Then we go back to the overhead and ask for a show of hands to see how many selected the various answers. I use the above question in my own class; the results are shown in Figure 1 (on next page).

Notice that 78% of the students got the correct answer before discussion and 88% after discussion. The pie charts show another benefit of the discussion periods: the fraction of students who are "pretty sure" of their answer increases from 56% to 81%. Of course, I did not have access to such detailed results in class, but the show of hands would have revealed an overwhelming majority of correct answers. I would therefore have spent only a few minutes explaining the correct answer before going on to the next topic.

The increase in the number of correct responses and the students" confidence is frequently much more pronounced than in the example shown in Figure 1.[8] Repolling the students" responses after the *Peer Instruction* periods systematically reveals a strikingly greater proportion of correct answers. It seems that the students are able to explain concepts to one another more efficiently than are their instructors, for whom such concepts are second nature. The students who understand the concepts when the question is posed, on the other hand, have only recently mastered the idea. Because of this, they are still aware of the difficulties one has in grasping that particular concept. Consequently, they know precisely what to emphasize in their explanation. Similarly, many seasoned lecturers know that their first presentation of a new course is often their best. Their initial presentation of the material is marked by a clarity and freshness often lacking in the more "polished" version. The underlying reason is the same: as time passes and one is continuously exposed to the material, the conceptual difficulties seem to disappear.

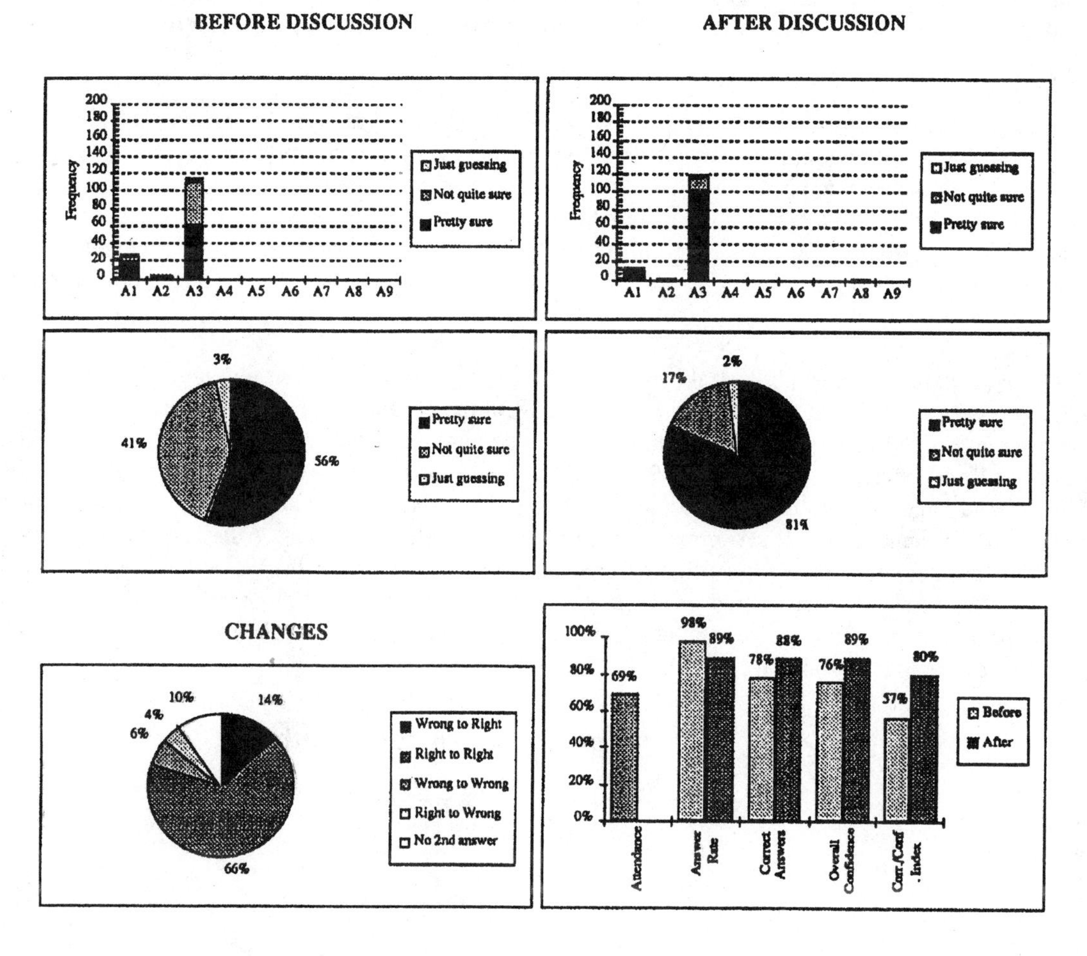

Figure 1. Data analysis of the ConcepTest about Archimedes principle shown in the text. The students" initial responses and confidence levels are displayed on the left; the students" revised post-discussion responses and confidence levels are displayed on the right. The lower left graph shows how students changed their mind as a result of the discussion.

In this new lecturing format, the *ConcepTests* take about one third of the total time in each lecture. This necessarily means that less time is available for straight lecturing. One therefore has two choices: (*a*) discuss only part of the material in the lectures, or (*b*) reduce the overall coverage of the material in the course. While (*b*) may eventually be the preferable choice, I have opted for choice (*a*). I do not cover all the material in class — after all, the details are always available in the book or in the notes. I start by throwing out nearly all derivations and *all* example problems (yes, that's right). As I have argued, students derive precious little benefit from seeing the instructor manipulate equations anyway. To make up for the omission of these more mechanical aspects of the course, I *require* the students to *read* the material ahead of the class. While this may sound surprising for a science course, students are accustomed to reading assignments in many other courses. In this way I can continue to cover the same amount of material as before. Moreover, the students" attention is focused more strongly on the underlying principles. The students still get the opportunity to learn problem solving in weekly sections, half of which are devoted to developing problem skills. In addition, the home assignments consist half of traditional problems, half of essay type questions.

Results

Before getting into the specifics and providing more detailed guidelines on the implementation of *Peer Instruction*, let me first summarize some of the results I have obtained with this method in my course. Results, I should emphasize, that are supported by findings from other institutions where *Peer Instruction* has been implemented.[9]

The advantages of *Peer Instruction* are numerous. The discussion periods break the unavoidable monotony of passive lecturing. Not only are the students kept alert, but they are actively involved in the lecture. I have found that the discussions among the students are always remarkably uninhibited and animated. Furthermore, the students do not merely assimilate the material presented to them; they must think for themselves and put their thoughts into words. The data collected over the past three years demonstrate that following the *Peer Instruction* periods, student confidence as well as the proportion of correct answers increases dramatically and systematically.

The long-term gains are even more striking. In the past few years I have used a diagnostic test, the *Force Concept Inventory* developed by Hestenes,[10] to test student understanding of the underlying concepts. This test has been utilized in a number of studies across the country to determine the effectiveness of physics instruction. Data obtained in my class in 1990 and 1991 allows one to compare the relative effectiveness of the *Peer Instruction* and the traditional approaches. The results are shown in Figures 2 and 3. Figure 2 shows a dramatic improvement in student performance using the *Peer Instruction* method. After instruction, only 4% of the students are below the cutoff identified by Hestenes as the threshold for the understanding of Newtonian mechanics. Notice also how the scores are strongly shifted towards full marks (29 out of 29). Conversely, with the traditional approach used the year before (1990), the improvement was much smaller, in agreement with what Hestenes has found at other institutions.

1991 results: peer instruction method

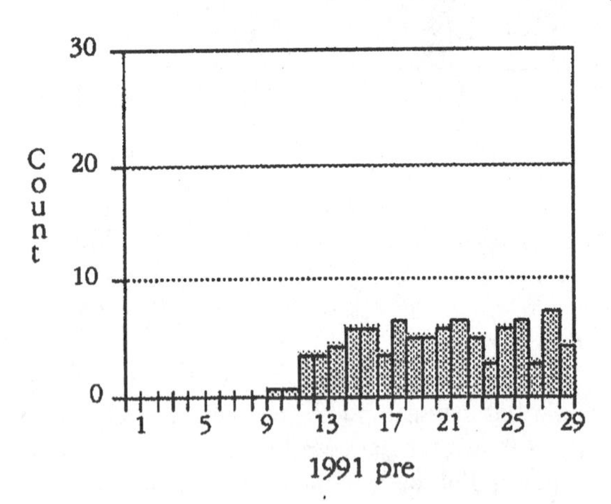

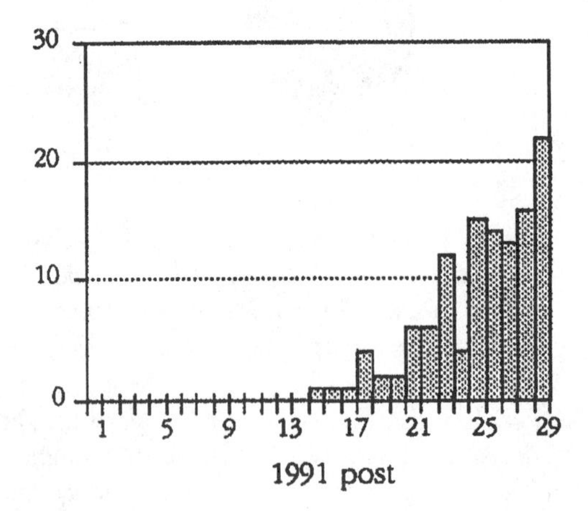

Figure 2. Histograms of scores on the Force Concept Inventory test obtained in 1991 on the first day of class (left) and after two months of instruction with the Peer Instruction method (right). The maximum score on the test is 29 (out of 29). The means of the distributions are 19.8 and 24.6.

1990 results: traditional instruction method

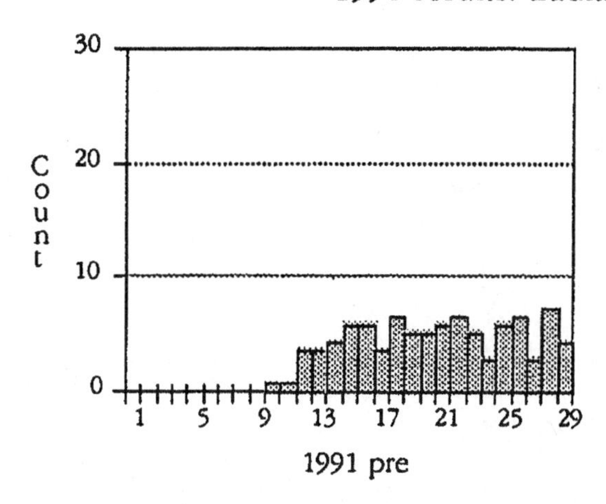

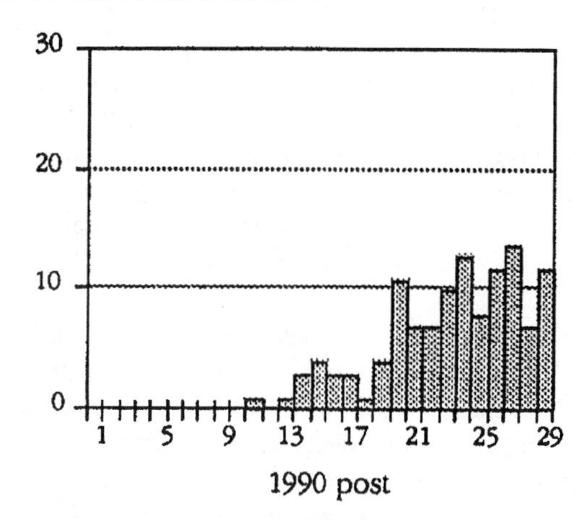

Figure 3. Histogram of scores on the Force Concept Inventory test obtained in 1990 after two months of traditional instruction (right). For comparison, data obtained on the first day of class in 1991 are shown on the left. The means of the distributions are 19.8 and 22.3.

While the improvement in conceptual understanding is undeniable, one might question how effective the new approach is in teaching the problem solving skills required on traditional examinations. After all, the restructuring of the lecture and its emphasis on conceptual material is achieved at the expense of time devoted to problem solving. To answer this question, I gave the *identical* final examination in 1991 as the one I gave in 1985. Figure 4 shows the distributions of final examination scores for the two years. Given the students" improvement in conceptual understanding, I would have been satisfied if the distributions were the same. Instead, there is actually a marked *improvement* in the mean, as well as a higher cut-off in the low-end tail.

conventional examination results

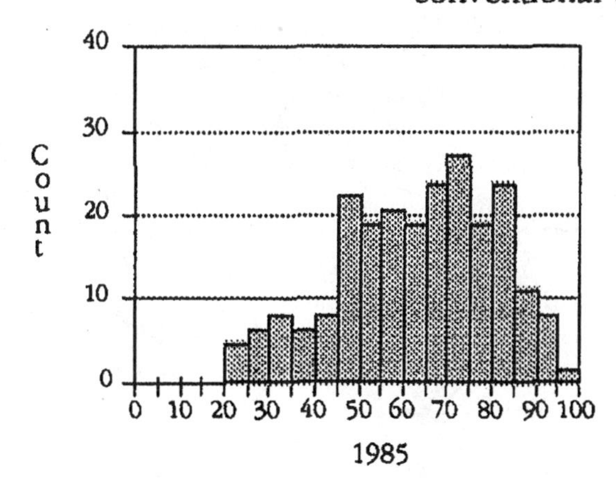

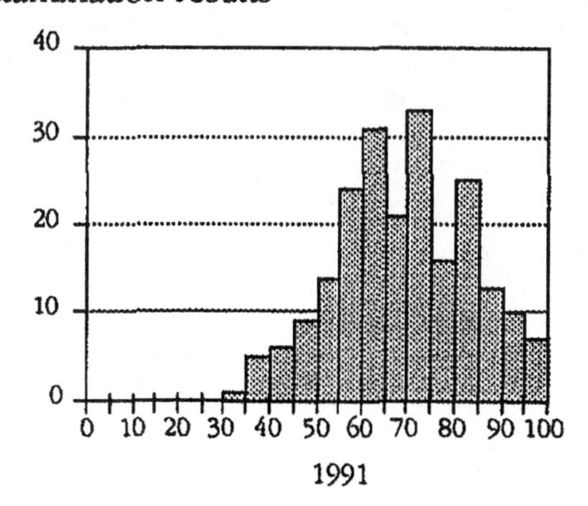

Figure 4. Histograms of final examination scores in 1985 (left) and 1991 (right). In both cases the examination was the same. In 1985 the course was taught in a traditional manner; in 1991 the method of Peer Instruction was used. The means of the distributions are 62.7 and 69.4 out of 100 points maximum in 1985 and 1991, respectively.

Converting Lectures From The Old To The New Format

Below I will attempt to describe what I have done with my own material over the past few years to change from a traditional lecturing style to *Peer Instruction*. I should stress that I still use my old lecture notes — it is not necessary to completely rewrite one's lecture notes! I hope this description will therefore serve as a guide for converting your own material for use with *Peer Instruction*.

1. *Reading assignments.* Since the *ConcepTests* take time away from the lecture, it will not be possible to devote as much time to straight lecturing as before. As I mentioned, I have *completely* eliminated all worked sample problems and many derivations from my lecture. Even though this may come as a surprise to many, the literature abounds with indications that students derive little or no benefit from seeing someone solve a problem. Besides, the results shown above indicate that the students" ability to solve problems is not affected by the omission of examples and/or derivations.

I tell the students on the first day of class that I will not lecture straight out of my notes or out of the textbook and that I *expect* them to read the relevant material (notes and book) in advance. To make sure they actually carry out their reading assignments, I provide them with some incentive.[11] As a result, I am still able to cover the same amount of material as before implementing *Peer Instruction*.

In fact, at the first lecture I distribute a schedule of lectures that provides the students with a reading assignment for the entire semester, and I stick *religiously* to the schedule — better than I ever was able to do before. If a certain lecture goes faster than anticipated (a rare event), the students get an early break; nobody is unhappy. If a certain lecture goes slower than planned (usually because a *ConcepTest* revealed some difficulty with the material), I skip the less important part(s) and rely on *a*) the students" reading, *b*) sections (weekly discussion sessions), and *c*) homework assignments. In some cases, I may use part of the next lecture to stress some important points or to give an extra *ConcepTest*. In any case, I always plan a single review lecture in the middle of the semester to allow for some slack in an otherwise very rigid schedule. So the flexibility is in the schedule of the individual lecture, not that of the semester.

A key point is thus to get the students to do part of the work ahead of time. Unfortunately most books are not ideal; they provide so much information that the student is usually unable to determine what is relevant and what is not. Part of the reading is therefore from the lecture notes.

2. *Key concepts.* In some lectures, taking away examples and derivations leaves surprisingly little material. This, however, is the "core" material of the lecture which contains the key concepts. After taking the derivations and worked example problems out of my presentation, the next thing I do is to determine what the four or five key points are that I want to get across to the students. I also frequently consult Arnold Arons" book *A Guide to Introductory Physics Teaching* for additional advice on where to expect the most difficulties. Eventually I am left with a skeletal lecture outline consisting of four or five key points (see Table 3).

1. definition of pressure
2. pressure as a function of depth
3. Archimedes" principle
4. Pascal's principle

Table 3. Outline for a lecture on fluid statics

3. *ConcepTests.* At this point, it is important to develop a number of good conceptual questions to test understanding of each of the key concepts in the above lecture outline. This constitutes perhaps the largest amount of work in converting the lecture. The importance of this task should not be underestimated — the success of the method depends to a large extent on the quality and the relevance of these questions. Sources for (inspiration for) such questions are listed in the next section.

While there are no hard and fast rules for the *ConcepTests*, they should at least satisfy a number of basic criteria. Specifically, they should:

1) focus on a single concept,
2) not be readily solvable by relying on equations,
3) have adequate multiple choice answers,
4) be unambiguously worded, and
5) be neither too easy nor too difficult.

The first three of these points are the most important because they directly affect the feedback of information to the instructor. If more than one concept is involved in the question, it will be more difficult to interpret the results of the question and correctly gauge the students" understanding. Similarly, if the students can derive the answer by relying on equations, then the students" responses will not adequately reflect their real understanding. The choice of answers provided is another important point. Ideally, the incorrect answers should reflect the students" most common misconceptions. At present, the incorrect answers to each *ConcepTest* have been formulated with this criterion in mind, but the ultimate source for the alternative responses ("detractors") should be the students themselves. For instance, by posing the question in a "fill-in" format and then tallying the most prevalent incorrect responses, a student-generated *ConcepTest* question accurately mirroring common misconceptions could be generated.

The last two points are harder to gauge beforehand, even though they may sound entirely unmistakable. I have been surprised time and again to see that questions that appeared to be completely straightforward and unambiguous to me, were misinterpreted by many students. These ambiguities can only be eliminated by class-testing questions. As for the level of difficulty, Figure 5 shows the percentage of correct answers after discussion versus that before discussion for all questions during a full semester. Notice that *all* points lie above a line of slope 1 (points on or below that line would correspond to an entirely useless discussion). As should be expected, the improvements are largest when the initial percentage of correct answers is around 50% (with jumps as large as from 40 to 90%). I consider an initial percentage of correct responses in the 50 to 80% range optimal.

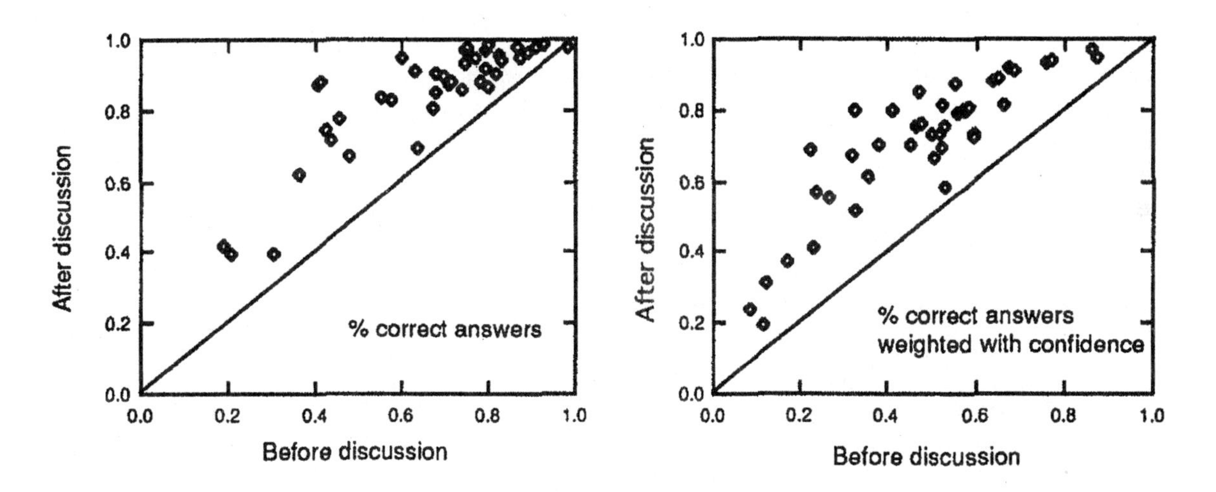

Figure 5. Percentages of correct responses after discussion versus that before discussion (left) and the same information weighted with the students" confidence.

4. *Lecture plan.* Once the questions are made up, I take a new look at my old lecture notes and decide at what point in the remaining material to put the newly made questions. At the same time, I plan which lecture demonstrations to give. Sometimes I may even combine a question with a demonstration, with one leading into the other.

5. *Lecture.* The actual lecture is much less "rigid" than before. It is necessary to keep a certain amount of flexibility to respond to the sometimes unexpected results of the *ConcepTests.* I find myself improvising more often than before. While this may seem like a disturbing prospect at first, I should say that the added flexibility

actually makes the teaching *easier* than before. During the periods of silence (when the students are thinking) I get a break — a minute or so to catch my breath and to reformulate my thoughts. During the periods of discussion, I usually participate in some of the discussion to get a feel for what goes on in the mind of the students. This helps me to focus better on the problems the students are facing and keeps me "in touch" with the class.

I should also mention that the new lecture format elicits more questions from the students than I have ever encountered before. Often these questions are very to the point and profound, and I attempt to address as many of them as I can.

Conclusion

So, with relatively little effort and no capital investment it is possible to greatly improve student performance in introductory science courses. To achieve the results reported above, I merely incorporated a number of conceptual questions into each lecture. For the remainder of my lectures I used my existing lecture notes. I omitted worked-out examples and derivations, assigning these as reading to the students. Despite the reduced time devoted to problem solving, the results convincingly show that conceptual understanding enhances student performance on traditional type examinations. Finally, student surveys show that student satisfaction — an important indicator of student success — is greatly increased.

Appendix 1: Sources for *ConcepTests*

Creating and compiling conceptual problems is an important task, since the *ConcepTests* are central to the success of the method. To make this task easier, I hope to develop an informal network of people sharing such questions with one another. Over the course of the past two years we have developed questions for all concepts covered in introductory physics.[12]

There are a number of good sources of both questions and inspiration. End-of-chapter *questions* (as opposed to "problems" or "exercises") in most standard introductory physics texts can be a useful starting point. The *American Journal of Physics* publishes many articles which may prove helpful in creating new *ConcepTests*. In addition, the books listed below emphasize fundamental concepts and contain numerous questions which are designed to both isolate these concepts, and to help students grasp them by exposing their most common misconceptions about the material. This list is not comprehensive, but represents the sources I have most frequently drawn from myself.

Arnold B. Arons, *A Guide to Introductory Physics Teaching*, John Wiley & Sons, New York (1990).
Lewis Carroll Epstein, *Thinking Physics*, Insight Press, San Francisco (1990).
Paul G. Hewitt, *Conceptual Physics*, Scott, Foresman and Company, Boston (1989).
Jearl Walker, *The Flying Circus of Physics*, John Wiley & Sons, New York (1977).

Appendix 2: Feedback Methods

One of the great advantages of *Peer Instruction* is that it provides immediate feedback on the level of student understanding; however, it requires that one keeps track of the students" answers to the *ConcepTests*. The tallying of these answers can be accomplished in a variety of ways, depending on the setting and purpose. Three methods that we have used are:

1. Show of hands
2. Scanning forms
3. Hand-held computer devices

The simplest method of data collection is a show of hands after the *Peer Instruction* periods. This method does not require any new technology or investment, but will still accomplish the goals of *Peer Instruction*. It will give a feel for the level of the class" understanding and allows the teacher to tailor the pace of the lecture accordingly. The only drawback is a certain loss of accuracy, in part because some students may hesitate to raise their hands, and in part because of the difficulty in estimating the distribution of answers. Another minor problem is the lack of a permanent record (unless one keeps data in class) and the lack of any data *before* the *Peer Instruction* discussion.

Since I was interested in actually quantifying the effectiveness of the *Peer Instruction* discussion both on the short and the long term, I have made extensive use of forms that were scanned after class. On these forms, which are reproduced on pages 50–51, the students mark their answers and the confidence they have in these answers, both before and after discussion. This method yields an enormous body of data on students" attendance, understanding, improvement, and on the short-term effectiveness of the *Peer Instruction* periods. The drawback of this method, however, is that it is labor intensive and that there is a delay in the feedback: the data are available only after the forms are scanned. For this reason, I always ask for a show of hands (in addition to requiring the students to mark their answers).

A year ago we installed an interactive computer response system called *ClassTalk*, produced by *Better Education, Inc.* The system allows the students to indicate their answers to the *ConcepTests*, as well as their confidence levels, on hand-held computers which they can share in small groups of three or four. Their responses are relayed to the instructor on a computer screen and can be projected so the students can see it too. The main advantage of the system is that the analysis of the results is available immediately. There are many additional features and advantages: student information (such as the students" name and seat location) is also available, making large classes more personal; the system can also handle numerical and non-multiple choice questions; we have also found that sharing of these handheld computers enhances the student interaction. Potential drawbacks are that the system requires a certain amount of capital investment and that it adds complexity to the lecture.

References

1. Sheila Tobias, *They"re not Dumb, They"re Different: Stalking the Second Tier* (Research Corporation, Tucson, AZ, 1989).
2. H. Richard Crane, *Am. J. Phys.* 36, 1137 (1968).
3. My ratings on the Harvard Committee on Undergraduate Education questionnaires have consistently been among the highest in the Physics Department at Harvard.
4. Ibrahim Abou Halloun and David Hestenes, *Am. J. Phys*, 53, 1043 (1985); ibid. 55, 455 (1987); Hestenes, David, *Am. J. Phys*, 55, 440 (1987).
5. See for example: Arnold Arons, *A Guide to Introductory Physics Teaching* (John Wiley & Sons: New York, NY, 1990); Richard P. Feynman, *The Feynman Lectures*, Vol. 1, (Addison Wesley, New York, NY, 1989) p. 1-1; Ken Wilson, *Phys. Today* **44:9** (1991) p. 71-73.
6. See *e.g.* Chapter 8 in *Revitalizing Undergraduate Science: Why Some Things Work and Most Don't* by Sheila Tobias (Research Corporation, Tucson, AZ, 1992).
7. This question is from Lewis Caroll Epstein, *Thinking Physics*, Insight Press, San Francisco (1990).
8. The improvement is usually largest when the initial percentage of correct answers is about 50%. If it is much higher (as in the case of Figure 1), there is little room lest for improvement. If it is much lower, there are too few students in the audience who are able to convince others of the correct answer. See also Figure 5.
9. See Sheila Tobias, *Revitalizing Undergraduate Science Education: Why Some Things Work and Most Don't*, (Research Corporation: Tucson, AZ, 1992).
10. D. Hestenes, M. Wells, G. Swackhamer, *The Physics Teacher* 30, 141 (1992).
11. At the beginning of the class, I give the students a special *ConcepTest* called a *Bonus Question* which allows the students to earn some credit towards their final grade. This question differs from the others in that the material is not first discussed and requires the students to have *read* the material before coming to class. In addition, the responses are collected immediately and the students do not discuss the answers with one another.
12. Typically between two and five fundamental concepts are introduced each lecture, for a total of about 150 for a standard introductory physics text. We have developed complete sets of questions and obtained detailed statistics on student performance for each of these. They are available upon request.

A New Order for Mechanics

Priscilla W. Laws, Department of Physics and Astronomy
Dickinson College

Abstract

A number of researchers have reported on conceptual difficulties students encounter in the study of Newton's Laws, including Newton's Third Law. This paper describes a project to restructure the introductory physics mechanics curriculum to present Newton's Laws in a more logical sequence. The curriculum is based on the use of direct experience coupled with Microcomputer-Based Laboratory (MBL) tools. Particular attention is given to the sequence of learning experiences developed to improve students' understanding of Third Law concepts applied to collision processes. The results of pre- and post-testing show significant gains in students' ability to apply the Third Law to different types of interactions.

Introduction

In the study of introductory mechanics, acquiring a conceptual understanding of Newton's Laws has proven to be one of the most difficult challenges faced by students. Recent surveys by Hestenes, et. al.,[1] of student conceptual gains before and after traditional instruction have been disappointing. Arons and Rothman[2] have observed that the treatment of Newtonian dynamics is logically inconsistent in many popular textbooks. In addition, science education researchers have discovered that many students begin the study of mechanics with misleading conceptions about the nature of motion, which are extremely hard to overcome[3].

Research has shown that microcomputer-based laboratory tools are effective in enhancing student learning in kinematics and dynamics.[4] Recent improvements in microcomputer-based laboratory systems for the study of force and motion[5] and the development of low friction dynamics carts[6] have made it possible to design new activities in which students can observe relationships between force and motion quickly and easily.

David Sokoloff, Ronald Thornton and the author outlined a sequence of laboratory activities for teaching mechanics concepts. These activities were designed for the Workshop Physics program[7] and the RealTime Physics project.[8] In the summer of 1992 a small conference was held at Tufts University for individuals active in physics education research and curriculum development.[9] Participants were asked to critique the ideas for the new mechanics sequence. Thus, the outcomes of research on student learning, insights offered by Arons and Rothman on logical development, new MBL tools, activities designed for Workshop Physics and Tools for Scientific Thinking programs, and ideas generated by participants in the new mechanics conference were integrated into a new activity-based mechanics curriculum. During the 1992-93 academic year this curriculum was tested in the Workshop Physics programs at Dickinson College and Gettysburg High School and in activity-centered RealTime Physics laboratories at the University of Oregon and Arizona State University.

The New Mechanics Sequence

The New Mechanics sequence differs from the traditional sequence in several ways:

(a) The order in which Newton's three laws are presented is based on the difficulty students appear to encounter in understanding them. Students begin with Second Law activities before they consider First

Law phenomena.[10] Finally they work with Third Law concepts which appear to be the hardest to master.[11]

(b) Activities using MBL force and motion sensors and low friction dynamics carts are designed to enable students to make direct observations of basic elements of Newtonian dynamics without recourse to textbooks.

(c) Extra efforts are made to help students look at the elements of Newton's laws and be able to distinguish definitions such as acceleration, force, and inertial mass from observed phenomena; for example, more "pull" causes more acceleration and more "stuff" causes less acceleration.

(d) Concepts in kinematics and dynamics are initially developed for one-dimensional horizontal motion with visible applied forces (pushes or pulls) with little friction present.

(e) Students are then asked to make additional observations which lead them to invent invisible forces (i.e., friction forces, gravitational interaction forces, normal forces, and tension forces) in order to maintain the viability of the Newtonian schema for predicting motions.

(f) The study of kinematics and dynamics is finally extended to two dimensional phenomena such as projectile motion, circular motion and motion on an incline.

(h) Students study the forces in collisions, the Law of Conservation of Momentum, and center-of-mass concepts before dealing with the conservation of energy.[12]

Although this paper focuses primarily on elements of the sequence designed to help students acquire an understanding of Newton's Third Law and collision processes, key elements of the New Mechanics Sequence are summarized in Table 1 below. A more detailed description of the activities developed to help students understand these elements will be published in the near future in an article currently being prepared by the author in collaboration with David Sokoloff and Ronald Thornton.

New Mechanics	**Traditional Mechanics**
I. *1D Kinematics w/ MBL*	I. *1D Kinematics*
–body motion	–lectures
–constant velocity	–textbook problems
–constant acceleration	–lab w/const.accel
II. *1D Dynamics*	II. *2D Kinematics*
(low friction w/ visible applied forces)	–projectiles
–define constant F_{app}	–centripetal acceleration
–observe $a \propto F_{app}$	
–define F scale	
–observe more stuff $\rightarrow < a$	
–define static mass scale	
–observe $a \propto (1/m)$	
–observe $F_{net} \equiv \Sigma F_{app} = ma$	
i.e. if $\Sigma F_{app} = 0$, then v=const.	
III. *1D Dynamics (invisible Fs)*	III. *1D and 2D Dynamics*
–observe friction w/ visible drag as a "passive force"	–state First Law
–postulate friction as force	–state Second Law F=ma
–observe vertical fall	–state superposition i.e $F_{net} \equiv \Sigma F$
	–describe action of forces

–postulate constant gravitational force –observe effects of strings and surfaces –postulate Newton's third law to explain a= 0 cases –postulate T and N as passive forces	due to gravity, friction, surfaces and strings –state third law as "action/reaction" –state an elaborate rules for problem solutions w/ free body diagrams
IV. 2D Dynamics –observe vector combination of forces –discover rules for freebody diagrams and 2D problem solving –apply Newton's laws to projectile, circular and inclined plane motion	*IV. Mechanical Energy Conservation*
V. Momentum Conservation and Collisions –Observe impulses and p-change –Derive $F\Delta t=\Delta p$ theorem from Newton's 2nd Law –Use MBL to verify $F\Delta t=\Delta p$ theorem experimentally –predict interaction forces during collisions if $m_1>m_1$ or $v_1>v_2$ –Use MBL to observe interaction forces ($F_{12}=-F_{21}$ always) –Combine dynamic Third Law w/ $F\Delta t=\Delta p$ to get 1D p-conservation –Center-of-mass, 1D collisions, 2D collisions, and particle systems	*V. Momentum Conservation and Collisions* –Impulse –p-conservation –1D collisions –2D collisions –Center-of-mass –Particle systems
VI. Mechanical Energy Conservation	

***Table 1**: Outline of the New Mechanics Sequence*

Helping Students Understand the Third Law

Overview

One of the most challenging and interesting parts of the New Mechanics sequence involves the application of Newton's Third Law to one-dimensional collision processes. Many students can apply Newton's Third Law to the construction of free body diagrams when two interacting objects are in equilibrium. Virtually all introductory physics students can recite Newton's Third Law in the form "for every action there is an equal and opposite reaction" or "forces are always equal and opposite." However, the majority of students who complete introductory mechanics either do not understand the meaning of these phrases or don't really believe them when considering contact forces in collisions. For example, traditional instruction in two high school classes in Arizona, one regular and one honors, reduced average error rates from 90% to only 72% on conceptual questions requiring an understanding of Third Law concepts.

The existence of common misconceptions about interaction forces in collisions is not surprising for two reasons. First, when students observe elastic collisions between a rapidly moving object (i.e. an active agent) and a stationary object having the same mass, a dramatic momentum transfer seems to take place. Pretest scores indicate that about 80% to 90% of students begin the study of introductory physics with the belief that in a collision there are circumstances under which one object exerts more force on another. Second, when students observe a head-on collision between a heavy object and a light one moving at the same speed, the light object undergoes a more dramatic acceleration than the heavy one. This leads to the belief that the object with the greatest mass exerts the most force in a collision. Only students with good physics training and intuition recognize that Newton's Second Law reveals that momentum changes are essential to determining relative magnitudes of interaction forces.

Arons asserts that an understanding of Newton's Third Law requires students to recognize that "*all* interacting objects exert equal and opposite forces on each other *instant by instant*, and this applies to widely separated gravitating bodies as well as to those exerting contact forces on each other, and that zero time elapses between a change occurring at one body and the effect of the change being felt by the other."[13] From the perspective of modern physics, we now understand that the requirement that zero time elapse between a change in one body and a change in another cannot be met. Thus, although Newton's Third Law does hold for mechanical contact forces between objects, modern physics ultimately gives primacy to Conservation of Momentum in the hierarchy of physical law. In fact, one of Newton's many brilliant insights was that the experimental fact that momentum (quantity of motion) is conserved in collisions implies that the interaction forces between two objects must have the same magnitude.

The Sequence of Activities on Collisions and the Third Law

A major goal of this sequence of activities is to help students understand how Newton's Laws lead naturally to the Law of Conservation of Momentum in the description of collision processes. As we explained in section 2 we decided to introduce momentum and its conservation before exposing students to energy concepts. The sequence of activities was designed to help students to: (1) understand the relationship between forces experienced by a single object and its change in momentum, (2) consider mutual interaction forces between two bodies undergoing a collision, and (3) realize that the Law of Conservation of Momentum is a consequence of Newton's Second and Third Laws. Students perform the following activities.

Recasting Newton's Second Law in Momentum Form

a. *Using a Thought Experiment to Define Momentum:* Students perform a thought experiment and try to predict at what speed, V, a small car of mass m must move in order to stop a truck of mass M moving at a slower speed, v. The outcome of this discussion is used as a basis for defining momentum as $p = mv$.

b. *Deriving Newton's Second Law in terms of momentum:* Students show mathematically that

$$\sum \bar{F} = m\bar{a} = \frac{d\bar{p}}{dt}$$

The Impulse-Momentum Theorem

a. *Reviewing the mathematical definition of p-change:* Students need help realizing that a super-ball undergoes more momentum *change* than a clay blob. They are asked to practice calculating momentum changes.

b. *Gaining intuition about impulse, average force and momentum change:* Students discuss why they tend to catch a raw egg more slowly than a ball. This discussion makes the definition of impulse as the time integral of force a bit more plausible.

c. *Measuring Impulses:* Quantitative data on forces during a collision is performed using the MBL system set up with motion software and a force probe. A force probe is attached to a cart and allowed to collide gently with a wall or another object. The data analysis feature of the motion software is used to determine the impulse resulting from the collision. A sketch of the apparatus is shown in Figure 1.

d. *Deriving the Impulse-Momentum Theorem from Newton's Second Law:* Students are asked to perform a mathematical derivation to show that

$$\int_{ti}^{tf} \Sigma\vec{F}\, dt = \int_{ti}^{tf} \frac{d\vec{p}}{dt}\, dt = (\vec{p}_f - \vec{p}_i) = \Delta\vec{p}$$

e. *Verifying experimentally that the impulse-momentum theorem holds:* A quantitative experiment is performed using the MBL system set up with motion software, a force probe, and a motion detector. A force probe is attached to a cart so that it can undergo a relatively slow collision with something soft such as a piece of foam rubber. Force readings are taken during the collision while a motion detector is used to determine the velocity of the cart just before and just after the collision. Students find that the quantitative verification of the impulse-momentum theorem is good to within 5 or 10% if they take careful measurements.

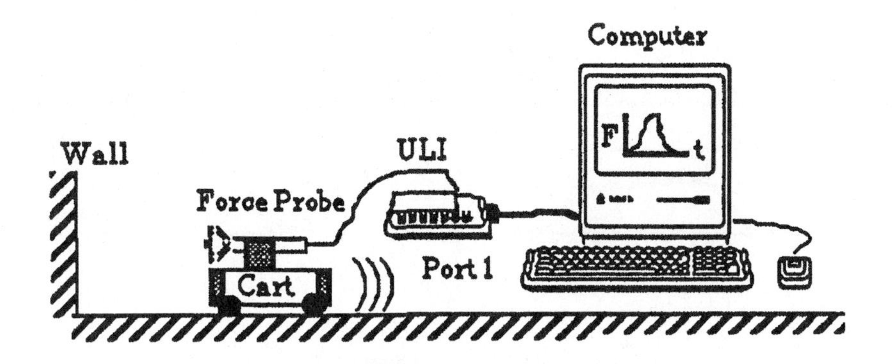

Figure 1: MBL apparatus set up for measuring collision forces on a force probe mounted on a cart. If a motion detector is placed to the right of the cart, the impulse-momentum theorem relating the change in momentum to the time integral of the collision force experienced by the cart can be easily verified.

Mutual interaction forces

a. *Predicting relative force magnitudes in a collision:* Students are presented with several collision scenarios such as two cars of equal mass undergoing a collision, a moving car hitting a stationary truck, and a school bus smashing a fleeing mosquito. They are asked to predict the relative forces and discuss the circumstances under which one object might exert a greater magnitude of force on another object.

b. *Observing interaction forces in "real time":* Students are given two low friction carts with force probes mounted on them and some extra masses. The force probes are hooked into an MBL system, and they are provided with a special version of the motion software which can record data from two probes simultaneously. Students are then asked to use this equipment to investigate the circumstances under which one object exerts more force on another. A typical setup is shown in Figure 2. When these observations are carefully done, students discover that contact forces of interaction are equal in magnitude and opposite in direction on an instant-by-instant basis for all circumstances including that of a heavily loaded cart bearing down on a light cart which is at rest. Many students are surprised to see the force vs. time graphs for the two force probes looking equal and opposite! Sample graphs from these types of experiments are shown in Figures 3 and 4 for slow and fast encounters respectively.

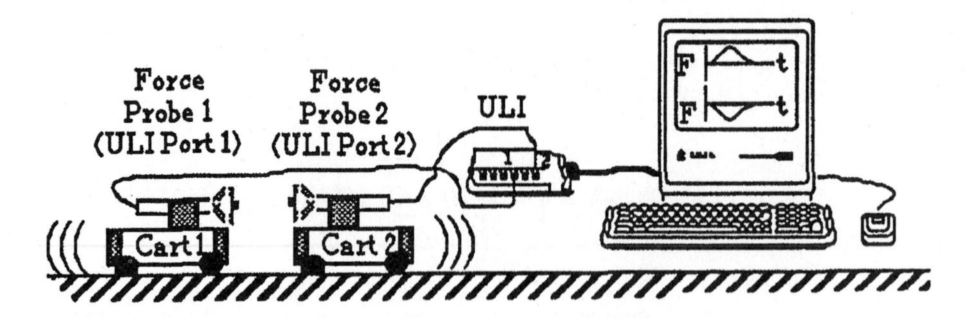

Figure 2: MBL set up for reading two forces at once during a gentle collision

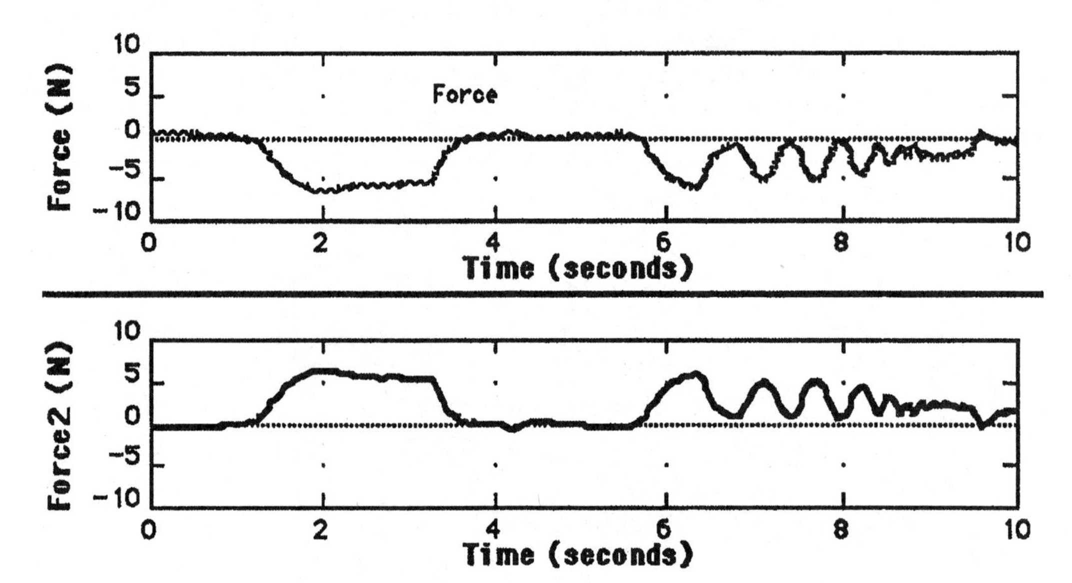

Figure 3: Two carts undergo slow collisions. Sometimes the first cart does the pushing and other times the second cart does the pushing. These graphs are made with MBL software, a ULI, and two force probes.

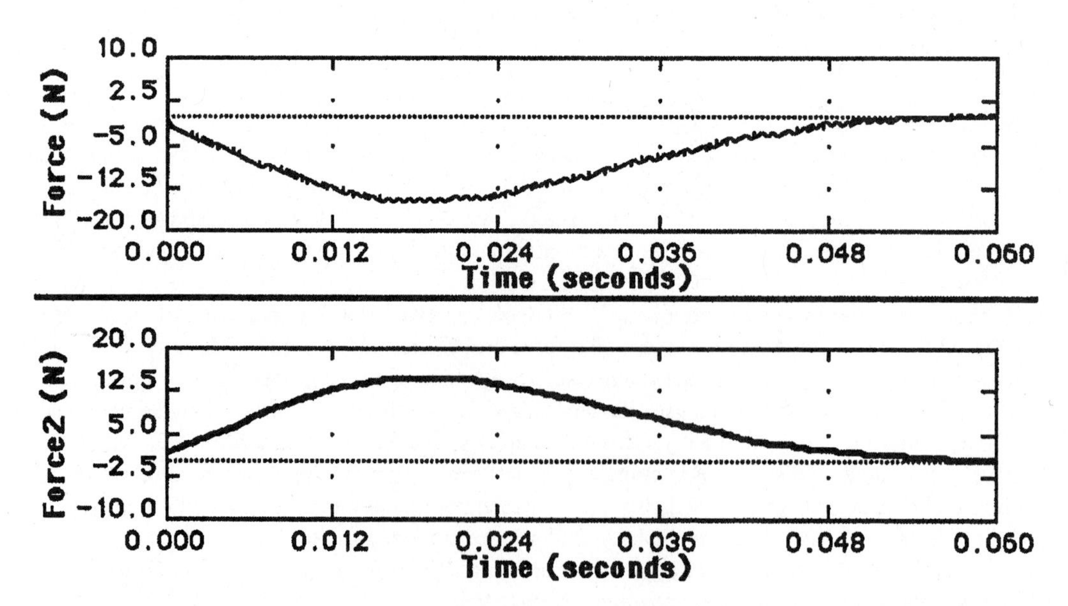

Figure 4: A 1.0 kg cart which is moving collides gently with a 0.5 kg cart which is at rest. These graphs are made with MBL software, a ULI, and two force probes. The data rate was set for 1000 readings per second.

Momentum conservation

a. *Deriving Momentum Conservation as a Consequence of the Second and Third Laws:* Students combine the impulse-momentum theorem which is a form of the Second Law and the Third Law to predict mathematically that momentum ought to be conserved for collision processes.

b. *Observing Momentum Conservation Qualitatively for Simple Situations:* Students watch carts of the same mass interact in elastic collisions involving both contact forces and magnetic action-at-a-distance forces. They also observe inelastic collisions and "explosions."

c. *Deriving the Equivalence between Momentum Conservation and Constant Center-of-Mass Motion:* The idea of center-of-mass is introduced and students show mathematically that the center of mass of an isolated system always moves at a constant velocity.

d. *Observing Center-of-Mass Motion in 1D and 2D Collisions:* Collisions between low friction carts having different masses and between pucks on an air table enable students to verify momentum conservation in 1D and 2D.

Assessing Learning Gains for Third Law Concepts

Initial use of the sequence at three institutions

Activity-based student worksheets using New Mechanics sequences were prepared in two slightly different formats, one for the Workshop Physics program and one for the RealTime Physics laboratory program. Preliminary versions of the New Mechanics curriculum were then tested at Dickinson College, the University of Oregon, and Arizona State University in the fall of 1992. The author introduced activities to three sections of the Workshop Physics calculus-based physics course at Dickinson College with a total enrollment of 71 students. These students had no formal lectures and met for three sessions of two hours in length each week for the semester. At the same time a RealTime Physics Laboratory program using the New Mechanics sequence was used under the direction of David Sokoloff in an algebra-based introductory laboratory course taken by 257 students. These students met in the laboratory once each week for three hours. In addition, the University of Oregon students were enrolled in a parallel lecture course which included recitation sessions in addition to lectures. Finally, Cheryl Claussen, a graduate teaching assistant at Arizona State University, introduced the new RealTime Physics Laboratory materials to students in one laboratory section of algebra-based physics at Arizona State University which met for two hours each week for a semester.

From an instructor's perspective the trials at the three institutions went as well as could be expected for a first time. However, many changes are being made as a result of the classroom testing. For example, the preparation for the activities involving the Third Law was quite labor intensive because the software was so new that there was insufficient time to test it and some of the older force probes didn't work properly with the new software. Thus, while many of the students made observations on colliding carts that convinced them that forces between carts were always "equal and opposite," some students encountered technical difficulties. We expect that with fully tested software and procedures the reliability of the MBL observations made throughout the sequence will be improved significantly.

Pre and post test results on Third Law concepts at several institutions

The Force Concepts Inventory examination was administered to students in the calculus-based sections of Workshop Physics in the fall of 1992 both before and after students worked with the New Mechanics activities. At the University of Oregon students were given a related Force and Motion Concepts Test[14] after completing the New Mechanics sequence as part of the RealTime Physics laboratory program. Since the testing of the curriculum at Arizona State University was done only on a pilot basis, no formal analysis of the ASU results was performed.

Three of the questions covering Third Law concepts on the Force and Motion Concepts Test were based on questions developed for the Force Concepts Inventory. Each of these three questions tests several important elements in student misconceptions about forces in collisions: (1) the notion that an object with a greater mass exerts a greater force even if the objects are moving at the same speed when they collide head on, (2) the notion that a larger active agent with more mass will exert more force on a smaller passive agent, and (3) the question of whether more force is exerted during contact by a small active agent or by a large passive agent. These three questions are reprinted in the Appendix to this paper. In this article the error rates are reported for these three questions at both Dickinson College and the University of Oregon after students completed the New Mechanics activities. In addition, Hestenes, et. al.[1] have reported results for those questions for some other groups including two high school classes, one honors and one regular, that had received traditional instruction and two high school honors sections that had received special instruction. The results are summarized in Table 2.

			Tradition Instruction	Special Instruction	RTP New Mechanics	WP New Mechanics
Misconception	**FCI**	**FMT**	**HS honors & regular % Errors**	**HS honors % Error**	**N=257 Univ of Oregon % Error**	**N=70 Dickinson College % Error**
1. Greater mass results in greater force when truck and car collide head on	Q2	Q36	65 (88)	08 (85)	11 (–)	14 (100)
2. Active agent w/ more mass exerts more force as a student pushes another	Q11	Q45	61 (89)	03 (86)	09 (–)	11 (73)
3. Active *or* most massive agent exerts more force when car pushes truck	Q13	Q42	89 (93)	22 (89)	39 (–)	30 (78)
Average Error Rate %			*72 (90)*	*11 (87)*	*20 (–)*	*18 (84)*

Table 2: Percentage Error on Post (Pre) test questions involving the application of Newton's Third Law to contact forces

Third Law concepts at Dickinson College before and after New Mechanics

Even though the number of students involved is not large, it is interesting to look at how changes in the curriculum can lead to progressive gains in the conceptual understanding of Newton's Third Law in the Workshop Physics program at Dickinson College. The average error rates have been recorded for questions 2, 11, and 13 for four different situations. These results are summarized in Figure 5 on the next page.

Pretest Results: The pretest results for the 70 students enrolled in Workshop Physics in the Fall of 1992 indicated that students that year came into the program with a somewhat better understanding of the concepts in questions 11 and 13 than the high school student groups listed in Table 1 above did.

Lecture Results: In the Fall of 1991 one of the sections of calculus-based physics at Dickinson was taught in a quasi traditional mode in which more lecturing was done and students did fewer Workshop Physics style activities. The error rates of these students was a bit higher that those reported in Table 2 above for high school students receiving traditional instruction. However, this result may not be statistically significant since there were only 14 students in the section.

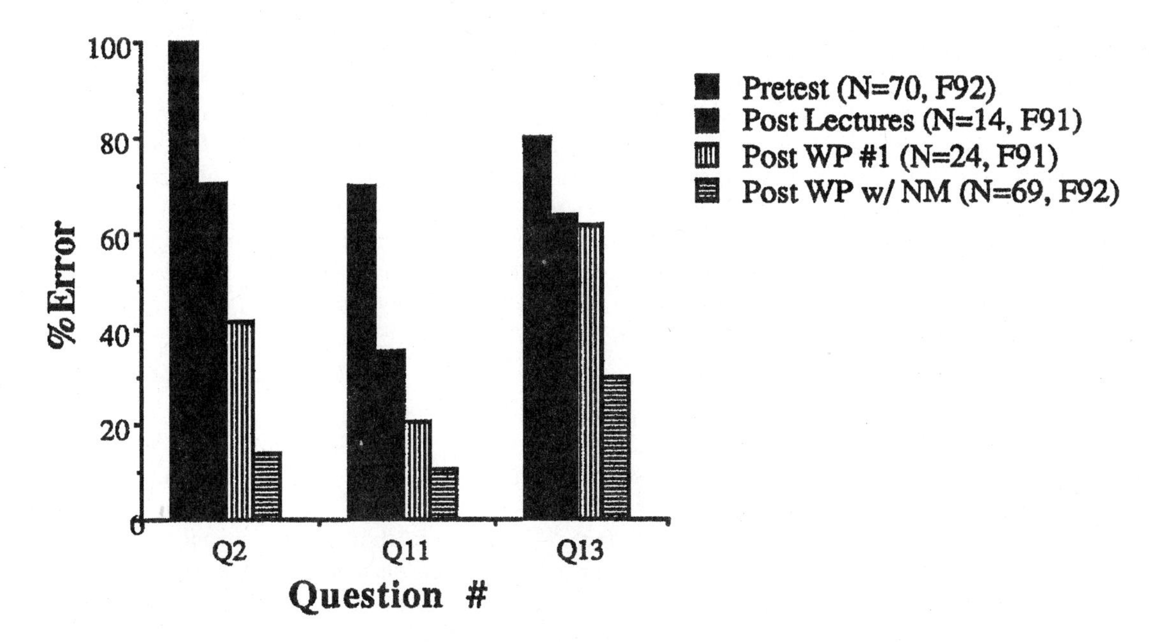

Figure 5: Average error rates on Newton's Third Law questions in the Dickinson College Workshop Physics program before instruction, after quasi-traditional instruction, after completion of Workshop Physics activities, and after completion of Workshop Physics activities using the New Mechanics sequence.

Workshop Physics Results (before New Mechanics): In the Fall of 1991 a section doing Workshop Physics activities that involved verifying the impulse-momentum theorem (but not performing force probe-force probe collisions) had a lower error rate than the group listening to lectures. Although the learning gains seemed significantly better for these student than for the students listening to lectures, the results were still disappointing.

Workshop Physics Results (after New Mechanics): In the Fall of 1992 all three sections of the calculus-based physics course completed Workshop Physics activities that used the New Mechanics sequence described earlier in this paper. They verified the impulse-momentum theorem and observed the outcomes of force probe-force probe collisions. The 69 students in these groups had lower error rates than any of the other groups tested at Dickinson even though there were some technical difficulties with the force probe-force probe collision observations.

Comments on the results

Some conclusions can be drawn from the data presented in Table 2 and Figure 5. Based on pretest error rates at Dickinson College and in the high school groups tested by Hestenes, et. al., between 80 and 90% of any class have significant misconceptions about interaction forces in collisions. After traditional instruction error rates are only reduced at best for the "easiest" of the questions to about 60%. The post instruction error rates on Third Law concepts for students using the New Mechanics sequence were very similar for students at Dickinson College and the University of Oregon. The lowest error rates were achieved for question 2 and were about 10% in each case with the average error rate on all three questions being about 20%.

Question 3 in which a small car was pushing a large truck with its engine turned off remained the hardest for all the classes tested. If students thought either that the more active agent exerts more force or that the more massive object exerts more force they could answer the question incorrectly. However, essentially all of the students at Dickinson and University of Oregon who still answered the question incorrectly after instruction did so because they believed that the small car as an active agent would exert more force on the large truck which was passive and pushed in the direction of the car.

This overall result for error reduction in Third Law concepts achieved using the New Mechanics sequence at Dickinson College and the University of Oregon, although quite good, are not quite as impressive as the 11% post test error rate achieved by a high school honors class taught by Malcolm Wells. No details are reported on how Wells achieved the learning gains in this particular class. There are several possible factors that might explain the difference between his results and those obtains by students completing the New Mechanics activities. It could be that difficulties (which we expect to overcome) in keeping the equipment calibrated and working smoothly in the MBL based force probe collisions prevented a few student groups from discovering the Third Law for themselves. Perhaps there are a larger proportion of students in classes at the University of Oregon and Dickinson College who were slow learners than in the high school honors class. It may be that differences of 10% in error rates are simply not statistically significant since the unreported sample size in the high school honors class is probably quite small.

Conclusions

As a result of the pilot testing in the fall of 1992, instructors generally agreed that the New Mechanics sequence shows promise in helping students develop a deeper conceptual understanding of Newton's Laws. The curriculum needs further refinement and more classroom testing. We must do much more careful analysis of learning gains for elements of all three of Newton's Laws before reaching firm conclusions about the impact of the new curriculum on learning. By examining the teaching of Third Law concepts in more detail, perhaps we have shed some light on the educational potential of the New Mechanics activities.

Acknowledgments

This work would not have been undertaken without the help of David Sokoloff and Ronald Thornton, who played a major role in the development and implementation of the New Mechanics sequence. As collaborators we owe a debt of gratitude Arnold Arons for consulting with us over a period of a year and a half to suggest better ways to present mechanics. Dewey Dykstra spent time early in the conception of the new sequence helping us refine our ideas and reminding us that teaching mechanics to students requires much work prior to the presentation of logical sequences and quantification. We would especially like to thank participants in the New Mechanics Conference who served as sounding boards for our ideas. Special mention should go to Cheryl Claussen who is cooperating with us in the testing of the activities at Arizona State University and to Bob Morse of St. Albans School who was usually several steps ahead of us in the conceptualization and classroom testing of a number of activities we have incorporated into the New Mechanics curricula.

Appendix

Conceptual questions on Newton's Third Law

Questions 36, 42, and 45 from Force and Motion Concepts Test developed by David Sokoloff and Ronald Thornton are reproduced below. These questions are adapted from Force Concepts Inventory[1] questions 2, 13, and 11 respectively.

Questions 36-40 refer to collisions between a car and a truck. For each description of a collision (36-40) below, choose the one answer from the possibilities **A** though **J** that best describes the forces between the car and the truck. You may use a choice more than once or not at all.

A. The truck exerts a greater amount of force on the car than the car exerts on the truck.

B. The car exerts a greater amount of force on the truck than the truck exerts on the car.

C. Neither exerts a force on the other; the car gets smashed simply because it is in the way of the truck.

D. The truck exerts a force on the car but the car doesn't exert a force on the truck.

E. The truck exerts the same amount of force on the car as the car exerts on the truck.

F. Not enough information is given to pick one of the answers above.

J. None of the answers above describes the situation correctly.

In questions 36 through 38 the truck is much heavier than the car.

→ **36.** They are both moving at the same speed when they collide. Which choice describes the forces?

37. The car is moving much faster than the heavier truck when they collide. Which choice describes the forces?

38. The heavier truck is standing still when the car hits it. Which choice describes the forces?

Questions 41-43 refer to a large truck which breaks down out on the road and receives a push back to town by a small compact car.

Pick one of the choices A through J below which correctly describes the forces between the car and the truck for each of the descriptions (34-36). You may use a choice more than once or not at all.

A. The force of the car pushing against the truck is equal to that of the truck pushing back against the car.

B. The force of the car pushing against the truck is less than that of the truck pushing back against the car.

C. The force of the car pushing against the truck is greater than that of the truck pushing back against the car.

D. The car's engine is running so it applies a force as it pushes against the truck, but the truck's engine isn't running so it can't push back with a force against the car.

E. Neither the car nor the truck exert any force on each other. The truck is pushed forward simply because it is in the way of the car.

J. None of these descriptions is correct.

41. The car is pushing on the truck, but not hard enough to make the truck move.

→ **42.** The car, still pushing the truck, is **speeding up** to get to cruising speed.

43. The car, still pushing the truck, is at cruising speed and continues to travel at the same speed.

→ **45.** Two students sit in identical office chairs facing each other. Bob has a mass of 95 kg, while Jim has a mass of 77 kg. Bob places his bare feet on Jim's knees, as shown to the right. Bob then suddenly pushes outward with his feet, causing both chairs to move. In this situation, while Bob's feet are in contact with Jim's knees,

A. Neither student exerts a force on the other.

B. Bob exerts a force on Jim, but Jim doesn't exert any force on Bob.

C. Each student exerts a force on the other, but Jim exerts the larger force.

D. Each student exerts a force on the other, but Bob exerts the larger force.

E. Each student exerts the same amount of force on the other.

J. None of these answers is correct.[15]

References

1. David Hestenes, Malcolm Wells, and Gregg Swackhamer, "Force Concept Inventory," *The Physics Teacher,* Vol. 30, 141-158. (March, 1992).
2. Arons, Arnold. *A Guide to Introductory Physics Teaching,* (John Wiley, New York, 1990) Chapter 3; and Rothman, Milton A. *Discovering the natural laws; the experimental basis of physics,* (Doubleday, New York, 1972) Chapter 2.
3. Lillian McDermott, "Research on conceptual understanding in mechanics," *Physics Today,* 2-10. (July, 1984); and David Hestenes, Malcolm Wells, and Gregg Swackhamer, "Force Concept Inventory," *The Physics Teacher,* Vol. 30, 141-158. (March, 1992).
4. c.f. The article in this volume by Ronald K. Thornton; and Ronald K. Thornton and David Sokoloff, "Learning Motion Concepts Using Real-Time Microcomputer-Based Laboratory Tools," *Am. J. Phys.* 58 (9) 858-867 (Sept., 1990).
5. A motion detector, force probe, interface and motion software for Macintosh and MS Dos computers can be purchased from Vernier Software Company, 2920 S.W. 89th Street, Portland, OR 97225.
6. Low friction dynamics carts can be purchased from PASCO Scientific Company, 10101 Foothills Blvd., PO Box 619011, Roseville, CA 95678-9011.
7. Priscilla Laws, "Calculus-Based Physics Without Lectures," *Physics Today,* Vol. 44, No. 12, (Dec. 1991); and Priscilla Laws, "Workshop Physics—Learning Introductory Physics by Doing It," Change, 20-27 (July/Aug. 1991); and Priscilla Laws, "Workshop Physics-Replacing Lectures with Real Experience," *Proceedings of the Conference on Computers in Physics Instruction,* Addison-Wesley, Reading, MA, 1989.
8. The RealTime Physics project, directed by David Sokoloff, is funded by the National Science Foundation ILI Laboratory Leadership program, #USE-9054224, at the University of Oregon. It involves the adaptation of curricular materials from the Workshop Physics and Tools for Scientific Thinking Programs to introductory laboratory sequences in Mechanics, Heat and Temperature, and Circuits.
9. The New Mechanics conference which was held on August 6-7, 1992 in Medford, MA was attended by Pat Cooney, Dewey Dykstra, David Hammer, David Hestenes, Priscilla Laws, Suzanne Lea, Lillian McDermott, Robert Morse, Hans Pfister, Edward F. Redish, David Sokoloff, and Ronald Thornton.
10. James Minstrell, "Teaching for the Development of Understanding of Ideas: Forces on Moving Objects" from the 1984 *Yearbook of the Association for Supervision and Curriculum Development,* Editor Charles W. Andersen, December 1984.
11. David Hestenes, Malcolm Wells, and Gregg Swackhamer, "Force Concept Inventory," *The Physics Teacher,* Vol. 30, 141-158. (March, 1992).
12. The inversion of momentum and energy topics was suggested by Arons on the basis that (1) the momentum concept is simpler than the energy concept, in both historical and modern contexts and (2) the study of momentum conservation entails development of the concept of center-of-mass which is needed for a proper development of energy concepts. Private Communication with Arnold Arons, "Preliminary Notes and Suggestions," August 19, 1990; and ________, *Development of Concepts of Physics* (Addison-Wesley, Reading MA, 1965)
13. Arons, Arnold. *A Guide to Introductory Physics Teaching,* (John Wiley, New York, 1990), p. 67.
14. For more information on the test, contact Ronald K. Thornton, Center for Science and Mathematics Teaching, Lincoln-Filence Building, Tufts University, Medford, MA 02155.

Part II

Contributed Poster Session I

Report Of The Workshop On The Teaching Of Thermodynamics In The IPC, A Particle Approach

Marcelo Alonso, Florida Institute of Technology
Edward J. Finn, Georgetown University

Workshop Participants:
Larry Josheno, Corning Community College
Robert Wallis, Baldwin-Wallace College
James Wheeler, Lock Haven University
Charles Whitney, Harvard Observatory

The purpose of the Workshop was to explore how to teach thermodynamics in the Introductory Physics Course (IPC) in a form more consistent and coherent with our current picture of the physical world. In this report we present the most relevant aspects of the discussions in a way we hope will serve to stimulate and orient those interested in the teaching of thermodynamics in the introductory courses. No effort has been made to refer to the extensive literature on the subject.

Introduction

Traditionally, and correctly, the IPC begins with mechanics. Although the order and extent may vary from one place to another, all courses basically deal with single particle dynamics (this term is not always used) and then discuss the rigid body, but not necessarily systems of particles. After a few other subjects, which in one way or another can be considered as part of mechanics, all IPCs normally proceed to what is designated as Heat, Thermal Physics or Thermodynamics. This part of the IPC is usually taught as an independent subject with a rather minimal connection to mechanics, except perhaps for a reference to work and energy, which are used in a limited sense. In addition, the emphasis is usually placed on thermal engines because, historically, thermodynamics evolved out of an interest in developing such devices. However, over time thermodynamics has become important for many physical and chemical processes, including the emission and absorption of radiation.

This situation requires a new focus on the approach to thermodynamics, an approach that has not as yet been fully incorporated into the IPC. In particular, although it is understood that all physical systems are aggregates of interacting particles, the ideas of Statistical Mechanics remain absent from the IPC, except for a minimal consideration of the kinetic theory of gases. Even in this case the interaction between the molecules is ignored and students get the idea that gases behave in haphazard ways. Students are thus deprived of a more profound and more easily visualized model of thermal phenomena. Fortunately there is strong interest in redressing this situation, an interest that has been recently encouraged by the Introductory University Physics Project (IUPP), and which was the motivation for the Workshop.

The main conclusion of the Workshop was to recognize the need to combine the *empirical approach* of classical thermodynamics, which evolved from macroscopic and sensor-ial perceptions and certain associated concepts (pressure, temperature, heat, etc.), with the *structural or particle approach* of statistical mechanics, which attempts to correlate the thermal properties of a system with the properties of the "particles" or constituent units of which it is composed. Another advantage of the statistical approach is that it can deal, in principle, with a system whether the system is in equilibrium or not (recall Boltzmann's H-theorem), while classical thermodynamics is practically restricted to systems in equilibrium or quasi-equilibrium. This has been put very clearly by one of the participants, Charles Whitney:

> With a suitable choice of descriptive parameters, statistical physics and its older sister thermodynamics permit us to discuss systems that we cannot know - and do not wish to know -in complete detail. We do not attempt a detailed accounting; we confine attention to statistical parameters and distribution functions that are approached, in the average, more and more closely as time moves on. From these distribution functions, we may derive important properties of the gas, such as temperature, pressure, density and stratification.
>
> The primary distinction between the two sisters is that thermodynamics deals with measurable, macroscopic parameters, such as pressure and temperature. The results of thermodynamics are often quite general so they can be applied to a wide range of phenomena. For example the "first law" of *thermodynamics*, which we describe later, is a general statement of energy conservation and it can be applied to the analysis of any type of engine, as long as the engine conforms to certain restrictions. *Statistical physics*, on the other hand, attempts to go deeper, by modeling the microscopic behavior of a system. The result is more powerful in some ways, and less powerful in other ways. Less powerful because statistical physics is more specific and cannot be applied with the sweeping generality of thermodynamics. More powerful because it does not impose severe restrictions on the systems it can discuss.
>
> Classical thermodynamics is, in fact, misnamed. It ought to be called "*thermostatics*," because it is confined to systems that are in equilibrium, which means that it only deals with systems that are isolated and not changing with time. But if a system changes slowly enough, it may be said to follow a concatenation of equilibrium states, and, in this way, thermodynamics can take slowly changing (*quasi-equilibrium*) states under its wing. Having done so, it can tell us much about the system without detailed modeling. But this limitation to slow changes necessarily eliminates a broad class of important phenomena: the so-called "irreversible" changes. While a system undergoes such changes, thermodynamics must briefly close her eyes. She may reopen them when the system has settled down once more to a near-equilibrium state. One example is the conduction of heat in a gas. If we add energy to the gas at one end of a tube, some of this energy will be carried to the other end. If the heating is gentle and gradual, thermodynamics can describe the entire process; if the heating is drastic and abrupt, it cannot cope and we must resort to statistical physics. We must proceed by constructing a model that permits us to specify the average behavior of the atoms in the gas.
>
> Thus the range of phenomena open to statistical physics is far greater. This increased range is purchased at the expense of our needing to build a specific model, but computers can help us cope with such models.[1]

There was a general consensus that from the pedagogical point of view it is desirable in teaching thermal physics in the IPC to first focus on the concepts students are acquainted with, using the empirical approach, and then proceed to elaborate on the concepts in more detail using the particle approach, all coupled with the proper laboratory experiments, demonstrations and simulations. The procedure is well established in the case of pressure, where it is customary to first talk about pressure phenomenologically, discussing how to measure it (barometers, manometers, gauges, Boyle's law, etc.) and then correlate the pressure of a gas with the mass and velocities of its molecules using the ideal gas model. The same applies to temperature: first we introduce the concept phenomenologically and indicate how to measure it with different thermometers. Afterwards, we relate temperature to the kinetic energy of the molecules.

However, the Workshop did not elaborate on how to approach each and every concept in the IPC and the appropriate experiments to be performed, since it was recognized that each instructor has his/her own methods and preferences. Instead, the Workshop concentrated on the conceptual aspects; that is, on how IPC students should understand each concept at the end of the course.

A point raised during the discussions was whether incoming students were sufficiently informed about the structure of matter to accept the particle approach to thermal physics without reservations. The consensus was that IPC students have at least been exposed to the ideas of atoms, molecules, electrons, etc. so they could accept the particle approach.

The Conceptual Basics of Thermodynamics in the IPC

At the beginning of the Workshop it was suggested that "thermodynamics" is too restrictive a term, dealing mostly with thermal engines, and perhaps the terms "thermal physics" or "thermophysics" ought to be used instead. However, to avoid a semantic discussion, it was agreed to consider "thermodynamics" as dealing with all thermal phenomena, and the word is used in this sense throughout this Report. Accordingly, if we look at thermal phenomena from an empirical, macroscopic and sensorial point of view, we may tell students that *thermodynamics deals with relations between temperature, heat, internal energy and work in systems with a large number of particles*. However, as one progresses in the study of thermal phenomena in the IPC, the above definition may be refined to emphasize the structural or particle point of view in that we tell students that *thermodynamics deals with energy (and momentum) exchanges between the particles of a system composed of a very large number($\sim 10^{24}$) of units (molecules, atoms, etc.)*[2] *and the particles in the surroundings; as a consequence, the internal energy of the system may change,* to which one may add that experimental evidence indicates that the energy exchanges can be grouped in three categories: *heat, work and/or radiation.*

In the discussion of thermal phenomena, the following basic quantities, among others, must be defined: pressure, temperature, internal energy, heat, work and entropy. The Workshop reviewed how those quantities are usually considered in the IPC and the following interpretations are offered, suggesting first a presentation more akin to the empirical approach and a second more refined consideration that emphasizes the particle approach.

TEMPERATURE (T)

All students are familiar with the notion of temperature and its measurement using some kind of thermometer. Therefore, after the necessary preliminaries, one may tell students that the *temperature of a body is a quantity that measures the degree of hotness or coldness of the body and that it is measured by using a thermometer on a temperature scale defined using prearranged reference points, such as the freezing and boiling points of water at standard pressure.* Next this definition should be criticized, showing that it is not related to any internal property of the body. Then one may indicate that *temperature is a measure of the average energy of the particles of a body.* It is important to emphasize that the relation between temperature and the particles' energy depends on the structure and composition of the body, being different for each kind of body (solid, liquid, gas). Only for an ideal gas in thermal equilibrium, where the particles have only kinetic energy, may we write: $E_{k,ave} = (3/2)kT$, where T is defined on an absolute scale.

The question of whether temperature is defined only for a system in thermal or statistical equilibrium was discussed at the Workshop. The consensus was that students should understand that temperature is really defined only for a system in thermal equilibrium, for which the average energy of the particles has a well defined value. However, in a system not in equilibrium, one can define local temperatures, which in general are not constant but fluctuate rapidly depending on the mobility of the particles composing the body.

INTERNAL ENERGY (U)

It is desirable that the notion of internal energy should be explained in the IPC when systems of particles are discussed, in which case its use in thermodynamics does not introduce any new ideas. If that is not the case, one may start by saying that *internal energy is the energy stored in a body* and that its change in a process is measured by the sum of the heat absorbed by the body and the work done on the body: $\Delta U = Q + W$, which is the First law of Thermodynamics. However, this definition of internal energy is rather imprecise and suffers from a lack of correlation with the structure of the body and does not indicate how U is measured, only its change. For that reason, it is better to explain that the particles in a system (body) are subject to internal and external forces. In addition, the particles are in motion relative to each other. Then the internal kinetic energy is the sum of the kinetic energies of the particles measured relative to the CM. It is essential to clarify this point because only in this case is the internal kinetic energy a property of the body, independent of the observer. The essential point is that *the internal energy of a body is the sum of the kinetic energy of the particles in a body, relative to its CM, and the internal potential energy due to their interactions (internal forces).* As shown in the dynamics of systems of particles, the change in internal energy in a process is given by $\Delta U = W_{ext}$, where W_{ext} is the work done by the external forces acting on the particles of the body.

For a body composed of a very large number (~ 10^{24}) of particles, one may tell students that the external work can be separated operationally into two parts of a collective nature: heat and mechanical work, $W_{ext} = Q + W$, a fact discussed below. Then one obtains the First law: $\Delta U = Q + W$. A point to be emphasized is that the internal energy is a property of the state of the system, whether in equilibrium or not, and that a given change ΔU can be obtained using several combinations of Q and W. At this point, one could make clear that the first law expresses a balance in the energy transfer between a system and its surroundings, which is the principle of the conservation of energy.

HEAT (Q)

Usually students are told that *heat is the energy transferred from a hotter body to a colder one as a result of their temperature difference, without necessarily doing any work.* Although this statement is basically correct, it fails to clarify the nature of heat or the mechanism by which it is transferred. For that reason, it is better to use the particle approach and explain that *heat is the process of kinetic energy transfer across the boundary (surface) of a body (system) by molecular collisions when the average molecular energy (temperature) of the body and of the environment are different.* This process of energy transfer involves a multitude of microscopic works done by force exerted by the external particles on the particles of the system and it thus part of W_{ext}, as indicated previously. It should also be added that heating is essentially a surface effect and the energy transferred as heat propagates into (out of) the body, resulting in a gradual heating (cooling) throughout the body. [This is why we get "rare" steaks when we broil.]

Terms frequently used are: heating, to heat, hot, etc., and should be defined properly. Thus, "heating" and "to heat" should refer to the act or process of energy transfer we have defined as heat, although by extension they may be used to refer to any other process of energy transfer that results in an increase in the temperature of a body. In that sense, it might be correct to say "heating of a transformer core by electromagnetic induction" when what one is actually doing is work on the core. On the other hand, "hot" and "cold" should be used as equivalent to "high" and "low" temperatures, without relation to heat or any other process.

WORK (W)

In the dynamics part of the IPC students are introduced to the notion of work on a particle. This notion can easily be extended to a system of particles. However, in thermodynamics it is usual to define *work* in a more restricted sense as *the energy transferred to a system from the surroundings (or the reverse) as a result of a change or modification of its configuration or shape by the action of external (either surface or body) forces.* This is basically *mechanical work* and is exemplified by the push of a piston as a result of a pressure difference. However, this interpretation is too restrictive and it is preferable to use the particle approach and say that *mechanical work is an energy exchange between a system and its surroundings as a result of external forces acting on the particles of the system, not necessarily involving molecular collisions, and usually resulting in a net macroscopic motion of the boundary.* This is the case when the forces correspond to particle collisions with the boundary (pressure). But there are many other mechanisms for doing external work. This is the case for long range interactions, such as electric and magnetic forces, that may act throughout the body (radiation pressure, magnetic compression of a plasma, magnetic induction heating, etc.), doing external work without necessarily modifying the boundaries.

RADIATION (R)

It is still customary in the IPC to use the expression "heat transfer" and say that it occurs as conduction, convection and radiation. This is an old fashioned, even incorrect, terminology that should no longer be used.

Heat, alone, is a process of energy transfer. Thermal conduction and convection are two types of energy transport phenomena by molecular collisions or by bulk motion due to pressure or density differences. Above all, radiation at the atomic and molecular level is a process of energy transfer by emission or absorption of photons, or electromagnetic waves, resulting in a change in the internal energy of the atom or molecule involved. When many atoms or molecules are involved, the result is a change in the internal energy of the system. In fact, students should be told that radiation is the most common energy transfer process in the universe. Microwave radiation is used for cooking in a microwave oven and infrared radiation in a conventional oven. [By extension, x-rays and γ-rays may produce even more profound changes (dissociation, ionization) in a system, but this will not be considered as part of thermodynamics as we have defined it.]

For the above reasons it is better to use the particle approach and tell students that *radiation refers to an energy exchange between a system and its surroundings through the emission or absorption of photons by the particles of the system and of the surroundings, resulting in a change of the internal energy of the system, a process that may take place throughout the entire system.* [For that reason there are no "rare" steaks in a microwave oven!] Therefore, one must modify the first law and write $\Delta U = Q + W + R$. This is a long overdue "modernization" in the IPC. Since, at this juncture of the IPC, students are not yet well acquainted with electromagnetic waves or photons, one should not elaborate on this mechanism of energy transfer, but it should be discussed later on.

ENTROPY (S)

There is general consensus that entropy is one of the more difficult concepts for IPC students to grasp. Traditionally it is introduced by telling students that *entropy is a quantity whose "change" during an infinitesimal reversible transformation of a system is* $dS = dQ/T$, *where* T *is the temperature of the system.* Then for a finite reversible transformation, $\Delta S = \int^{2}_{(\text{rev})} dQ/T$. It should be carefully explained that the calculation applies only to a system that remains in thermal equilibrium during the process (reversible transformation); otherwise temperature is not well defined for the system. One should also tell the students that to obtain ΔS when the transformation is irreversible, a reversible transformation that takes the system from the initial to the final state must be imagined. This is hard to visualize and accept without first proving that S is a state variable. However, there is no simple way in the IPC to provide that proof using classical thermodynamics. The method of dividing a cycle into a series of infinitesimal Carnot cycles appears artificial and not at all clear to IPC students. Summarizing, the three difficult aspects IPC students encounter when entropy is presented this way are: first, acceptance that it is defined by a "change"; second, to be able to understand that it is a state variable; third, to accept that for an isolated system $\Delta S \geq 0$ (Second law of Thermodynamics). All these obstacles are easily overcome in the particle approach. Using the particle approach, *entropy can be defined as a state variable given by* $S = k \ln P$, *where* P *is the "thermodynamic" probability of the state of a system.* The definition applies whether the system is in equilibrium or not, but P is hard to calculate when the system is not in equilibrium. A new problem then appears: What is meant by thermodynamic probability, particularly for a system of interacting particles? Another difficulty is to correlate S with other variables used in thermodynamics, although it is easy to carry it out for an ideal gas, and then to prove that the empirical and particle definitions of entropy are fully equivalent without going into the details of how to calculate P. The fact that the use of the log transforms probabilities, which are multiplicative, into an additive quantity, which is a property of entropy, should be explained. These matters were discussed in the Workshop and our collective response appears in Section 3 of this Report.

Two terms that should be used with great care are "entropy production" and "entropy flow". Entropy is not something that exists by itself but is a property of the state of a body. If in a process the entropy of a system increases, one could perhaps say that entropy has been produced. Also, considering the expansion of the universe, one could say that entropy is produced continuously, but this is a very special case. Further, because entropy is not a conserved quantity, it is not "exchanged". Obviously a body in motion, say a fluid, carries its entropy with it; but that does not mean that there is a flow of entropy. On the other hand, energy is a conserved quantity and one may say that can be transferred or may flow from one body to another.

THE SECOND LAW

The traditional statements of the second law are those of Kelvin-Planck: "It is impossible for a reversible cyclical thermal engine to absorb heat from a body and perform mechanical work without releasing some heat to a body at a lower temperature." and of Clausius: "It is impossible for a reversible cyclical thermal engine to transfer heat from one body to another at a higher temperature without the input of external work." Usually, no connection is made between these statements and the rest of thermodynamics and they sound too mechanistic, lacking physical generality. Perhaps it is better to follow a method originally proposed by Fermi and state that *in an isolated system composed of subsystems 1, 2, 3, ... at temperatures T_1, T_2, T_3, ..., if there are transfers of energy among themselves as heat, the relation $\Delta S = \sum_i Q_i/T_i \geq 0$ holds.* This statement is easy to accept by students, even if it is difficult to explain how to verify it experimentally, except in some particular examples. Using the particle approach, students may be told *experience shows that an isolated system tends to evolve toward statistical equilibrium, eventually reaching the most probable configuration or state compatible with the structure and internal forces of the system; therefore, since the only processes that can occur in an isolated system are those that tend to bring the system to a state of higher probability and $S = k \ln P$, then necessarily $\Delta S_{syst.} \geq 0$.* In this way, the second law appears as something obvious. The Kelvin-Planck and Clausius statements are easily obtained as corollaries, once the equivalence of $dS = dQ/T$ and $S = k \ln P$ is established (see Note 2 for a method to obtain this equivalence that is adequate for IPC students). A point to be emphasized in discussing the second law is the role of internal forces in the trend toward equilibrium, a point sometimes ignored.

THERMAL ENERGY

There was agreement that this is a frequently used, but poorly defined and, perhaps, superfluous term. Some use it to refer to the energy of a body associated with its temperature, or, which is the same, to the energy of a body associated with the motion of its particles (kinetic energy); others consider it as equivalent to the internal energy. If the term is to be used, it must be defined uniquely. The general consensus was that the term should not be used any more or, at best, used in the first sense.

THERMAL EQUILIBRIUM

This term is used frequently but its meaning is not always stated precisely. From the empirical point of view, *a system is in thermal equilibrium when all its macroscopic parameters (pressure, volume, temperature, density, etc.) are well defined and do not change nor fluctuate appreciably with time.* One may also add that two bodies are in thermal equilibrium when there is no net exchange of energy as heat or radiation; that is, they have the same temperature. However, in the particle approach, *thermal equilibrium corresponds to a state where the distribution of the particles among the possible energy states is fixed statistically (i.e., fluctuates slightly) around values determined by the nature of the particles and their interactions (i.e., Boltzmann distribution law, etc.) and this is reflected in the system having well defined macroscopic parameters.* It this respect "thermal" and "statistical" equilibrium can be considered as equivalent. This procedure helps IPC students recognize that thermal equilibrium is dynamical and quite different from the equilibrium of a body they studied in statics.

REVERSIBLE TRANSFORMATION

A process during which the external parameters vary gradually so that the system remains close to a state of thermal (statistical) equilibrium at all steps. Therefore, the macroscopic parameters of the system are well defined throughout the process and empirical thermodynamics applies. Students must understand that a reversible transformation must be slow enough to allow the system to continuously adjust its microscopic states to the equilibrium partition corresponding to the new and changing macroscopic parameters as the process goes on.

IRREVERSIBLE TRANSFORMATION

A process during which the external parameters suffer sudden large changes so that the system is not in thermal (statistical) equilibrium. Therefore, the macroscopic parameters of the system are not always well defined and empirical thermodynamics does not apply to the process and the microscopic configuration is irregular and, in general, far from the equilibrium partition. The behavior of such a system is very difficult to predict. In general, transformations that are fast are irreversible because the system does not have time to adjust its microscopic states to the rapidly changing equilibrium partition as the process proceeds.

Methodology for the Particle Approach

The following considerations made at the Workshop describe a possible methodology or model for using the particle approach in Thermodynamics, keeping in mind that the subject is only part of a more general course and not an isolated course. Emphasis is placed on the formulation of the two laws. In the IPC, Thermodynamics (or Heat) usually follows Mechanics. Therefore, the particle approach to thermodynamics should draw on the mechanics part of the course, assuming students are familiar with the basic facts of the structure of atoms and molecules and the forces among them, which is usually the case. This procedure can serve to display the "unity" of physics, a goal we are always striving for.

In Mechanics students study the dynamics of a particle where momentum, force, work, kinetic energy and potential energy are all introduced (see Table 1 below). The student should also learn that, at the fundamental level, anything that happens in the world can ultimately be reduced to exchanges of momentum and energy. Next, the more realistic problem of the dynamics of a system of particles interacting, both among themselves through internal forces and also with the surroundings by means of external forces, should be considered (Table 1). The ideas to be discussed in this section of the IPC are the motion of the CM, internal and external forces, internal and external potential energy, internal and orbital kinetic energy, internal energy and work done on the system by external forces. The basic relation is $\Delta U = W_{ext}$, where both quantities are measured relative to the CM frame of reference.

The transition to thermodynamics (Table 1) occurs when the system is composed of a very large number ($\sim 10^{24}$) particles. Then, based on empirical evidence, new concepts are introduced and related to the molecular structure of the body. Internal energy remains as defined before. External work, W_{ext}, is separated into two parts: "heat" Q and "mechanical work" W, corresponding to two different mechanisms of energy exchange between body and surroundings, as explained before. (A more elaborate distinction between heat and work is analyzed in Note 1.) Other new concepts are also introduced: "temperature", related to the average energy of the molecules of the body, and "entropy", related to the arrangement of the molecules in the body. As mentioned earlier, a third mechanism of energy transfer, which is usually ignored, is radiation (mostly infrared and microwave on Earth); that is, emission and absorption of photons with exchange of energy and momentum. Then a modern or extended statement of the first law should read $\Delta U = Q + W + R$.

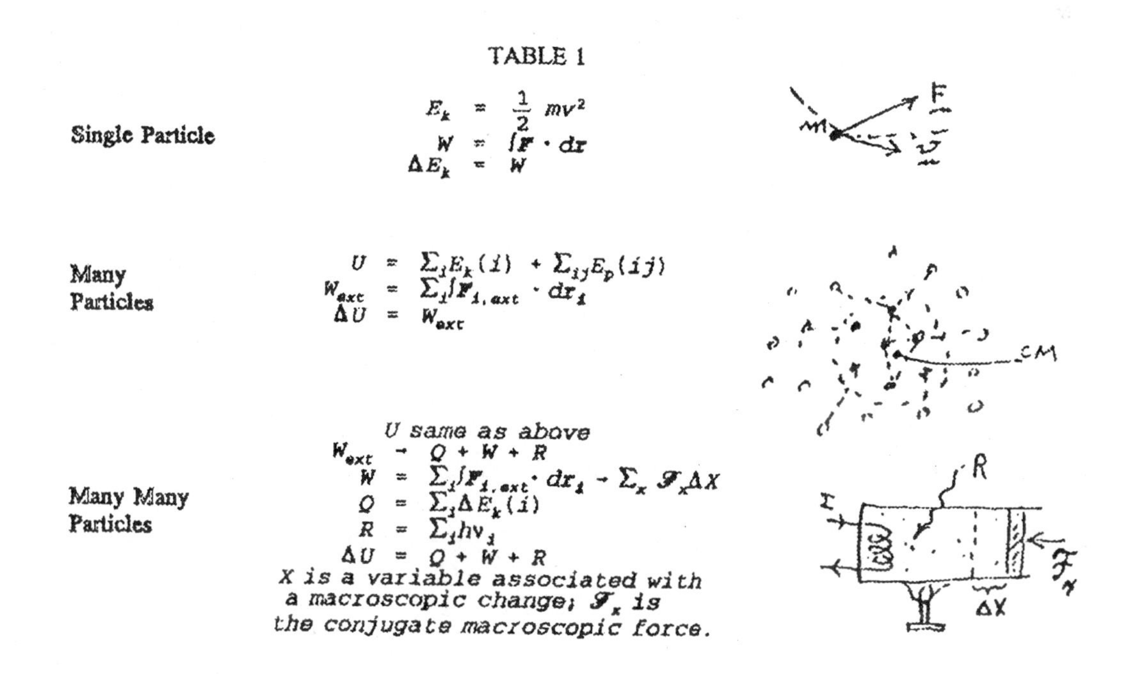

TABLE 1

Single Particle	$E_k = \frac{1}{2}mv^2$ $W = \int \mathbf{F}\cdot d\mathbf{r}$ $\Delta E_k = W$	
Many Particles	$U = \Sigma_i E_k(i) + \Sigma_{ij} E_p(ij)$ $W_{ext} = \Sigma_i \int \mathbf{F}_{i,ext}\cdot d\mathbf{r}_i$ $\Delta U = W_{ext}$	CM
Many Many Particles	U same as above $W_{ext} \rightarrow Q + W + R$ $W = \Sigma_i \int \mathbf{F}_{i,ext}\cdot d\mathbf{r}_i - \Sigma_x \mathcal{F}_x \Delta X$ $Q = \Sigma_i \Delta E_k(i)$ $R = \Sigma_i h\nu_i$ $\Delta U = Q + W + R$ X is a variable associated with a macroscopic change; $\mathcal{F}_x$ is the conjugate macroscopic force.	R, $\mathcal{F}_x$, ΔX

Discussing entropy and the second law in the particle approach requires some elaborate considerations that can be carried out at various levels of detail and sophistication within the simplicity required by the IPC, but none are beyond the capability of students who take this course. We suggest a sequence of analysis in the particle approach as follows:

A. Establish that a physical system (body) is composed of a number N of units or particles having well defined properties (mass, charge, spin, etc.) and subject to their mutual interactions (gravitational, electric, magnetic, etc.).

B. Define what is meant by "state" of a system, distinguishing between macrostates and microstates of a system of particles. When a system is in equilibrium, a *macrostate* is characterized by macroscopic, experimentally measurable, variables: pressure, temperature, volume, density, etc.. Otherwise, the macrostate is not well defined. A *microstate* is characterized by the dynamical state of each particle, whether or not the state is in equilibrium. However, if the system is not in equilibrium, the microstate is changing continuously and, consequently, the macrostate variables cannot be defined.

C. The internal energy of the system, relative to its Center of Mass (CM), is:

$$U = \sum_i E_k(i) + \sum_{ij} E_p(ij).$$

As such, the energy of each individual particle is not well defined; only that of the whole system. For weakly interacting particles (i.e., an ideal gas), the potential energy $E_p(ij)$ may be ignored or students can be told that [if a self-consistent field approximation is valid] $\sum_j E_p(ij)$ can be replaced by $E_{p,av}(i)$. Then the internal energy of the system can be written as the sum of single particle energies:

$$U = \sum_i \{E_k(i) + \sum_{p,av}(i)\} = \sum_i E_i.$$

However, this approximation cannot be valid for phenomena such as phase transitions, which involve strong interactions.

D. Next, assume that more than one particle may have the same energy, in which case $U = \sum_i n_i E_i$, with $N = \sum_i n_i$. At this point it is appropriate to mention that for certain systems the single particle energies E_i may be limited or constrained (i.e., quantized) to certain values and that the number of particles in each energy state may also be limited. The numbers n_i are called "occupation numbers" and the set $(n_1, n_2, n_3, ...)$ is called a "partition" or "complexion". Under these definitions a microstate of the system is characterized by its partition.

E. The two conditions involving the n_i's ($U = \sum_i n_i E_i$ and $N = \sum_i n_i$) are not sufficient to uniquely determine a partition and other assumptions are necessary to fix a partition. In any case, there may be several microstates (partitions or complexions) compatible with a given macrostate. This may be called the multiplicity of the macrostate. It is safe to assume that the state of statistical equilibrium is a mix of the partitions that are more favored by the structure of the system and its internal forces, as well as by external parameters.

F. A very important assumption (*ansatz*) must now be made: Given the external parameters, the internal energy of the system, its number of particles and the internal forces, a system always evolves toward a state of statistical equilibrium. When statistical equilibrium is reached, the system's internal and external parameters (density, pressure, temperature,...) are well defined, regardless of its initial conditions and the system is in a steady macrostate, although there may be minor fluctuations. In this state there is one partition (or possibly a group of partitions) that is more favored than any other. The microstate of the system fluctuates among those partitions close to the most favored one.

G. When a system experiences a transformation, exchanging energy with its surroundings, the occupation numbers n_i and the particle energy states E_i change. This allows for a precise definition of work and heat, which students understand without difficulty (see Note 1). When the process is reversible, the occupation numbers n_i adjust at each instant to those corresponding to the equilibrium partition or state at each stage of the process. (This is the basic definition of a reversible transformation.)

H. The problem at the heart of statistical mechanics is to determine the state of statistical equilibrium and the corresponding most favored partition or microscopic configuration. We may tell students that this is the state more "compatible with" or "determined by" the forces acting on the particles and the external parameters. The customary expression is that it is the most "probable" state. However, this term evokes in the students the idea of random or accidental events (coin tossing, for example). But, the equilibrium state is not the result of erratic or haphazard behavior. Only for systems of very weakly interacting particles is the usual concept of probability more or less applicable. As already mentioned, a state of equilibrium does not correspond to a unique microstate and the system may fluctuate among several more or less equivalent microstates, which gives some meaning to the concept of thermodynamic probability, designated P. The best example of a system in statistical equilibrium is a gas whose molecules obey Maxwell's velocity distribution law. (This law is determined by the availability of states in phase space, but this is difficult to explain in the IPC.) However, the concept of thermodynamic probability is perhaps the most difficult to explain to IPC students in the particle approach, particularly if internal forces are important. The method of ensembles is not recommended for the IPC.

After reaching this stage in the course, one of several possible paths may be chosen, differing in detail and complexity.

I. The simplest option is to introduce the Boltzmann factor for the equilibrium partition: $n_i \sim g_i e^{-(Ei/kT)}$, justify its form by telling students that the higher the energy, the lower the occupation and illustrate it with its application to Maxwell's velocity distribution law in a gas. At this point, a simulation even with a reasonably small number of particles would be of extreme assistance. Next define entropy as $S = k \ln P$, explaining that because probabilities are multiplicative entropy is an additive property. From there it is easy to explain the three fundamental properties of entropy that are a direct consequence of its statistical definition: (i) S is a state variable, (ii) $\Delta S \geq 0$ for an isolated system (2nd law) and (iii) $dS = dQ/T$ for reversible transformations. The third consequence is difficult to prove in general terms at the IPC level with the particle approach, but it is relatively simple to justify it in an intuitive way for an ideal gas, as suggested in Note 2 at the end of this discussion paper.

J. Once the connection between $S = k \ln P$ and $dS = dQ/T$ has been established, the relation
$dU = TdS + dW$ for a reversible process, thermal engines, the Carnot cycle (which is better represented using a T-S diagram), the Kelvin-Planck and Clausius statements, in fact all the traditional aspects considered in thermodynamics, can now be readily analyzed. Relations such as $(\partial U/\partial S)_V = T$, where IPC students would find it hard to see their physical or operational meaning, should be avoided.

K. A more elaborate presentation, still within the possibilities of the IPC, may include (i) a discussion of how to calculate P for weakly interacting identical particles in Maxwell-Boltzmann statistics:
function, $Z = \sum_i g_i e^{-Ei/kT}$ and (iii) how that function is related to the internal energy and entropy of the system. We do not feel it is necessary in the IPC to discuss the mathematical details of how to calculate the maximum of P or how to derive the Boltzmann factor.

Great caution should be exercised when considering the terms "order" and "disorder" in connection with entropy and the second law. In the first place, "order" and "disorder" are hard to define and express quantitatively in connection with a system of particles and even harder to measure. The fact is that there are differing points of view of these concepts when applied to a physical system. (Are "order" and "disorder" additive quantities? Is a uniform state ordered or not?) Entropy is not a measure of the degree of disorder of a system and the second law should not be presented as a trend toward disorder in an isolated system. The structure, or organization, of a system in thermal equilibrium is the result of a balance between energy of motion (kinetic energy) and the potential energy due to the internal forces between the particles of the system. In an isolated system, conservation of energy requires that the average energy of the particles in the system be constant. Furthermore, in thermal equilibrium the Principle of Detailed Balance requires that the distribution of energy among the particles remains essentially constant. At low temperatures, when the kinetic energy of the particles is small, intermolecular forces are the dominant factor and certain structures that we call "ordered" are possible, or favored. On the other hand, at high temperatures, when the kinetic energy of the particles dominates their internal potential energy, other structures or organizations, which we say are "less ordered" or "disordered", become favored. Nevertheless, some kind of "order" still exists. Often a gas is presented as an example of a disordered system; however, a gas in thermal equilibrium displays an almost perfect spherical symmetry at each point (homogeneity and isotropy) and its molecular velocities are distributed according to Maxwell's law and could be considered as a highly ordered system.

In a system far from equilibrium, energy is continuously exchanged among its constituents in an irreversible manner and the energy of the particles may suffer large variations. In some instances there may appear, under the action of internal forces, highly organized or "ordered" subsystems if the local conditions are appropriate. This rather common phenomenon is called "self-organization". The emergence of structure in the universe is the best example of this process. Because of the cooling of the universe due to its expansion, more and more "ordered" or organized structures have appeared, even if there is no apparent overall organization. First the strong force gave rise to structures called hadrons and, shortly after, to nucleons. Later on the electromagnetic force gave rise to atoms, molecules and small bodies, a process that is still going on. Finally, gravitation has allowed the emergence of large structures, such as planets, solar systems and galaxies, that may show less organization but are highly stable. The most sophisticated and complex case of "self-organization" is that associated with life, which is a consequence of the local conditions on Earth (and perhaps elsewhere in the universe), which is *not* an isolated subsystem of the universe, as well as the nature of the molecules involved and which lend to a very particular structural and functional organization.

As a second example, suppose we place a certain mass of ice at a very low temperature, say 10 kg at 223 K, in contact with a smaller mass of water, say 2 kg at 275 K. After thermal equilibrium is established, the water is frozen and the system is at an intermediate temperature, about 258 K. The process is irreversible and the final state has been the result of the interplay between the molecular energies and the intermolecular forces. It definitely has more entropy than the initial state, but it exhibits a regularity similar to that of the initial ice crystal at 223 K, and certainly less irregularity than the initial mass of water at 275 K. Can we say the final system has more disorder? No. What has actually happened is that the water molecules in the solid state (ice) at 258 K are in slightly higher vibrational levels than at 223 K and that can hardly be considered as a more disordered state.

Thus, one may tell students that the emergence of local "order" in an isolated system initially far from equilibrium is not a violation of the second law, but rather the natural consequence of irreversible processes occurring in the system that allow for the appearance of certain local conditions, coupled with the action of the internal forces. The prediction of the appearance of such local conditions is very difficult due to the non-linear nature of the relations regulating the behavior of the system, making the outcome of a local process very sensitive to small fluctuations in local conditions.

Under no circumstances should one tell students that an isolated system tends to the state of maximum disorder. Unfortunately, this idea has become popular because most of the examples and considerations given in textbooks deal with gases, where the intermolecular forces are weak and the mobility of the molecules gives the appearance of a disordered system. Therefore, for the case of gases, simple probabilistic considerations related to the occurrence of independent events may be invoked. This situation has given rise to a distortion in how students understand how things really happen in the universe. Of course, probabilistic considerations are necessary to analyze the macroscopic properties of matter because of our limitations in dealing with systems composed of a very large number of particles, and this is the reason for statistical physics. This situation should, however, be discussed with great care. The fact is that an isolated system tends to its "attractor" in phase space and this is the essence of the second law [this may be a bit too abstract for IPC students]. We must tell IPC students the truth, but not necessarily the whole truth, and this should be a good reason for using the particle approach to thermodynamics in the IPC.

NOTE 1: Statistical analysis of work and heat.

It is relatively easy to analyze the concepts of work and energy from a statistical point of view. Consider a system where the particles can have energies E_1, E_2, E_3,.... (we assume weakly interacting particles). When the numbers of particles in each energy level are n_1, n_2, n_3,..., the total internal energy of the system is $U = \sum n_i E_i$. If the system is not isolated, it may exchange energy with the surroundings. As a result of the energy exchange two effects may occur. First, the particles will shift between energy levels, resulting in changes dn_1, dn_2, dn_3,.. in the occupation numbers to adjust to the new internal energy; this modifies the partition. Secondly, there may be changes dE_1, dE_2, dE_3,... in the energy levels if the structure or size of the system changes, which also modifies the partition. For example, consider a gas in a box of side a; if the size of the box changes because its sides increase by a small amount da, the energy levels of the molecules change as illustrated in Fig. 1.

The change in internal energy can then be expressed as

$$dU = \underbrace{\sum_i E_i \, dn_i}_{\text{redistribution of particles}} + \underbrace{\sum_i n_i \, dE_i}_{\text{change in structure or size of system}} .$$

The first sum, $\sum_i E_i \, dn_i$, corresponds to a change in internal energy due to a redistribution of the molecules among the available energy levels. The second sum, $\sum_i n_i dE_i$, corresponds to a change in internal energy due to a change in the energy levels. Let us examine the second term first. For the case of the expansion of a gas, the change in energy $\sum_i n_i dE_i$ is due to the change in the dimensions (i.e., volume) of the container under the action of external forces. This energy change corresponds to what we have called external work. Thus, we conclude that the external work done on the system is $dW = \sum_i n_i \, dE_i$. We may also conclude that the first term corresponds to the heat absorbed by the system and write $dQ = \sum_i E_i \, dn_i$. That is, the statistical quantity we have defined as heat corresponds to a change in energy of the system due to molecular jumps between energy levels. These jumps result from energy exchanges with the surroundings. When radiation energy is absorbed by the system, it also results in a redistribution of particles among energy levels; that is, it contributes to the term $\sum_i E_i \, dn_i$.

This analysis puts the statistical definitions of the work done and the heat absorbed by a system on an understandable basis.

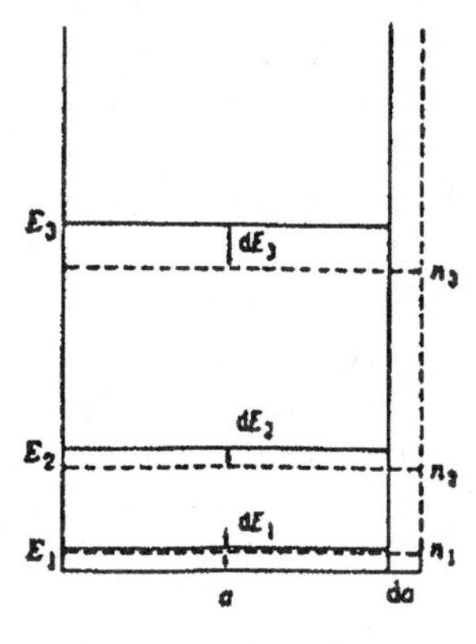

Figure 1. Change in energy levels of a gas in a box when the width of the box is changed.

NOTE 2: Entropy of an Ideal Gas in Statistical Equilibrium.

The calculation of the thermodynamic probability of an equilibrium state of an ideal gas in terms that can be understood by IPC students is relatively simple since the definition of an ideal gas is one composed of molecules that move independently of each other. In that case, we can apply the same logic that is used to calculate the probability of occurrence of independent events, such as coin tossing or throwing a die.

Suppose we have a gas composed of N molecules occupying a volume V. If we divide the volume into a number of small cells, each with volume τ (Figure 2), the number of possible different places where a molecule can be found is V/τ. The number of possible simultaneous occupations for two molecules is $(V/\tau)^2$ and for N molecules the number of occupations is $(V/\tau)^N$, which is then the number of possible microscopic configurations of the gas. Since τ is arbitrary, we conclude that the thermodynamic probability of a gas is proportional to V^N. Now consider a gas that expands isothermally from volume V_1 to volume V_2 (Figure 3). Then the thermodynamic probabilities of the initial and final states of the gas at the same temperature are in the ratio $P_2/P_1 = (V_2/V_1)^N$. Note that the volume τ disappears. Therefore, the change in entropy of the gas in an isothermal expansion, assuming it is in equilibrium in both the initial and final states, is

$$\Delta S = S_2 - S_1 = k \ln(P_2/P_1) = k \ln(V_2/V_1)^N = kN \ln(V_2/V_1) \qquad [1]$$

This expression gives the change in entropy of an ideal gas in an isothermal process. Incidentally, once the gas has expanded to volume V_2, the possibility of finding all N molecules simultaneously in volume V_1 is $(V_1/V_2)^N$, which is an extremely small number if N is very large ($\sim 10^{24}$), and in fact is never observed.

Next we must take the effect of a change of temperature on the entropy of a gas in statistical equilibrium into account (recall that the temperature of a gas is not well defined unless the system is in equilibrium). In this case, the molecular velocities obey Maxwell's distribution law, illustrated in Figure 4. We note that for each temperature there is a most probable velocity v_{mp}, corresponding to the maximum in the curve, irrespective of the direction of motion of the molecule and of the volume occupied by the gas. That is, when the gas is in statistical equilibrium, the molecular velocities tend to group around v_{mp}; but the spread of velocities around v_{mp} increases with an increase of temperature. Expressed in other terms, as the gas temperature increases, the number of energy states significantly occupied by the molecules also increases, thus increasing the thermodynamic probability of the equilibrium macrostate. It might then be reasonable to assume that the thermodynamic probability of the equilibrium state of the gas composed of N molecules at temperature T is proportional to v_{mp}^N. Actually, it is proportional to $(v^3{}_{mp})^N$, or $v_{mp}{}^{3N}$,in order to take into account the three independent directions of motion of the molecules. From Maxwell's distribution law we know that v_{mp} is proportional to $T^{1/2}$. Thus we conclude that the thermodynamic probability of the equilibrium state of an ideal gas at temperature T is proportional to $(T^{1/2})^{3N}$; that is, $P \sim T^{3N/2}$. Therefore, if the temperature of an ideal gas changes from T_1 to T_2 without a change in volume, the ratio of the thermodynamic probabilities of both equilibrium states is $P_2/P_1 = (T_2/T_1)^{3N/2}$ and the change in entropy of an ideal gas due to a change in temperature at constant volume is

$$\Delta S = S_2 - S_1 = k\ln(P_2/P_1) = k\ln(T_2/T_1)^{(3N/2)} = (3/2)kN\ln(T_2/T_1) \qquad [2]$$

Combining this result with Equation [1] we have that the change in entropy of an ideal gas when it goes from equilibrium state (V_1,T_1) to equilibrium state (V_2,T_2) is

$$\Delta S = kN\ln(V_2/V_1) + (3/2)kN\ln(T_2/T_1) \qquad [3]$$

which suggests that the entropy of an ideal gas in statistical equilibrium is

$$S = kN \ln(VT^{3/2}) + \qquad [4]$$

where S_o is a constant to be determined using a more precise method.

The final step is to show that for an ideal gas the definitions $S = k \ln P$ and $dS = dQ/T$ for entropy are equivalent. From Equation [4] we have that, for an infinitesimal reversible transformation,

$$dS = kN[dV/V + (3/2)dT/T] \qquad [5]$$

From the gas law $pV = kNT$, we have that

$$kN(dV/V) = p\,dV/T = -dW/T$$

where $-dW$ is the work done <u>on</u> the gas. Also, for an ideal gas $U = (3/2)kNT$, and

$dU = (3/2)kNdT$. Therefore, Equation [5] becomes

$$dS = -dW/T + dU/T = (dU - dW)/T = dQ/T \qquad [6]$$

where the relation $dU = dQ + dW$ has been used.

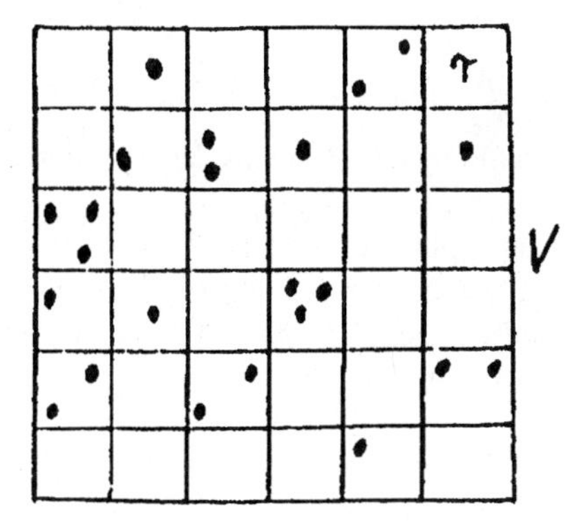

Figure 2.

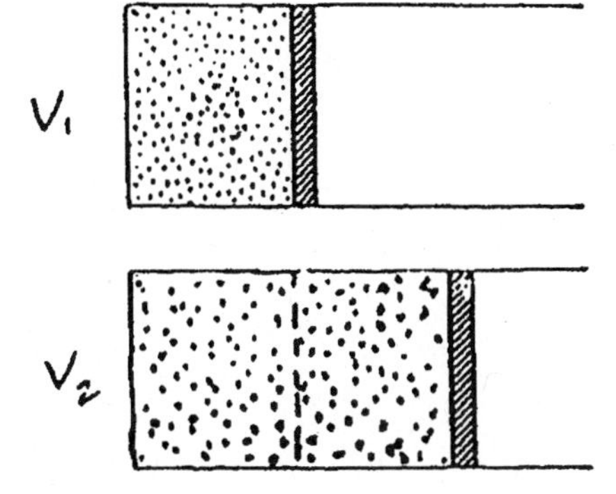

Figure 3.

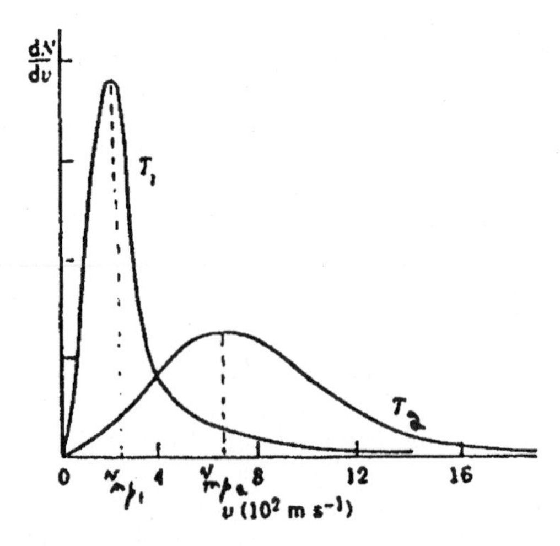

Figure 4. Molecular velocity distribution for a gas at two temperatures T_1 and T_2, with $T_2 > T_1$.

References

1. C. A. Whitney, Random Processes in Physical Systems, Wiley-Interscience Publications, New York, 1990. Section 7.1.
2. Statistical considerations have been applied to systems of a small number of particles, such as to the structure of atoms and large nuclei. However, the problem of the minimum number of particles to which statistrical considerations may be applied, though important, was not discussed during the Workshop because we felt it too sophisticated a topic to raise with IPC students.

MODERN INTRODUCTORY PHYSICS AT COLGATE

J. C. Amato, E. J. Galvez, H. Helm, C. H. Holbrow, D. F. Holcomb, J. N. Loyd & V. N. Mansfield
Department of Physics, Colgate University

Introduction

Eight years ago, Colgate's Physics and Astronomy Department embarked on a long term campaign to revitalize its calculus-based introductory physics curriculum. The traditional two-semester survey of classical mechanics and electromagnetism had ceased to be an effective teaching medium or an inspiration for further study in physics, astronomy, and engineering. Students were discouraged by their inability to master the rapid-fire treatment of physical and mathematical concepts, and faculty members were equally discouraged by low class morale, low enrollments, and low retention rates. Since then, we have replaced these courses with a new three semester sequence, the first two semesters of which are novel and unique to Colgate.

Physics 120, the lead-off course of our new curriculum, plays an important double role as an interface between high school and college level science, and, equally important, as a strong enticement for continuing study in physics and astronomy. Because of this mission, the course content is most unusual. We use what has proven to be a compelling theme, "Why do we believe in atoms?" to ignite student interest and to illustrate the excitement and concerns of present day physics research. Modern physics has the leading role, and classical concepts are summoned only when necessary to advance this main theme. The subject matter is well suited to our mathematically unsophisticated first year student audience. Most of the key ideas of atomic and sub-atomic physics can be derived via conservation principles, obviating differential and integral calculus as the primary mathematical tools in this first exposure to college physics. To wean our calculator-empowered but calculator-bound students from their dependence on rote plug-in methods for problem solving, we stress understanding over fact and formula, promoting basic mathematical strategies such as reasoning by ratios, order of magnitude estimation, and graphical representation. "How do we know?" replaces "What do we know?" as the primary question addressed in lecture, lab, discussion section and homework.

Physics 121 is the second course of the new curriculum. Like its predecessor, it is theme-based in order to capture student interest and to motivate the heady brew of classical physics topics which must be mastered. The conservation principles of momentum, energy, and angular momentum are derived and used to address the questions "Where in the Universe are we?" and, of course, "How do we know?" Physics 121 develops, in a spare and concise fashion, the fundamentals of classical mechanics (including electromagnetic forces), introducing and using vector algebra and calculus as resources to simplify theoretical presentations, proofs, and problem solving. Together, Physics 120 and 121 illustrate the broader theme of unity in physics: the same conservation principles which govern motion on the lab or human scale operate throughout the Universe on all scales from the sub-atomic to the extra-galactic. Using these principles, plus the quantum concepts encountered in Physics 120, we can probe and understand the far outposts of space-time.

The third course of the introductory curriculum, Physics 122, presently remains a traditional survey of electric and magnetic fields and phenomena. When we are satisfied with the development of Physics 121 course materials, we intend to adopt the same theme-based strategy to generate a fresh approach to electromagnetism. With its emphasis on contemporary topics, the three course introductory sequence is expected to provide focus and direction for the entire concentration program. The capstone of Colgate's program in physics and astronomy is a required senior research project. Ideally, our curriculum should foreshadow the research areas available to our students, and prepare them to understand, investigate, and carry out an independent research assignment.

Physics 120

Table 1 summarizes the first semester syllabus. As can be seen, we place great emphasis on the experimental underpinnings of our knowledge of atomic structure. Original sources and data are used whenever possible and appropriate. Students are surprised to learn that careful study of the systematics of chemical reactions (Dalton, Gay-Lussac, Avogadro) provide the first strong evidence for the existence of atoms and molecules: Pan balances and pipettes were the first atomic probes. Using a hard-sphere model and the experimentally derived ideal gas law (familiar from high school physics), we develop a simple microscopic picture of gases and obtain a crude estimate of atomic size from measurements of mean free path. After a concise review of electric and magnetic forces and fields, we encounter Faraday's electrolysis experiments (to obtain N_Ae), Thomson's discovery of the electron (e/m), and Millikan's oil drop experiment (e) to arrive at values for Avogadro's number N_A and values of atomic and electronic masses. The study of atoms is well underway.

Table 1: Physics 120 Syllabus

Early Evidence For Atoms: B Dalton, Gay-Lussac and Avogadro
Hard Sphere Atoms and the Ideal Gas Law: Atomic Sizes and Energies
Electric and Magnetic Forces
Electrical Nature of Atoms: Faraday, Thomson, and Millikan
Wave Phenomena: Interference, Diffraction, Resolution; X-rays and the Bragg Law
Special Relativity
The Granularity of Light: Photoelectric Effect and X-ray Cutoff Wavelength
Wave-Particle Duality: deBroglie, Davisson-Germer, Compton and Heisenberg
Bohr-Rutherford Model of the Atom: Rutherford Scattering, Balmer Spectrum, Moseley's Law

To peer within our hard sphere atoms, we need more precise probes. Students are introduced to atomic spectra, and to the wave phenomena of interference, diffraction, and resolution in order to interpret spectroscopic data and to understand the fundamental limits of such measurements. The class learns to analyze interference patterns - not just to measure wavelength - but also to determine the structure of the slit assembly responsible for the pattern. Having identified interference as a structural probe with resolution $\sim \lambda$, we then discuss X-rays and Bragg diffraction to study the crystalline structure of solids, and conversely, to illustrate how X-ray wavelengths are measured.

After a brief interlude on special relativity, in which we emphasize the relativistic generalizations of momentum and mass-energy, we encounter the excitement and confusion associated with the photoelectric effect, and its controversial explanation by Einstein. The particle nature of photons is made more explicit by the Compton effect. We then ask, along with deBroglie, that if waves can have particle properties, can particles then have wave properties? Students answer this question by studying the experiments of Davisson and Germer, and Thomson, but are then confronted with the bewildering non-classical behavior of matter (e.g. interpretation of electron double-slit experiments, delayed choice experiments) dictated by the Heisenberg Uncertainty Principle.

To dig deeper, we once again need shorter wavelength probes. These are provided by high momentum particles in accordance with the deBroglie relation. We briefly describe the anatomy of a scattering experiment before analyzing Rutherford's data to discover the nucleus. The Bohr-Rutherford planetary model of the atom is then derived and tested against the experimental Balmer spectrum. The course ends on a breathtaking note: Moseley's association of the wavelength of an element's Kα radiation with its nuclear charge, and the unambiguous placement of each element in the periodic table according to its nuclear charge.

Physics 120 has evolved over a period of eight years. The course is supported by a polished set of notes (written by Holbrow, Lloyd and Amato), a large repertoire of exercises, quizzes and activities, and a well-integrated laboratory (Table 2) boasting several novel experiments and associated apparatus. Two supplementary texts (Segre's *From X-Rays to Quarks* and Shamos' *Great Experiments in Physics*) have been used to provide historical background. The course is presently offered in a two lecture, two recitation format, and has been well received by our students. Student retention is high (80% to 90% enroll in the second semester), and participating faculty members express enthusiasm and satisfaction with their teaching efforts.

Table 2: Physics 120 Labs

Molecular Velocities
Electrolysis
Lorentz Force
Electronic charge/mass
Wave Superposition
Interference and Diffraction of Light
Bragg Diffraction - Microwaves
Bubble Chamber Collisions
Half-Life of Indium-116
Spectroscopy (Take-home Lab)
Balmer Spectrum

Physics 121

The content of Physics 121 is outlined in Table 3 (see next page). Like Physics 120, the course begins historically, tracing the origins of modern astronomy from Aristotle to Copernicus. There are three important reasons for doing so. First, the course theme occupies center stage at the outset of the semester, and foreshadows the emergence of modern science during the 16th and 17th centuries. Second, students are forced to activate their dormant understanding of geometry and trigonometry in order to comprehend the clever deductions of the early astronomers. Third, the use of vector algebra is motivated to simplify the mathematical representation of motion. In this historical overview, we follow man's displacement from the center of the Universe to our present, undignified location in the outback of the Milky Way Galaxy. Students determine the distance to the Moon, Sun, planets and stars by a variety of methods including triangulation, parallax, and apparent luminosity. The Doppler effect is explained and demonstrated for sound and light, and Hubble's Law is used to extend distance measurements to objects far beyond our own galaxy. To understand the *motion* of the planets, as summarized by Kepler's Laws, we begin our formal development of mechanics.

We start with a study of two-body collisions. It is straightforward and satisfying to introduce momentum conservation as an experimentally observable fact, and then to define the force on either body as the rate of change of its momentum. Newton's second and third "laws" follow naturally and without complication. Recalling Kepler's Laws, the force guiding the Earth, Moon, and planets is determined to be an inverse square central force, and, following Newton, it is found to be identical to terrestrial gravity. The Cavendish experiment is studied to find G, allowing us to determine the mass of the Sun, Earth, Jupiter, and any other body with a visible orbiting satellite. Other applications include the modeling of stellar structure, and the determination of stellar mass in visible and spectroscopic binary systems.

Table 3: Physics 121 Syllabus

Early Astronomy: Measuring Solar System Distances by Triangulation
Vectors and Relative Motion
Greater Distances: Using Parallax, Apparent Luminosity and the Doppler Shift; Hubble's Law and the Age of the Universe
Copernican Revolution: Tycho, Galileo, and Kepler
Collisions, Momentum and Force: Newton's Laws
The Law of Universal Gravitation; Cavendish
Work, Energy and Potential Energy: Gravitational, Escape Velocities, Nuclear Binding Energies
Rotations and Angular Momentum: Pulsars, Shape of Galaxies
Gravitational Collapse and Black Holes

Energy conservation and the concept of potential energy are introduced in the standard ways, but then are applied to a wide spectrum of problems including gravitational self-energy and nuclear binding energy. We discuss and criticize Kelvin's model of solar evolution, and then identify nuclear reactions as the source of the Sun's energy. Rotation energetics, angular momentum and its conservation are then introduced and applied to explain stellar collapse, pulsars, and the flattening of rotating galaxies. We end the course at the frontiers of our knowledge: black holes and the ultimate fate of the Universe.

The new version of Physics 121 has been offered only twice, and so its course content and materials are still far from finished. We are reasonably satisfied that we have cut a straight and direct path through the heart of mechanics, guided by a clear story line. This year, six new laboratories (Table 4, on next page) were designed and developed, along with four numerical spreadsheet exercises. Three of the labs (Doppler Effect, Magnetic Torsional Pendulum, and Mystery Moments of Inertia) are profoundly different from the usual mechanics laboratories. The numerical exercises emphasized vector algebra (Copernican vs. Ptolemaic systems of the world) and integration of the equations of motion (via spreadsheets) for several important physical systems (constant acceleration, oscillators). Computer literacy was further encouraged by the use of electronic mail and bulletin board conferences to communicate among students and faculty members. Once again, we found that small recitation classes were very effective forums for active learning. Four students, whose identities were not known to the course instructors, were hired as journal-keepers, and asked to record their personal impressions of Physics 121 throughout the semester. Overall, the journalists were sympathetic to the course theme and structure, but expressed some frustration with the lack of a well-matched text (we used French's *Newtonian Mechanics*) and the unavoidable oversights in our first edition notes and assignments. There is still much development work to be done, but we are confident that our theme-based strategy is paying off.

Where Do We Go from Here?

Development of the first semester course is complete. Although we are continually finding new examples and exercises, the written materials are in a polished state ready for export. Likewise, the laboratories have been refined and tested over the last eight years and are now in a stable state. We are only half-finished with Physics 121. While a robust theme has been established and supporting materials produced, we must still improve the integration of the two semesters, and also capitalize more fully on our astronomical theme. Laboratories and lab apparatus need further development to better support the course theme and to increase student satisfaction. Above all, we must solve the problem of a text. The past eight years have taught us that a well-written text, carefully matched to the course theme and syllabus, is vital for student success. In the long term, our diverse materials, activities, problem sets, laboratory instructions, and numerical exercises will need to be absorbed under one cover. Once this is accomplished, there's always the third semester!

Table 4: Physics 121 Labs

Vector Algebra
Finding Distance by Parallax
Luminosity and $1/r^2$
Doppler Effect
Collisions
Oscillators
Beam Balance
Torsional Pendulum
Rotational Dynamics
"Black Box" Moment of Inertia

PRECONCEPTIONS IN INTRODUCTORY PHYSICS COURSES

Robert G. Arns, Department of Physics
University of Vermont

Abstract

Various studies have shown that students' deep-seated beliefs, based on early learning, must be taken into account in the teaching of introductory physics courses if the understanding of physical concepts is an important objective of such courses. Other research is in progress to design and test instructional methods and materials that are sensitive to this problem and to the general need for improvement of physics learning. However, much remains to be done and various resource constraints may delay widespread adoption of such results. As an interim measure, this investigation reviews assumptions and principles that need to be considered in order to improve learning through traditional lecture-recitation-laboratory formats in those parts of undergraduate physics courses in which preconceptions are both firmly-held and incompatible with the degree of understanding sought.

Students come to their first physics class with many beliefs about the physical world which are rooted in early childhood learning and everyday experience.[1] They use these beliefs in attempts to understand new phenomena; hence they function as theories which are already in place when formal instruction begins. Although these prototheories are often incomplete and inconsistent, they are based on beliefs which are firmly established and not easily reinterpreted by the learner. Some of the beliefs are quite close to physics as we know it, others quite far from it. But because they depend on context, even those beliefs which are close to the physicist's beliefs cannot always be used as a reliable starting point for instruction.[2]

To illustrate the teaching and learning challenge posed by preconceptions, consider that babies learn to pull themselves up, to crawl, to walk, and to throw their food in a world in which friction is dominant. Through these experiences children develop a concept of force and a theory of dynamics which they use successfully every day for seventeen years before they enter a physics classroom for the first time. Before they hear of Newton's Laws, they have personally demonstrated:[3]

> Prenewton's First Law: Motion implies a force.
> Force is a property of objects that makes them move; Force may be external or internal.
> Constant external force produces constant velocity.
> An internal force maintains motion of an object independent of external agents.

A child never experiences the isolated interaction of two bodies. In a tug-of-war with a sibling, both are in contact with the earth. When an object is pushed, friction and the floor are also present. Hence, the beginning student believes:

> Prenewton's Third Law (The Principle of Overcoming): In an interaction of two objects, the larger or more active exerts the greater force.[4]

The preconception corresponding to acceleration is not crisp; acceleration is associated with speeding up due to *increasing* force. This does not lead to a quantitative Prenewton's second law of motion. However:

"...In learning Newtonian mechanics, the terms `mass' and `force' must be acquired together, and Newton's Second Law must play a role in their acquisition. One cannot, that is, learn `mass' and `force' independently and then empirically discover that force equals mass times acceleration...all three must be learned together, parts of a whole new...way of doing mechanics....one cannot learn to recognize forces without simultaneously learning to pick out masses and without recourse to the Second Law. That is why Newtonian `force' and `mass' are not translatable into the language of a physical theory ([Prenewtonian] or Einsteinian, for example) in which Newton's version of the Second Law does not apply. To learn any of these three ways of doing mechanics, the interrelated terms in some local part of the web of language must be learned or relearned together and then laid down on nature whole. They cannot simply be rendered individually by translation."[5]

This example, involving elementary dynamics, is only one of the patterns of belief that students bring to the first class in physics. Others are found elsewhere in mechanics; also with regard to heat and temperature, current electricity, etc.

Incommensurability

The foregoing quote is taken from Thomas Kuhn's elaboration of the concept of "local incommensurability" in the history of science. This term has been used to describe an important characteristic of scientific revolutions, namely the partially incompatible languages (both concepts and the relationships among concepts) of scientists on either side of the revolutionary divide and the severe failures of communication and persuasion that result. In these cases, there is no common language into which all concepts and relationships of the competing theories can be translated without loss or residue, or without significant interpretation. The gap between a present-day physics teacher and the students in the learning of the Newton's Second Law is of the same nature as that which characterized the understanding of dynamics before and after Newton.

Students bring a very large number of preconceptions to the first class in physics. Some, such as those involving force and elementary dynamics described here, warrant special attention. Others represent less of a barrier to learning and probably do not. The presence or absence of incommensurability may be a useful gauge in identifying the most serious problems posed by students' preconceptions, those cases in which learning is characterized by a wrenching novice → expert shift like that seen in periods of major upheaval in the history of science.

Incommensurability can also be induced in the classroom. For example, most beginning students do not possess an understanding of acceleration. Concepts of distance, velocity, and acceleration are not clearly separated. The concepts of "time interval" and "instant of time" are not differentiated; average velocity and instantaneous velocity also overlap.[6] However, even in the absence of strong preconceptions, incommensurability can still arise if textbooks and teachers introduce the variables such as time, velocity, and acceleration and the kinematic equations linking them *as if* the student had an understanding of these quantities based on experience of physical reality. The teacher assumes a web of concepts and a theoretical structure which the student does not possess. The same kind of communication breakdown occurs. When this happens, the student may learn kinematics as a separate formal structure without being able to connect it to the physical quantities which appear as variables in the equations.

The criterion of incommensurability can also be used to identify preconceptions that are less troublesome. Incommensurability does not result when the word used to express the teacher's concept can be easily and fully translated into the language in which the students' (differing) preconception is expressed. For example, the words `energy' and `work' have meanings in physics that are both more precise and different from their meanings in everyday language. However, the vernacular preconception does not depend much on physical learning; no incommensurability occurs. Careful definition should be sufficient to set the stage for using these concepts in the physics classroom.

Issues and Assumptions

There is convincing evidence that traditional instruction, with emphasis on lecture presentation and problem solving, has not been effective. Several efforts are in progress to develop and test ways of changing physics teaching in order to improve learning. Applications of new technologies, especially computing and multimedia tools such as real-time microcomputer-aided data acquisition in the introductory laboratory[7] and in lecture-demonstrations, and the use of the interactive videodisc, are particularly exciting. Older tools, such as the use of writing to promote conceptual understanding, are getting a new look.[8] Considerable effort is also being directed at finding ways of overcoming problematical preconceptions.[9] And the *Physics by Inquiry* modules being developed by the Physics Education Group of the University of Washington, and other efforts aimed at finding new ways of teaching,[10] offer promise for the future.

However, only a few physics teachers have the opportunity to be part of a special project to improve physics instruction and none of these efforts has as yet produced a complete course, with materials, to assist the physics instructor who must face 150 freshmen in a traditional lecture-recitation-laboratory format on short notice. Until such materials become available, conventional textbooks and traditional formats will continue to be used. What can the individual physics teacher do to improve learning in the interim?

First, with only a little time available for trying new things, there is a need to set priorities. Traditional courses are particularly weak at those places in the subject matter at which incommensurability, as described above, is encountered. This criterion may be useful in limiting the domain of special attention. Second, there is a significant research base to guide efforts at instructional improvement;[11] the individual instructor is not alone. Some items worthy of attention are:

1. *Course objectives.* Behind the litanies of weaknesses in traditional physics courses are assumptions concerning what students should know and be able to do when they complete such a course. If the student is to develop a useful understanding of certain important concepts and of the relationships and principles that make them important, and to develop a limited proficiency in scientific reasoning, and if we want the student to be able to use formal representations (diagrams, graphs, and equations), then these goals must be reflected in what is taught, how it is taught, and how students are examined. Introductory courses tend to be focused on problem solving, which cannot fully support these other objectives. Course objectives need periodic affirmation; and instructional methods need periodic alignment with objectives.

2. *Constructivist View.* Successful instruction is geared to the recognition that: (1) students construct their own beliefs, and (2) what they already know - or think they know - plays a big part in what they can learn.

3. *Prior Knowledge and Belief Patterns.* Physics teachers need to be aware of what students know and of what they do not know. Included here are: (1) the particular topics in which students' preconceptions are incommensurable with the teacher's view; (2) conceptual gaps, i.e., understanding which the teacher should not assume; and (3) unfounded beliefs, e.g., concepts which students believe they know, perhaps from high school physics, but have only heard about. Several of the references cited above provide information about students' prior knowledge.

4. *Design of Instruction.* General advice[12] is supplemented here by attention to potential islands of incommensurability within the subject matter of the introductory course.

   Usually those preconceptions that are described above as incommensurable are in clear conflict with the canonical view of the subject matter that we expect students to learn. If change is to occur, i.e., if the student is to see the preconception in a new light, the student must be made aware of the conflict, be dissatisfied with the preconception, and find the new concept intelligible, plausible, and fruitful (i.e., the new concept must lead to useful new insights).[13] Various mechanisms can be used, even in the lecture format, to help expose and resolve conceptual conflict. For example, demonstrations and simple quizzes or pretest results can be presented and discussed. Handouts such as the *Active learning Problem Sheets* described by

Van Heuvelen[14] may also be useful in challenging preconceptions in the lecture setting. In addition, Brown and Clement have described the use of analogies to help students over preconceptions.[15]

The avoidance of teacher-induced incommensurability is not simply a matter of being aware of what the student does not know. In addition, the student must be given an opportunity to learn the basic concepts in a meaningful way. For example, in the case of introductory kinematics described above, laboratory can play a key role in the development of clear concepts.[16]

Research has also shown that one pass at the material is not enough when profound preconceptions are involved.[17] The use of recitation or tutorial meetings for revisiting these topics may be especially helpful.

Fortunately, the number of potential incommensurabilities is not very great. The teacher can countenance special attention to these topics within a more-or-less traditional course setting. Still, teaching solely by talking does not work in these cases. Some additional time will be required, especially at the outset of the first course, to overcome profound preconceptions and to lay the foundations of kinematics. And this additional allocation of time means that some of the topics usually "covered" in the traditional course must be omitted.

Significant research is underway aimed at radical transformation of the teaching of introductory physics. However, hundreds of thousands of students will be taught in lecture-recitation-laboratory formats using conventional textbooks before that is accomplished. In the meantime, existing research results can be used to improve learning. In addition, more research is needed on how to improve traditional instruction, not as an alternative to the radical transformations being sought but in practical recognition of the fact that radical change takes a very long time.

References

1. For example, Clement, John, "Students' preconceptions in introductory mechanics," *Am. J. Phys.* 50 (1982) pp. 66-71; the ERIC database listed 178 papers on "misconceptions" in "physics" as of December, 1992.
2. Clement, J., D. E. Brown, and A. Zietsman, "Not all preconceptions are misconceptions: Finding `anchoring conceptions' for grounding instruction on students' intuitions," *Int. J. of Science Ed.* 11 (1989) pp. 554-565.
3. The following brief statements of the naive force concept and of the antecedents of Newton's laws can scarcely do justice to the careful mapping carried out by Hestenes and his coworkers. See, for example: Halloun, Ibrahim A. and D. Hestenes, "The initial knowledge state of college physics students," *Am. J. Phys.* 53 (1985) pp. 1043-1055, and "Common sense concepts about motion," Am. J. Phys. 53 (1985) pp. 1056-1065; Hestenes, David, M. Wells, and G. Swackhamer, "Force Concept Inventory," *Phys. Teacher* 30 (1992) pp. 141-158.
4. Adapted from p. 742 of Hestenes, David, "Modeling games in the Newtonian World," *Am. J. Phys.* 60 (1992) pp. 732-748.
5. Page 677 of Kuhn, Thomas S., "Commensurability, Comparability, Communicability," in *PSA* 1982 [Proceedings of the 1982 Biennial Meeting of the Philosophy of Science Association], Peter D. Asquith and Thomas Nickles, Eds. (Phil. of Sci. Ass'n; East Lansing, MI; 1982) Vol II, pp. 669-688.
6. Halloun, Ibrahim A., and D. Hestenes, *Am. J. Phys.* 53 (1985) pp. 1063-1064.
7. For example, Thornton, R. K. and D. R. Sokoloff, "Learning motion concepts using real-time microcomputer-based laboratory tools," *Am. J. Phys.* 58 (1990) pp. 858-870.
8. Mullin, William J., "Writing in Physics," *Phys. Teacher* 30 (May, 1989) pp. 112-117.
9. For example, Brown, David E. and J. Clement, "Classroom teaching experiments in mechanics," in *Research in Physics Learning: Theoretical Issues and Empirical Studies,* R. Duit, F. Goldberg, and H. Niedderer, Eds. (Inst. für die Pädogogik...; Kiel; 1992) pp. 380-397; also Brown, D. E., "Using examples and analogies to remediate misconceptions in physics: Factors influencing conceptual change," *J. of Res. in Science Teaching* 29 (1992) pp. 17-34.

10. For example, McDermott, Lillian C., "What we teach and what is learned — Closing the gap," *Am. J. Phys.* 59 (1991) pp. 301-315; Van Heuvelen, Alan, "Overview, Case Study Physics," *Am. J. Phys.* 59 (1991) pp. 898-907; Rigden, John S., D. F. Holcomb, and R. DiStefano "The University Physics Project," *Physics Today* (April, 1993) pp. 32-37.
11. See, for example, Arons, Arnold B., *A Guide to Introductory Physics Teaching* (John Wiley & Sons; New York; 1990)
12. See, for example, Arons, reference 11, and Van Heuvelen, Alan, "Learning to think like a physicist: A review of research-based instructional strategies," *Am. J. Phys.* 59 (1991) pp. 891-897.
13. Posner, G. J., et al., "Accommodation of a scientific conception: Toward a theory of conceptual change," *Science Education* 66 (1982) pp. 211-227.
14. Reference 12, p. 895.
15. References 2 and 9.
16. Rosenquist, Mark L. and L. C. McDermott, "A conceptual approach to teaching kinematics," *Am. J. Phys.* 55 (1987) pp. 407-415; see also Thornton, Ronald K. and D. R. Sokoloff, reference 7.
17. Brown, David E., and J. Clement, reference 9.

MODELS OF REALITY: PHYSICS BY EXAMPLE†

Gordon J. Aubrecht, II, Department of Physics
Ohio State University

Abstract

The current University Physics course, as IUPP groups have noted, includes little knowledge from physics education research about how students learn, little from contemporary physics topics, and does not generally use the microcomputer effectively. These 3 "*C*"s, existing *C*ognitive studies in physics pedagogy, more *C*ontemporary physics content, and use of [micro]*C*omputers in teaching, have been incorporated as an integral part of the project. This FIPSE-supported project at Ohio State is paying particular attention to teaching by example. Students in the introductory course for engineers have a strong tendency to focus on "hot formulas" and plug-and-chug computation, and to consider this the essence of physics. We wish our students to be able to describe physical situations both qualitatively and numerically. Several examples developed to help students think in "physical" ways and our use of them in the course is discussed.

Introduction

As one among many physicists inspired by use of the first edition of *Physics for students of science and engineering*, I have been working on a FIPSE project to carry on that tradition into the future entitled "Retooling University Physics for the 21st Century". The physics community has recognized for some time that the University Physics course is in serious trouble, as evidenced by invited talks at physics meetings, articles in physics journals,[1-5] and by the existence of a conference such as this one.

Dissatisfaction reaches beyond that of the physicists who teach and the students who take the course; some engineering colleges have expressed the view that their students are not being taught the physics concepts that are widely used in modern engineering practice. Difficulties include:

1) Research in physics education during the last decade and a half has begun to provide solid evidence that our traditional teaching techniques are not nearly as effective as we had thought. Such work[6] is providing insights on ways to improve the situation.
2) The increasing length of textbooks, most well over 1000 pages, and the overwhelming similarity among them has led to a situation with too much material being "covered" too rapidly for assimilation and with too little connective tissue among the ideas. The course lacks unity and purpose.
3) Despite the increasing length of the textbooks, what 20th century physics is taught in the course has been added to the end, "extended". As we all know, the end is never reached. While this might have been less of a concern 25 or 50 years ago, relativity, quantum physics, and nonlinear phenomena are now a significant part of human thought and engineering technology. Their omission from the remainder of University Physics is not defensible.
4) The present University Physics was conceived and developed during the era of slide rules and trig tables. It has not really been changed as the student has been able to use a pocket calculator, and in many places a desktop microcomputer. The computer has a major role to play in physics education.[4,5] I discuss one of my approaches to this aspect in a separate contribution to the conference.

†*Work supported in part by the U.S. Department of Education under FIPSE grant #P116B01237.*

Efforts to revise and revitalize University Physics have been organized under the auspices of the Introductory University Physics Project (IUPP). I have participated in IUPP as a member of the phenomenology subgroup. The material considered for inclusion in the IUPP phenomenology curriculum was on a "need-to-know" basis. The phenomena to be treated should be observed by students or demonstrated in lectures. There would follow an analysis of the details of the phenomenon. I have worked to incorporate this approach in my text.

In addition, the department at Ohio State has revised the laboratory recently, and we are knitting the laboratory experience more firmly into the course to provide the basis in student experience for the theory discussed in lecture and in the text. The enhanced integration of labs should give students a better idea of exactly what quantities can be measured, and of how the measurements can be used.

The goals of my work on the introductory text are:

1) To streamline the course while simultaneously adding a significant component of 20th century physics. A reorganization of the entire course structure is necessary, but it does end up with many topics in the current list. Students are interested in the topics that are causing excitement in any field. This was done in *Models of Reality*.
2) To reorganize the course structure and format with a spiral approach, returning to the same topic a number of times at different levels. This will explicitly recognize the nature of student learning while emphasizing the unity of concepts and techniques in physics. These insights are being incorporated in *Models of Reality*.
3) To introduce computers as a fundamental tool for model building, problem solving, and data acquisition. My spreadsheets, keyed to the text, are discussed elsewhere.

The Basis of the Course

There are three important facets to consider about the way students learn:

1) *Student understanding is greatly enhanced by direct experience, to the extent possible, of the phenomena of physics prior to the presentation of the theory.* This experiential basis for the curriculum emphasizes the importance of well-designed labs and computer simulations, both purposefully integrated into the course text rather than, as now, used simply as an adjunct to the course.
2) *Learning by the student is not a sequential, linear process. A subject must be seen repeatedly, in different contexts, before assimilation can be complete.* When students first learn something, their reaction is often to think how implausible it sounds. By the third time a concept is seen, the students may realize that the idea is obvious. The spiral approach follows from this consideration.
3) *Emphasis should be on transferable skills, those useful to the student in many areas of physics as well as elsewhere in school and after graduation.* Examples include graphing and recognition of functional relationships, dealing with error and uncertainty, and use of scaling and symmetry arguments. The current University Physics assumes that students will pick up these ideas on their own, whereas experience has shown that these skills must be taught explicitly. In addition, the sorts of problems we will treat in the course will be expanded to include nontraditional problems involving techniques of estimation and approximation discussed in class.

It is important to place the physics in perspective, and we have attempted to organize the text around five major unifying concepts (and their ramifications) and one subtheme common to all five. They are:

I Particles and their behavior
- A objects are reducible to particles
- B vibrations of particles
- C energy

II Forces and interactions
- A how particles interact
- B forces in Nature

III The field
- A kinds of fields
- B particle interaction through the field
- C wave propagation of a disturbance in a field

IV Wave behavior
- A waves in materials
- B wave properties
- C particles and waves together

V Wave-particle duality
- A wave behavior
- B waves can be described as particles
- C particles can be described as waves
- D particles or waves?

VI Conservation laws and symmetry
- A quantization
- B conservation laws

Significant components of modern physics is being interwoven into the development of the story throughout the entire course, as the list above suggests. We have tried to have a course organized around a coherent "story line," a study of how the one facet of the physical world can be understood better with the increasing knowledge the students can bring to bear. The physics is integrated into the story of increased understanding of some facet of nature. This seems to hold student interest better than the current approach.

First Quarter

Introduction ($\frac{1}{3}$ week)
Forces ($2\frac{1}{3}$ weeks)
Displacement, velocity, acceleration (2 weeks)
Momentum and reference frames ($1\frac{1}{3}$ weeks)
SHM & waves (incl. bandwidth theorem→Uncertainty Principle) (2 weeks)
Work & energy (2 weeks)

Second Quarter

Electricity, gravitation, & magnetism ($3\frac{2}{3}$ weeks)
Capacitance ($\frac{2}{3}$ week)
DC circuits ($1\frac{1}{3}$ weeks)
Magnetism (Faraday & Ampère) ($\frac{2}{3}$ week)
Electromagnetic oscillation (incl. relativity) (1 week)
Waves and particles (1 week)
Applied electromagnetism (scattering, acceleration, detection of particles) ($1\frac{1}{3}$ weeks)

Third quarter

Waves (classical & quantum) (1 week)
Rigid body motion ($1\frac{1}{3}$ weeks)
Hydrogen & multielectron atoms ($1\frac{1}{3}$ weeks)

CONDENSED MATTER PATH
- Thermal & statistical physics ($1\frac{2}{3}$) weeks)
- Quantum mechanics in solids ($3\frac{1}{3}$ weeks)
- Lasers (1 week)

PARTICLE STRUCTURE PATH
- Nuclear physics ($2\frac{1}{3}$ weeks)
- Particle physics ($2\frac{1}{3}$ weeks)
- Cosmology ($1\frac{1}{3}$) weeks)

Table 1

Wave phenomena are found in all contextsÆboth classical and quantum mechanical. Waves can form the bridge between classical and quantum mechanical interpretations of nature. An additional advantage of an emphasis on waves is that the topic lends itself to the spiral approach. One may introduce waves near the beginning of the course, then return to it when oscillations are considered, again in electromagnetism, and again at another level of sophistication in the discussion of quantum mechanics. This return to the topic reinforces student confidence in the techniques learned and the familiarity helps prepare the student to go beyond what has already been understood. Table 1 shows the outline of the course that follows the text.

Problem Solving

Problem solving is emphasized in the current University Physics course, but techniques of problem solving used by experts are not usually taught or utilized. It is important to emphasize the methodology of problem-solving in the recitations in the proposed course revision. Our multitudinous detailed examples (an average of about 17 per chapter) emphasize the approach of the expert in first thinking globally, in breaking a problem into its components, in making errors, and the like. The examples often are qualitative, in contrast to current practice elsewhere. Also, great care is taken with definitions and derivations so as not to obscure the physics.

While the space here is not sufficient for detailed examples of all the categories shown, let me summarize the approach taken.

i) There are some examples in which the solution is reached by two or more alternate paths.

Example 6.14. The original velocity of a ball [at point A] is 10 m/s at 45° above the horizontal. Assume that air resistance is negligible. List the velocities and accelerations at points A, B, and C shown in Figure 6.16.

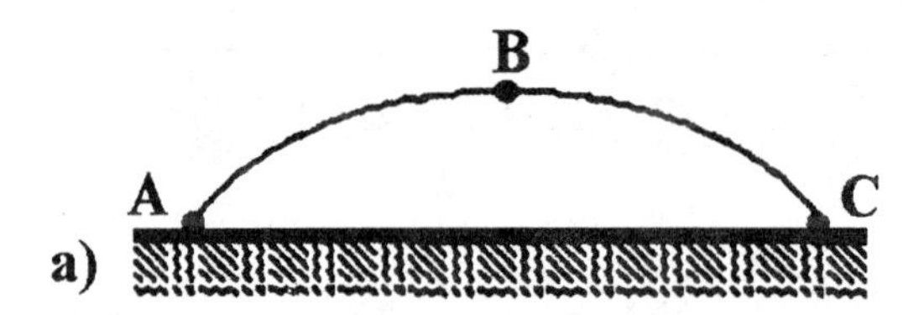

Figure 6.16. The trajectory of a ball subject to no air resistance.

The velocity at point A is 10 m/s **at** 45° **above the horizontal, which means that the x-component is**

$$v_{A,x} = (10 \text{ m/s}) \cos 45° = 7.1 \text{ m/s}$$

and the y-component is

$$v_{A,y} = (10 \text{ m/s}) \cos 45° = 7.1 \text{ m/s}$$

also. Since there is no air resistance, the x-component of velocity does not change, and

$$v_{B,x} = v_{C,x} = v_{A,x} = 7.1 \text{ m/s}.$$

The y-component of velocity does change. At point B, it is zero. By the symmetry of the diagram, it may easily be seen that the y-component of velocity at point C is the same in magnitude but opposite in direction from that at A:

$$v_{B,y} = 0,$$

$$v_{C,y} = -7.1 \text{ m/s}.$$

Alternatively, we could solve the equations for the velocity and elevation,

$$v_y(t) = v_{y,0} + a_y t = v_{A,y} - gt$$

and

$$y(t) = y_0 + v_{y,0}t + \frac{1}{2} a_y t^2 = v_{A,y}t - \frac{1}{2} gt^2.$$

The point B is found from $v_y(t_B) = 0$, which gives

$$t_B = -\frac{v_{y0}}{a_y} = -\frac{v_{Ay}}{g}$$

The point C is determined by $y(t_C) = 0$, which leads to

$$0 = v_{y,0}t_C - \frac{1}{2}gt_C^2 = t_C(v_{y,0} - \frac{1}{2}gt_C),$$

or,

$$t_C = 0 \quad \text{and} \quad t_C = 2\frac{v_{Ay}}{g}$$

The first solution corresponds to point A, not point C, so that the y-velocity at point C is

$$v_{C,y} = v_{A,y} - gt_C = v_{A,y} - g2\frac{v_{Ay}}{g} = v_{A,y} - 2v_{Ay} = -v_{A,y} = -7.1 \text{ m/s}.$$

The acceleration is the gravitational acceleration, and so is 9.8 m/s^2, downward everywhere.

ii) There are many examples that are totally qualitative (perhaps half of them).

iii) In some cases, two examples in sequence do the problem qualitatively, then quantitatively.

iv) In many examples, the solution is graphical rather than numerical (both are, of course, equally quantitative).

v) The spreadsheets are used in the examples as well as in the text and problems.

Summary

The work so far has generated about 2200 pages of typescript organized into 525 sections, 891 worked examples, 2235 problems at the end of the chapter, and 343 spreadsheet problems. We are tackling a persistent problem, and have some small indications that the students are learning as much as other students. Additionally, they do not react negatively to the text. We hope the result will be a useful addition to the texts now in use.

References

1. J. F. Rigden, "The current challenge: Introductory Physics," *Am. J. Phys.* **54,** 1067 (1986).
2. D. F. Holcomb, R. Resnick, and J. F. Rigden, "New approaches to introductory physics," *Phys. Today* **40** (5), 87 (1987).
3. E. F. Redish, "The coming revolution in physics instruction," in L. A. Steen, ed., *Calculus for a New Century,* National Academy Press, 1988. See also the article in Academic Computing, Nov. 1988, page 18: "From here to the future: The impact of the computer on College Physics Teaching."
4. W. M. MacDonald, E. F. Redish, and J. M. Wilson, "The M.U.P.P.E.T. Manifesto," *Computers in Phys.* **2**(4), 23 (1988).
5. J. M. Wilson and E. F. Redish, "Using Computers in Teaching Physics," *Phys. Today* **42** (1), 34 (1989).
6. G. Aubrecht, "Using Spreadsheets in the University Physics Course," in K. Johnston and J. M. Wilson, eds., *Proceedings of the Conference on the Introductory Physics Course.*

COMPUTER-BASED INTERACTIVE DEMONSTRATIONS IN KINEMATICS AND DYNAMICS

Antonino Carnevali & Anthony Blose
Department of Physics & Earth Sciences,
University of North Alabama

Recent research has shown that classroom lectures are ineffective vehicles for teaching basic concepts in kinematics and dynamics. At the University of North Alabama, we are exploring the potential of computer-based interactive classroom demonstrations as a more effective way to promote understanding. A motion detector and force probe are interfaced to a computer; real-time graphs of the actual motion of a cart are produced and projected onto a screen. An interactive protocol, developed by David R. Sokoloff and Ronald K. Thornton, is used to foster students' understanding. Standardized pre-tests and post-tests were administered to monitor students' progress and evaluate the effectiveness of the method.

Introduction

Recent research in science education has shown that students come to the introductory physics course with a well defined set of "alternative conceptions" (e.g., that a force is needed to keep an object in uniform motion) and that traditional style lectures are very ineffective at replacing those preconceptions with appropriate (e.g., Newtonian) physical concepts [1,2,3]. On the other hand, it has been established that interactive methods using computers in a physics laboratory setting are successful in teaching those concepts to introductory physics students [4,5]. At the University of North Alabama, we are exploring the potential of computer-based interactive classroom demonstrations as an alternative to a full program of microcomputer-based laboratory (MBL) activities. This alternative would be particularly valuable for institutions that can not afford the implementation of MBL programs or are otherwise not ready to adopt such programs. We believe that it would also be useful as a complement to programs with ongoing MBL activities to introduce basic concepts and leave more time for independent investigation of more complex situations.

Configuration and Methods

The system configuration is shown in Figure 1. We use a Macintosh LC-II (without monitor), but other models are also satisfactory, such as Macintosh SE or Classic. An ultrasonic motion detector and a Hall-effect force probe are interfaced to the computer, and various graphs of the actual motion of the cart are produced essentially in real-time and projected onto a screen using a Liquid Crystal Display (LCD) panel - a Sharp QA75, in our case. The ULI data acquisition interface and related software are very easy to set up and they are user-friendly.

The interactive motion demonstrations consist of a series of 7 demonstrations in kinematics, which take approximately 45 minutes, and 6 demonstrations in dynamics (conducted one or two weeks later), which take another 45 minutes. The demonstrations we use are modifications of those developed by David Sokoloff and Ron Thornton (see also Ron Thornton's paper in these Proceedings). At the beginning of class, students are given an Interactive Motion Demonstration Prediction Sheet, in which they write their name, sketch their predictions, and draw the actual results displayed on the screen. The Prediction Sheets are collected at the end of class. Sokoloff/ Thornton's protocol, with which we concur, is as follows:

1. Describe the first demonstration and carry it out without MBL tools.
2. Students discuss the demonstration in groups of 2 or 3, and then, with a dashed line, sketch their predictions of the velocity-time graph, and, when appropriate, the acceleration-time and force-time graphs, on the Prediction Sheet.
3. Do the demonstration using MBL tools.
4. Ask a few students to describe the results displayed on the screen, and discuss the results in the context of the demonstration and analogous physical situations.
5. The students draw, with a solid line next to their predictions, the graphical results of the actual motion as displayed on the screen.
6. Then, and only then, go on to the next demonstration.

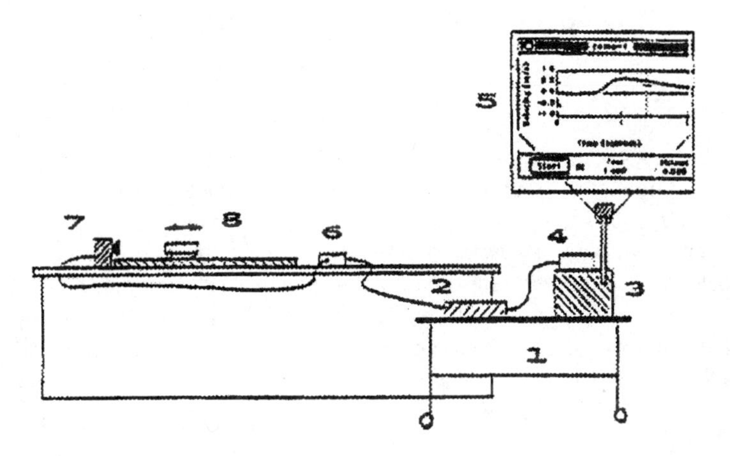

1. Movable cart
2. Computer
3. Overhead projector
4. LCD panbel
5. Screen
6. Interface (ULI)
7. Motion detector
8. Cart on frictionless track

Figure 1. System Configuration for Computer-based Interactive Classroom Demonstrations.

Results

We used the kinematics demonstrations in the Fall 1992, and both the kinematics and dynamics demonstrations in the Spring 1993, in several sections of Physics 101, an introductory physics course for non-science majors. About one half of the students in this course are in the Elementary Education program and take it because it is required; most of the remaining students take the course to satisfy the General Studies science requirement.

Appropriate pre- and post-testing procedures were followed in all cases, administering the same standardized, multiple choice questions used by other investigators (e.g., Ron Thornton). Figure 2 shows the error rates in 4 questions on kinematics, in each of which the students were asked to select the velocity-time graph that corresponds to a particular type of motion, respectively as follows:

1. Constant velocity to the right
2. Reversing direction
3. Constant velocity to the left
4. Increasing speed at a constant rate

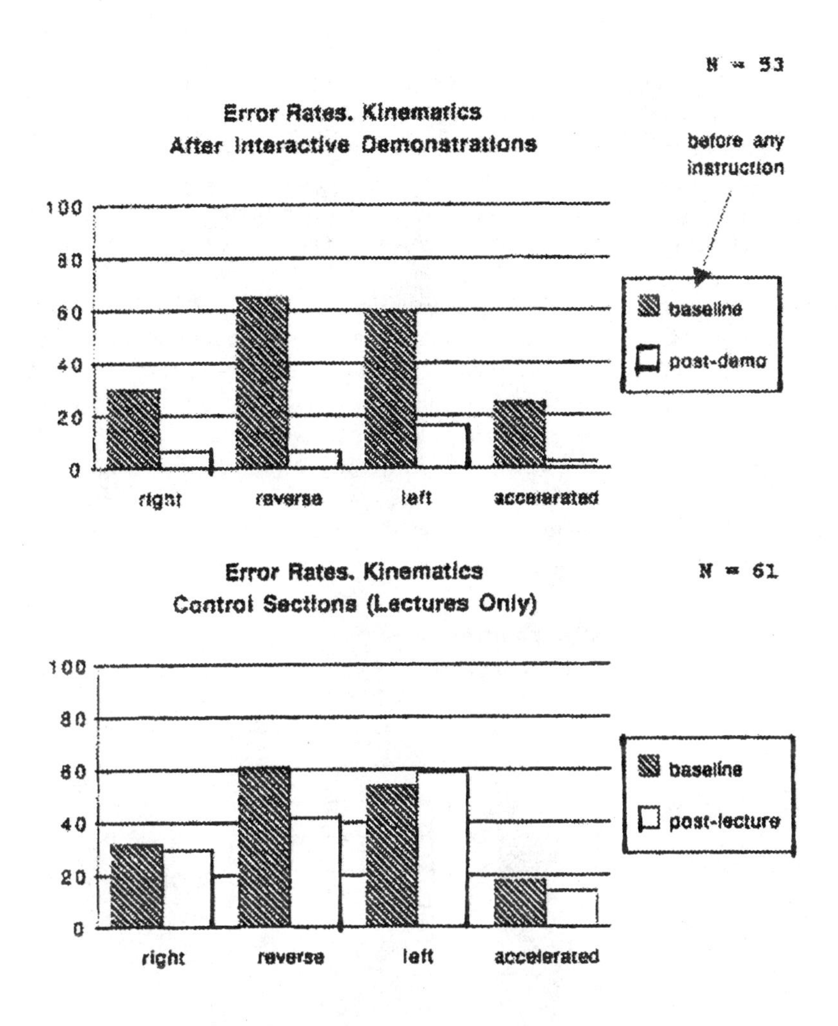

Figure 2. Comparison of error rates on 4 kinematics questions for students who participated in the interactive demonstrations and students who attended traditional style lectures.

The same 4 questions were asked in the pre-test and the post-test (which contained other types of questions as well), with the graph choices shuffled. The same amount of time (5 class periods) was allotted to kinematics in both the "demonstration sections" and the "lectures only sections". The pre-test, which was administered before any instruction (and was not for grade), shows that students in the various sections of the course started out with similar error rates. The error rates decreased quite significantly after the interactive demonstrations, whereas in the control sections, with lectures only on the same material, they did not. The results on this particular test after the interactive demonstrations are close to those obtained at other universities using a full program of MBL activities.

In the Spring 1993, after administering the kinematics post-test comparison, we also tested the effectiveness of the interactive demonstrations in dynamics, as compared to lectures only, in the same course for non-science major. The student sample size for this comparison was N=42. The time sequence was the following:

- dynamics pre-test (baseline, before any instruction)
- 2 weeks of lectures on dynamics
- pre-demo (post-lecture) test on dynamics
- demonstrations (kinematics + dynamics in control section)
- post-demo test on dynamics

The baseline, post-lecture, and post-demo tests consisted partly of identical questions, with the various choices of answers shuffled. The standardized questions were the same as those used, for example, by Ron Thornton. The tests contained 7 force questions, in each of which the students were asked to select, from among 7 choices of direction (right or left) and magnitude (i.e., whether constant, increasing, or decreasing strength), the force which would produce a particular type of motion of a sled on ice. As an illustration, one question read: "Which force would keep the sled moving toward the left and slowing down at a steady rate (constant acceleration)?"

Before any instruction, the error rates on the 7 questions ranged from 92.1% to 100%, with an average error rate on the 7 questions of 95.9%. Not only did most students answer incorrectly all the questions, but, most significantly, 86.8% of the incorrect answers were exactly the same, and they corresponded to the preconceptions that a constant force is needed to keep an object moving at constant velocity and an increasing force is needed to produce constant acceleration. On the post-lecture test, after two weeks of lectures, the error rates on the same 7 questions ranged from 28.6% to 90.5%, with an average of 56.1%. One single demonstration set (45 minutes) brought these error rates down to a range from 7.1% to 66.7 %, with an average of 26.2% over the 7 questions. This was the first time that we used the demonstrations in dynamics, and we believe that we will be able to achieve even better results next time.

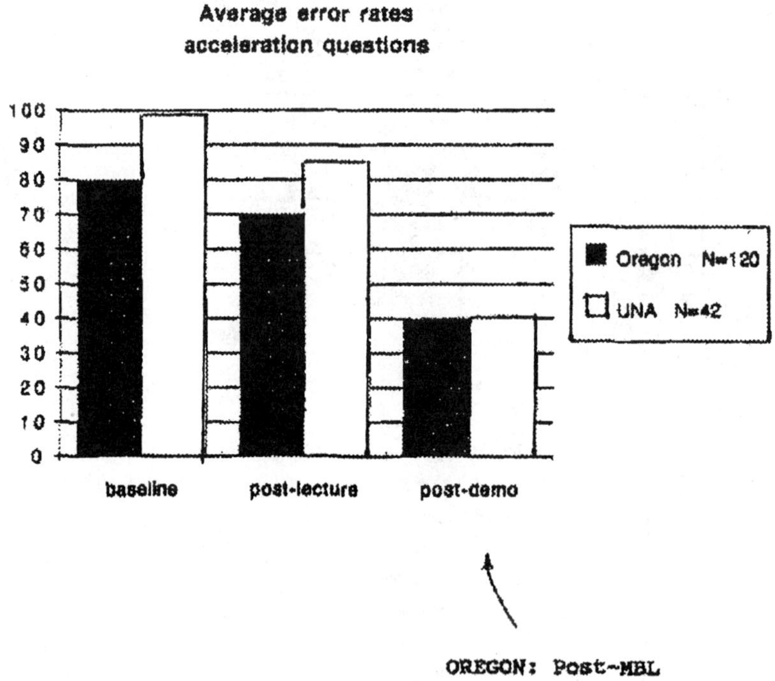

Average error rates
acceleration questions

Oregon N=120
UNA N=42

baseline post-lecture post-demo

OREGON: Post-MBL

Figure 3. Performance comparison of interactive demonstrations and early MBL program (Oregon, 1988).

The tests also contained 5 identical but reshuffled acceleration questions (the same standardized questions used by other investigators), in which the students were asked to select the correct acceleration-time graph corresponding to a particular motion of a toy car, such as "the car moves toward the right (away from the origin), speeding up at a steady rate."

Before any instruction on dynamics and acceleration-time graphs, the error rates on the 5 questions ranged from 94.7% to 100%, with an average error rate of 97.9% over the 5 questions. It is interesting to note again that the incorrect answers were not random, rather revealed a definite pattern: 79.8% of the incorrect responses were the same, and they would have been the correct answers if the questions had asked for velocity-time graphs! Clearly, the students could not distinguish the concepts of velocity and acceleration (this conclusion was supported by their written explanations of the reasons why they picked those particular choices). After the two weeks of lectures in dynamics, the error rates ranged from 76.2% to 95.2%, with an average of 83.3%. After the single Interactive Demonstration set in dynamics, the error rates decreased to a range from 26.2% to 61.9%, with an average over the 5 questions of 37.6%. Again, we believe that we will be able to do better as we gain experience with conducting the demonstrations. Nonetheless, our initial results are already similar to those obtained after full MBL programs at an early stage of their development. For example, our results are compared in Figure 3 to those obtained in an introductory General Physics course for non-calculus students at the University of Oregon in the Fall 1988, averaging the error rates over the same set of 4 acceleration questions (question # 1 was not asked that time at Oregon). Although our students started from behind before any instruction was imparted, and after lectures the error rates decreased by approximately only 10% for both groups, the error rates decreased significantly after the interactive demonstrations and went down to the same level as they did after the full MBL program at the University of Oregon.

Conclusions

Although computer-based interactive classroom demonstrations by no means offer all the benefits of a full MBL program or a workshop style course a la Priscilla Laws, they do provide a cost effective way to promote students' understanding of basic kinematics and dynamics concepts, which can not be achieved by traditional style lecturing. They also result in a livelier environment for learning, as the students enjoy the demonstrations and are willing to make predictions and discuss their reasoning and their surprise at the results. Finally, we believe that these interactive demonstrations would be a valuable complement to MBL activities where those programs already exist.

Acknowledgments

Work supported in part by an NSF grant administered by Dickinson College, Carlisle, PA, and by a research grant from the University of North Alabama.

We wish to thank Priscilla Laws, David Sokoloff and Ronald Thornton for promoting this research at the workshop held at Dickinson College in June, 1992.

References

1. Lillian C. McDermott, "Research on conceptual understanding in mechanics", *Physics Today*, July 1984.
2. David Hestenes, Malcolm Wells, and Gregg Swackhamer, "Force concept inventory", The Physics Teacher, Vol. 30, March 1992.
3. David Hestenes and Malcolm Wells, "A mechanics baseline test", *The Physics Teacher*, Vol. 30, March 1992.
4. Ronald K. Thornton and David R. Sokoloff, "Learning motion concepts using real-time microcomputer-based laboratory tools", *Am. J. Phys*. 58 (9), September 1990.
5. Priscilla W. Laws, "Calculus-based physics without lectures", *Physics Today*, December 1991.

A SURVEY OF THE INTRODUCTORY PHYSICS COURSE

Bradley W. Carroll, Department of Physics
Weber State University

Introduction

In May of 1993, more than three hundred people converged upon Rensselaer Polytechnic Institute to attend the Conference on the Introductory Physics Course, held on the occasion of the retirement of Robert Resnick. This provided an ideal opportunity to collect information on how instructors were handling their courses. The questionnaire reproduced in Appendix A was conceived, written, distributed, and collected during the three days of the meeting.

Obviously, there is no such thing as *the* introductory physics course. There are many versions of the calculus-based course, usually called something like "Physics for Scientists and Engineers." Everyone involved with such a course faces the dilemma of covering an ever-expanding amount of material in a limited time. Some material must be sacrificed, even in courses lasting three semesters. One of the objectives of the questionnaire was to survey the different strategies employed to balance the conflicting demands of time and content. Other objectives were to determine how class time is used, and to what extent computers and video, demonstrations, and conceptual writing are being utilized. The results of the survey give a portrait of "the introductory physics course" as it is today, the culmination of the many compromises that every instructor is forced to make.

Of the 247 questionnaires distributed, 145 usable responses (59 percent) were obtained.[1] Most of the respondents provided additional comments to clarify their answers when the questions seemed ambiguous.[2] These proved invaluable in interpreting the survey's results.

Results

Physics majors and physics courses

Figure 1 (on next page) shows the number of physics majors graduated per year vs. the institution's undergraduate enrollment. Obviously there is no simple scaling law that connects the two, although there is a "1 percent ceiling" that is rarely exceeded. Of the 101 institutions represented on the figure, half (51) report that between 0.1 percent and 1 percent, inclusive, of their undergraduate students graduate each year as physics majors.

Table 1 (on next page) shows the types of physics courses offered. Many respondents listed additional courses offered at their institutions, although the table entries are restricted to the three general categories of descriptive physics (requiring minimal math), college physics (non-calculus), and the calculus-based physics for scientists and engineers course. The latter two are almost always accompanied by a laboratory, while just over one-third of the descriptive physics courses do not include a lab.

Course	# with Lab	# without Lab	No Response
Descriptive physics	56	30	1
College physics	107	1	0
Physics for scientists and engineers	129	3	1

Table 1: Types of physics courses, with and without labs.

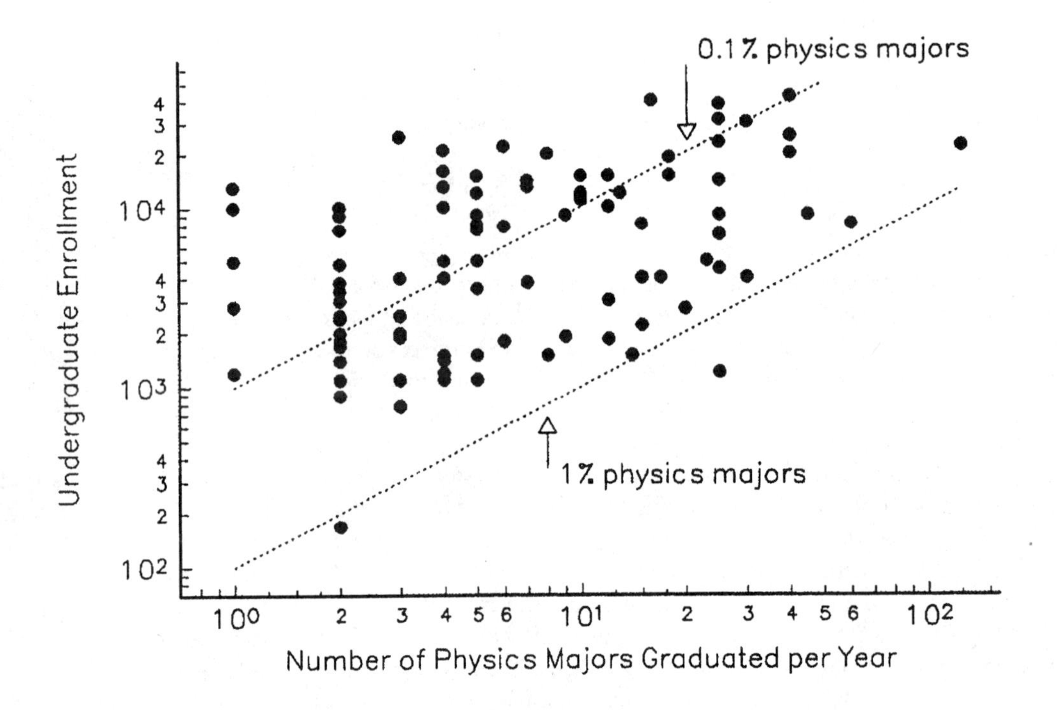

Figure 1: Number of physics majors graduated per year vs. each institution's undergraduate enrollment (N = 101). Thirty-one responses of zero physics majors per year have been omitted.

Calculus-based physics for scientists and engineers

Surprisingly, the size of the college or university bears little relation to the number of students in the calculus-based physics classroom. Figure 2 (on next page) shows that small physics classes are found at institutions of all sizes. About 63 percent (86/137) of the calculus-based physics courses require calculus as a corequisite for the first semester or quarter of the course, while the rest require calculus as a prerequisite. The type of calculus requirement has no discernible influence on the number of physics majors graduated per year, either in absolute numbers or as a fraction of the institution's undergraduate enrollment.

Table 2 (on next page) indicates that most courses are taught in the traditional lecture/recitation format, although ten respondents reported completely non-traditional approaches. Typically there are three lectures accompanied by zero or one recitation or tutorial session per week. Recitations are utilized more by larger classes, as shown in Table 3, so individual students are not always lost in the crowd of a large lecture hall.

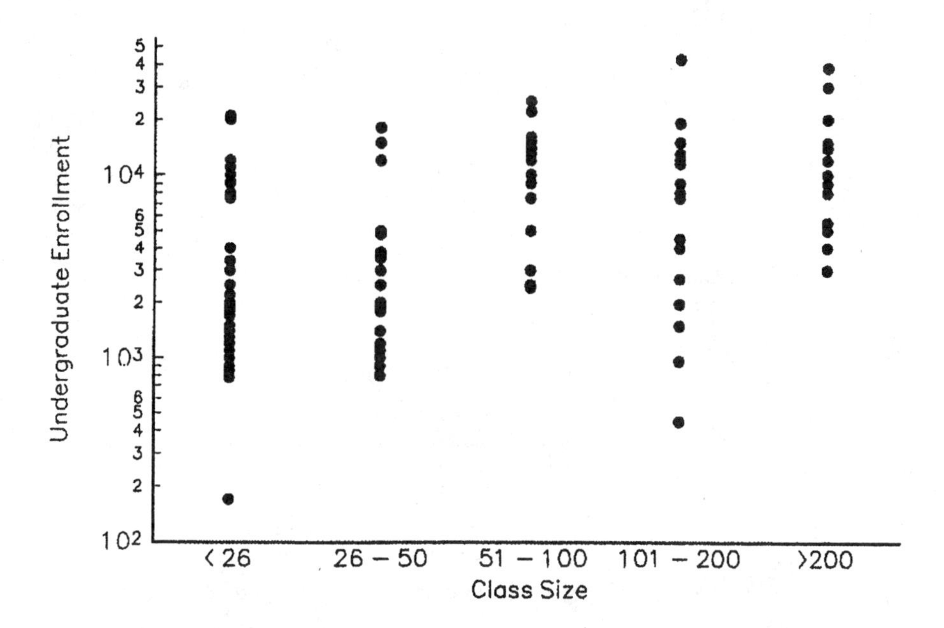

Figure 2: Binned class size vs. each institution's undergraduate enrollment (N = 123).

# of Recitations on Tutorials per Week	# of Lectures per Week: 0	1	2	3	4	5	non-trad
0	0	0	1	40	11	1	0
1	0	0	4	38	2	0	1
2	0	0	6	8	1	0	0
3	0	3	0	0	0	0	0
4	1	0	0	0	0	0	0
5	0	0	0	0	0	0	0
non-trad	0	0	0	1	0	0	10

Table 2: Use of lectures and recitations or tutorials (N = 128).

Class Size	# of Recitations or Tutorials per Week: 0	1	2	3	4
< 26	25	4	1	3	0
26 - 50	26	9	1	0	0
51 - 100	5	7	2	0	0
101 - 200	5	7	4	0	0
< 200	2	11	4	1	0

Table 3: Number of recitations or tutorials vs. class size (N = 103).

The use of computers and video, demonstrations, and writing stressing conceptualization

Although there were frequent discussions during the conference of innovative ways to use computers and video equipment, Tables 4 - 6 show that they are most commonly used in physics labs for analyzing data and for graphing.

Use of Computers and/or Video:

	1 (*Never*)	2	3	4	5 (*Frequently*)	ave	std dev	N
During lectures	32	65	26	12	2	2.2	0.9	137
During lab for data acquisition	43	30	24	26	17	2.6	1.4	140
During lab for data analysis/graphing	23	25	19	48	25	3.2	1.4	140

Table 4: Use of computers and/or video during lectures and labs.

Homework Assignments Using:

	1 (*Never*)	2	3	4	5 (*Frequently*)	ave	std dev	N
Mathematica / Maple	118	11	5	2	0	1.2	0.6	136
Spreadsheets	91	28	11	6	2	1.6	0.9	138
Programming language	105	17	9	2	0	1.3	0.7	133

Table 5: Use of computers with homework assignments.

Use of Computers and/or Video Outside Class:

	1 (*Never*)	2	3	4	5 (*Frequently*)	ave	std dev	N
Supplementing lecture material	65	41	16	5	4	1.8	1.0	131
Introducing new material	91	25	8	5	2	1.5	0.9	131

Table 6: Use of computers and/or video outside the classroom.

Demonstrations are widely used, at least by the instructor. Table 7 indicates that demonstrations which involve students are less common. Students are infrequently asked to express theirä understanding of concepts in writing, whether on homework or on exams; see Table 8.

Use of Demonstrations:

	1 (*Never*)	2	3	4	5 (*Frequently*)	ave	std dev	N
By instructor	4	16	42	44	34	3.6	1.1	140
Involving students	26	74	20	10	8	2.3	1.0	138

Table 7: Use of demonstrations.

Writing Stressing Conceptualization:

	1 (*Never*)	2	3	4	5 (*Frequently*)	ave	std dev	N
On homework	30	57	28	13	10	2.4	1.1	138
On exams	19	53	32	22	14	2.7	1.2	140

Table 8: Use of writing stressing conceptualization.

Coverage of topics in calculus-based physics courses

The tyranny of time demands that certain topics be covered less thoroughly than others. As more time becomes available, the choice must be made whether to increase the coverage of neglectedä subjects, or to use the time to introduce new material. The questionnaire asked instructors to rate the attention they gave to fifteen topics on a scale of 1 (no coverage) to 5 (complete coverage). It is instructive to look at the results as a function of the time span of the course. Most institutions offer either a two-semester (N = 81) or a three-semester (N = 26) course. (Concerning the latter, several respondents indicated that their third semester is not required of all students.) Most of the remaining courses last for three quarters (N = 19), although courses spanning one semester (N = 1), four semesters (N = 7), two quarters (N = 3), four quarters (N = 1), and five quarters (N = 1) were also reported.

The results for courses lasting two semesters, three quarters, and three semesters are tabulated in Appendix B. For many topics, the values of the average coverage are nearly the same for courses of all lengths, typically within ±0.2 on the scale of 1 - 5. The standard deviation of 1.0 - 1.5 may be interpreted as a natural spread in the amount of coverage given to each topic.

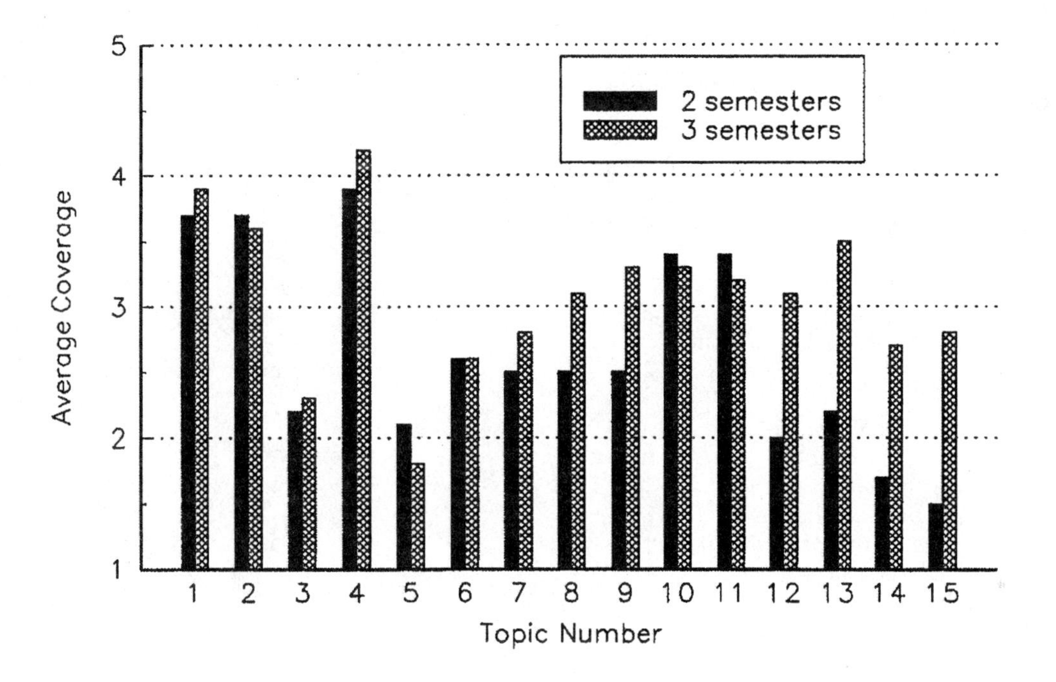

Figure 3: Average coverage of topics for two and three-semester courses. The topic numbers correspondto: 1) rotational motion; 2) statics; 3) propertiesof solids; 4) Newtonian gravity; 5) fluid flow;6) calorimetry; 7) heat engines; 8) kinetic theory; 9) ac circuits; 10) geometric optics; 11) sound waves; 12) relativity; 13) Bohr atom; 14) nuclear physics; 15) wave mechanics (e.g., particle in a box).

Figure 3 compares the average coverage given to the fifteen topics for courses lasting two and three semesters. The coverage given to the first seven topics, and to topics 10 and 11, is about the same regardless of the length of the course. Of these, properties of solids (3) and fluid flow (5) are dealt with most cursorily, averaging just 2.2 and 2.1, respectively, for all courses. Calorimetry (6) and heat engines (7) are also covered less thoroughly than the others, averaging 2.7 and 2.6, respectively, for all courses.

The "modern physics" topics 12 - 15 (relativity, the Bohr atom, nuclear physics, and wave mechanics) are among the most neglected topics in the two-semester courses. Despite efforts to include more modern physics into the introductory physics curriculum, most students see little of these topics in a one-year course.[3] With more time available in a three-semester course, a significant increase in coverage is seen for the modern physics material. The time devoted to kinetic theory (8) and ac circuits (9) also increases. However, even with a 50 percent increase in the time allotted for the course, both the properties of solids (3) and fluid flow (5) continue to be de-emphasized.

Appendix A. The Questionnaire

1. Institution: ______________________________ Quarter / Semester (**circle**)
2. Institution's undergraduate enrollment: ______________
   # Physics majors graduated per year: ______________
   % women: __________ % minorities: ______________

3. Number of different introductory physics courses: ______
   Check:

	# students per class	lab component?
[ ] Descriptive Physics (minimum math)	____	Y / N
[ ] College Physics (non-calculus)	____	Y / N
[ ] Physics for Scientists and Engineers	____	Y / N
[ ] Other ______________	____	Y / N
[ ] Other ______________	____	Y / N

THE REMAINING QUESTIONS CONCERN YOUR PHYSICS FOR SCIENTISTS AND ENGINEERS COURSE (CALCULUS - BASED)

4. Length of course: ______ Quarters / Semesters

5. Calculus prerequisite / corequisite (**circle**) for first quarter or semester?

6. # Lectures per week (circle):
   0 1 2 3 4 5 non-traditional
   # Recitations or tutorials per week (circle):
   0 1 2 3 4 5 non-traditional

7. Use of computers and/or video:

	never			**frequently**	
During lectures: (e.g., simulations, graphing)	1	2	3	4	5
During lab for data acquisition	1	2	3	4	5
During lab for data analysis and graphing	1	2	3	4	5

7. Use of computers and/or video (continued):

Homework assignments using:	**never**			**frequently**	
Mathematica / Maple	1	2	3	4	5
spreadsheets	1	2	3	4	5
programming language (Basic/FORTRAN/C/Pascal)	1	2	3	4	5
Outside class: supplementing lecture material	1	2	3	4	5
introducing new material	1	2	3	4	5

8. Use of demonstrations:

	never			**frequently**	
by instructor	1	2	3	4	5
involving students	1	2	3	4	5

9. Writing stressing conceptualization

	never			**frequently**	
on homework	1	2	3	4	5
on exams	1	2	3	4	5

10. Does your course cover:

	not at all			**completely**	
rotational motion	1	2	3	4	5
statics	1	2	3	4	5
properties of solids	1	2	3	4	5
Newtonian gravity	1	2	3	4	5

fluid flow	1	2	3	4	5
calorimetry	1	2	3	4	5
heat engines	1	2	3	4	5
kinetic theory	1	2	3	4	5
ac circuits	1	2	3	4	5
geometric optics	1	2	3	4	5
sound waves	1	2	3	4	5
relativity	1	2	3	4	5
Bohr atom	1	2	3	4	5
nuclear physics	1	2	3	4	5
wave mechanics (e.g., particle in a box)	1	2	3	4	5

11. Other topics omitted: ______________________________

Appendix B. Coverage of Topics

Two Semester Courses:

Does your course cover:	1 *Not at all*	2	3	4	5 *Completely*	ave	std dev	N
1 rotational motion	3	6	21	30	19	3.7	1.0	79
2 statics	5	9	14	25	26	3.7	1.2	79
3 properties of solids	20	31	19	6	2	2.2	1.0	78
4 Newtonian gravity	1	5	15	35	23	3.9	0.9	79
5 fluid flow	32	22	11	10	3	2.1	1.2	78
6 calorimetry	23	18	14	14	10	2.6	1.4	79
7 heat engines	23	17	21	13	5	2.5	1.2	79
8 kinetic theory	20	19	22	13	5	2.5	1.2	79
9 ac circuits	24	17	19	9	10	2.5	1.4	79
10 geometric optics	11	7	18	23	19	3.4	1.3	78
11 sound waves	9	7	20	30	13	3.4	1.2	79
12 relativity	38	19	7	8	6	2.0	1.3	78
13 Bohr atom	39	13	13	4	10	2.2	1.4	79
14 nuclear physics	52	6	15	3	2	1.7	1.1	78
15 wave mechanics	61	7	2	3	5	1.5	1.1	78

Table 9: Coverage of topics in two-semester courses.

Three Quarter Courses:

Does your course cover:	1 *Not at all*	2	3	4	5 *Completely*	ave	std dev	N
1 rotational motion	0	0	6	10	3	3.8	0.7	19
2 statics	0	2	6	5	5	3.7	1.0	18
3 properties of solids	7	4	1	3	1	2.2	1.4	16
4 Newtonian gravity	0	5	3	5	6	3.6	1.2	19
5 fluid flow	9	2	4	1	1	2.0	1.3	17
6 calorimetry	5	3	5	2	3	2.7	1.4	18
7 heat engines	5	3	5	3	2	2.7	1.4	18
8 kinetic theory	5	1	3	6	2	2.9	1.5	17
9 ac circuits	4	3	5	5	1	2.8	1.3	18
10 geometric optics	1	2	5	8	3	3.5	1.1	19
11 sound waves	2	4	4	4	3	3.1	1.3	17
12 relativity	6	3	2	5	2	2.7	1.5	18
13 Bohr atom	8	1	4	3	3	2.6	1.6	19
14 nuclear physics	10	1	4	1	2	2.1	1.5	18
15 wave mechanics	9	4	2	1	2	2.1	1.4	18

Table 10: Coverage of topics in three-quarter courses.

Three Semester Courses:

Does your course cover:	1 *Not at all*	2	3	4	5 *Completely*	ave	std dev	N
1 rotational motion	0	1	6	13	5	3.9	0.8	25
2 statics	2	3	5	8	7	3.6	1.3	25
3 properties of solids	6	10	5	4	0	2.3	1.0	25
4 Newtonian gravity	0	2	3	9	11	4.2	0.9	25
5 fluid flow	10	12	1	1	1	1.8	1.0	25
6 calorimetry	8	5	3	7	2	2.6	1.4	25
7 heat engines	6	4	6	7	2	2.8	1.3	25
8 kinetic theory	5	1	7	11	1	3.1	1.2	25
9 ac circuits	1	6	5	11	2	3.3	1.1	25
10 geometric optics	3	4	5	8	5	3.3	1.3	25
11 sound waves	3	6	2	10	4	3.2	1.3	25
12 relativity	2	7	7	5	4	3.1	1.2	25
13 Bohr atom	2	3	6	6	7	3.5	1.3	24
14 nuclear physics	5	8	3	6	2	2.7	1.3	24
15 wave mechanics	7	4	5	6	3	2.8	1.4	25

Table 11: Coverage of topics in three-semester courses.

Acknowledgments

I would like to thank everyone who took the time to complete the questionnaire. I am grateful to Jack Wilson for his encouragement and assistance, and especially to Linda Kramarchyk for her help in completing this whirlwind project. This work was funded by the Department of Physics at Weber State University.

References

1. Duplicate information was not used. Several questionnaires contained information about high school physics courses which, however interesting, did not address the objectives of the survey.
2. Unfortunately, the responses for the percentage of women and minorities under item #2 of the questionnaire had to be discarded because (as was pointed out by several respondents) it was not clear whether the question pertained only to physics majors, as was intended, or to the entire institution.
3. Of the nineteen respondents who listed other topics omitted from their courses, all but one teach a course lasting a year or less. The most frequently mentioned omission was all of thermodynamics, followed closely by certain aspects of electricity and magnetism (mutual induction, electromagnetic waves, magnetic properties of materials).

A MULTIMEDIA INTRODUCTORY PHYSICS LABORATORY

P. Cassabella/Y.-F. Liew/T.-M. Lu/T. Shannon/B. R. Schnell/K. Ware/J. Wilson
Department of Physics, Rensselaer Polytechnic Institute

Introduction

Traditionally, the physics laboratory provides the students with an environment to verify physical laws and to experience the physical phenomena they learn in their classrooms. It is also a place where they can learn from each other through interaction with their peers and through hands-on experience. More importantly, the laboratory should be used to discover and build concepts or fundamental laws of physics, to build general laboratory and instrumentation skills, and to practice the scientific method. A good physics laboratory should thus include most of the tools, if not all, that are used by the physicists everyday. We choose to implement our introductory physics laboratory around multimedia workstations, taking advantage of the inexpensive but technological advanced and powerful personal computers and peripherals. Each of these multimedia workstations is centered around an IBM PS/2 personal computer equipped with Audio/Video accessories and a Universal Laboratory Interface board. The laboratory instruction, written using a popular hypertext application, allows the student to access hypermedia linked database of text, graphics, video and data. The students can go through the laboratory instruction at their own pace and depth and also watch a video demonstration of a particular experiment on the computer monitor. Coupled with the software tools provided by the Comprehensive and Unified Physics Learning Environment (CUPLE), these multimedia workstations are also capable of fast and real-time data acquisition via either a video system or transducers. CUPLE also provides the students with data analysis and visualization softwares. The students can also model the physical phenomenon using the modeling and simulation tools provided. The modularity of the CUPLE software, and the bagful of developer tools and students exploratory tools, makes the development of laboratory instruction and access to the multimedia equipment and hypertext database easy and flexible.

A Hypertext Computer Book

A page from a computer laboratory book. It shows the richness and flexibility of a hypertext book. Shown above is (a) a **hotword**, (b) a **video icon**, and (c) a **WINPHYS module icon**. When the mouse cursor is placed over the hotword, it changes shape. If the user clicks on that word, the text or page linked to this hotword is displayed. Double clicking on the video icon would open up a video window with the associated remote video control panel and video tool for the display and manipulation of video clips. The WINPHYS module icon when activated would open up a graph window which allows the user to interact with the Universal Laboratory Interface (ULI) and to activate the ULI for data acquisition. The data is collected in real-time and displayed in a graph window. (See Figure 1 on page 2.)

Graph Window

The graph windows of the MOTION program. MOTION is the program that controls the ULI to acquire data in real-time and plotting the data simultaneously into several different windows. The above shows the data collected from a spring-mass system using an ultrasonic ranging problem connected to the ULI. The position, x, the velocity, v, and the acceleration, a, of the mass are plotted as a function of time. In addition a phase plot is also shown. From the simple x-t data, other useful physical information can be derived and displayed. (See Figure 2 on page 2.)

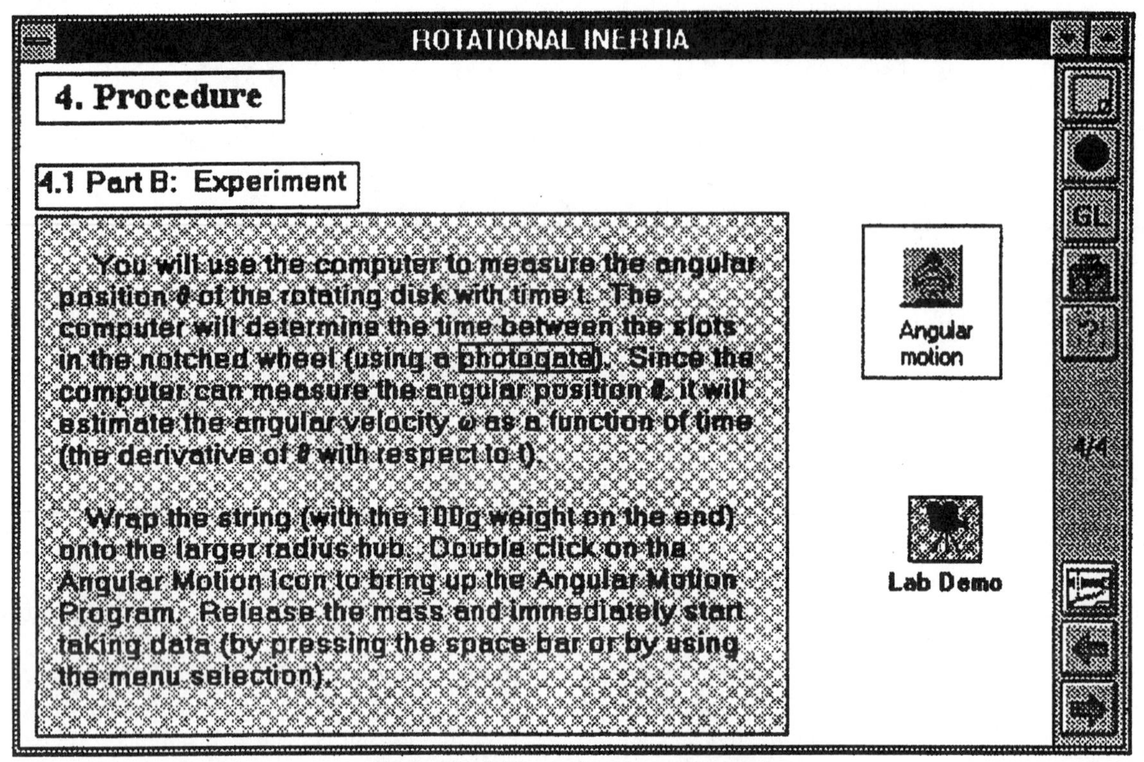

Figure 1. A Hypertext Computer Book

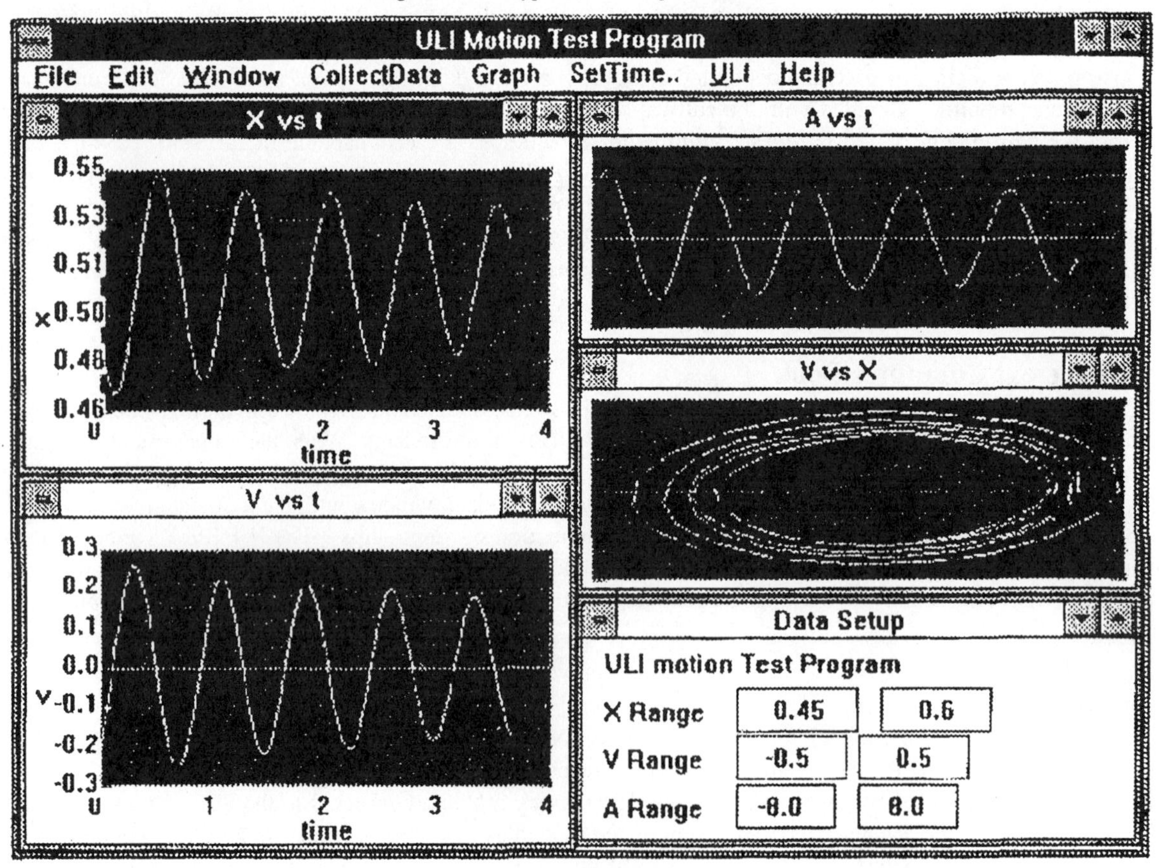

Figure 2. Graph Window

Schematic of the Multimedia Laboratory

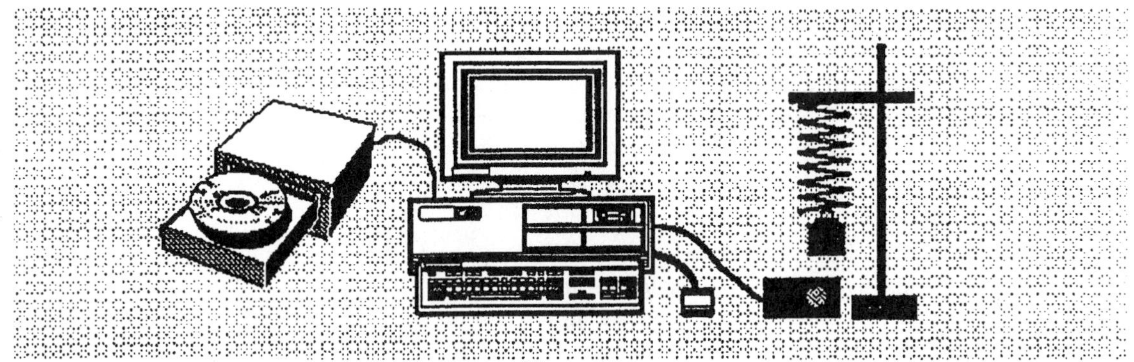

Figure 3

Hardwares:
(a) Personal Computer with M-Motion Adapter,
(b) Audio-Video player,
(c) Universal Laboratory Interface (ULI) board,
(d) transducers, and the necessary apparatus for the equipment.

CUPLE Software

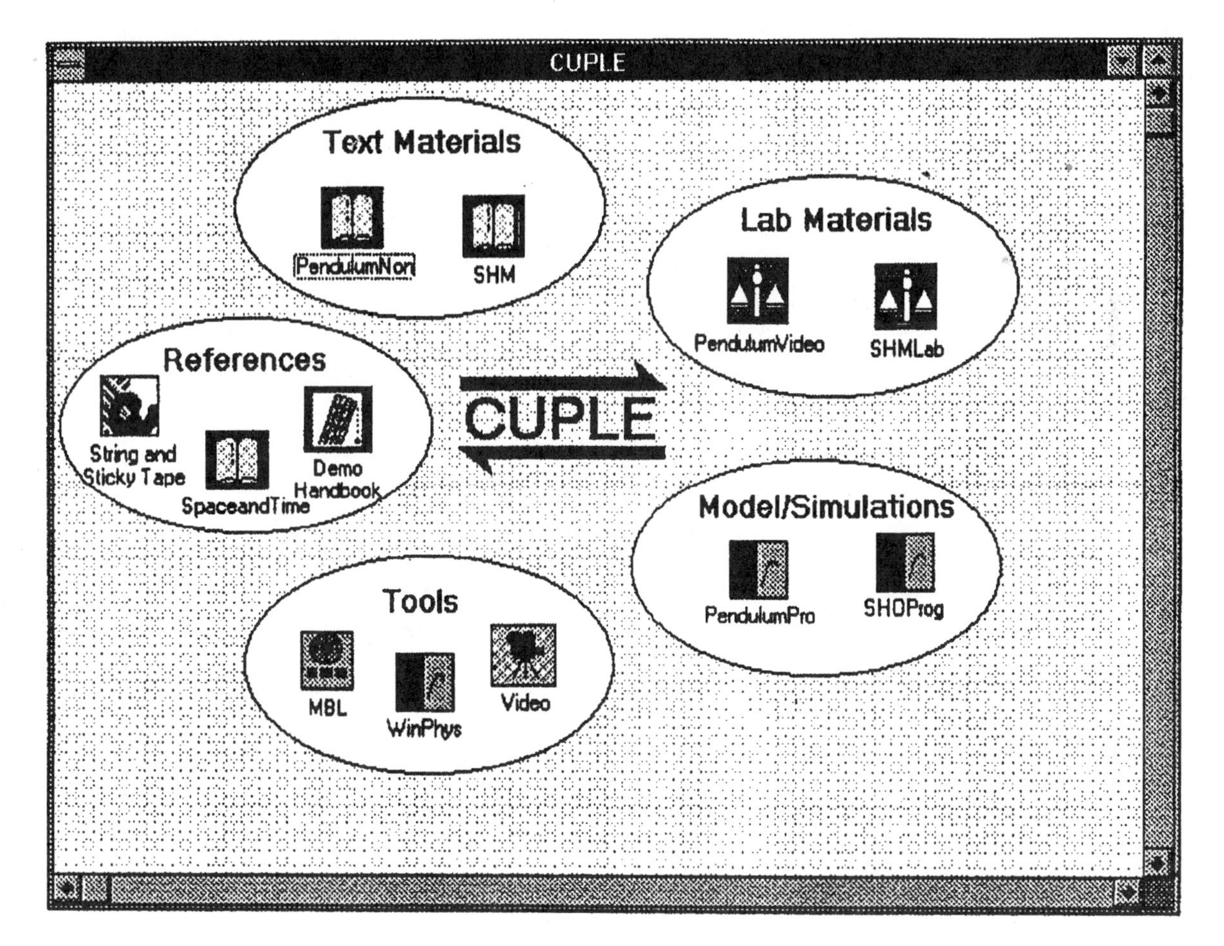

Figure 4

The multimedia laboratory computer is installed with the CUPLE softwares for (a) **data acquisition**, (b) **data analysis**, (c) **data visualization**, and **programming language** compilers. In addition the students also have access to some **hypertext reference materials**, links to **spreadsheets**, and **programming language** compilers. A video demonstration of the laboratory could be viewed on the computer monitor. **Video tools** are also available to step through the recorded video clips of an experiment frame by frame. This way, the time dependence position of moving objects could be measured. (See Figure 4 on page 3.)

Some Advantages

- reduced the dependence on TA for laboratory instruction.
- promote peer-to-peer interaction and group learning.
- reduced the tedium of data taking, freeing the students to think more about the physics.
- made the lab more interesting and information more easily retrieved.
- the analysis, modeling, simulation, and visualization tools allowed more physics to be extracted from the data.

Implementing the Results of Physics-Education Research without Expensive Technology

Paul D'Alessandris, Department of Physics
Monroe Community College

In an effort to integrate results from physics-education research into the introductory physics sequence, without the monies available for microcomputers, a local adaptation of Alan Van Heuvelen's *Overview, Case Study (OCS) Physics* has been developed and implemented at Monroe Community College (MCC). In this OCS-based curricula, students actively construct a knowledge hierarchy on a foundation of qualitative understanding and receive repeated exposure to fundamental concepts in a variety of contexts through a flexible, spiral format. This approach has produced impressive gains in student performance on a variety of assessment tools, comparable to those obtained through more technology-intensive curricula.

Research in physics education reveals serious weaknesses in conventional instruction for students in introductory college physics courses.[1] Students who perform satisfactorily on tests are nonetheless found to have little gain in qualitative understanding,[2] disorganized knowledge hierarchies,[3] and primitive problem solving skills.[4] Furthermore, students enter physics courses with strongly held beliefs on the way the physical world behaves, which, at best, are only partially consistent with the proper perception, and leave with the majority of these views intact.[5]

Recent cognitive research in how, and how well or poorly, students learn physics has been applied in a new generation of curricula. One of these is *Overview, Case Study Physics*[6] developed by Alan Van Heuvelen. The Overview, Case Study method of instruction combines the results of physics-education research in a flexible, spiral format that helps students construct a knowledge hierarchy on a foundation of qualitative, conceptual understanding. The program places considerable emphasis on student construction of conceptual knowledge, on the use of that knowledge to reason qualitatively about physical processes, and on the development of analytic techniques needed to use knowledge to analyze complex problems. Students receive repeated exposure to concepts over an extended time interval and in a variety of contexts. They are actively involved in their instruction. Student interest is motivated by developing understanding and by applying knowledge to interesting phenomenon. Preliminary trials of the OCS method have produced promising gains in student qualitative understanding, problem solving ability, and in the number of students that successfully complete their study. Although other research-based curricula have produced similar gains in conceptual understanding, most notably the microcomputer-based laboratory approach of *Workshop Physics*[7] and the *Tools for Scientific Thinking,*[8] developed by Priscilla Laws and Ron Thornton, respectively, the capital required to implement such schemes can be prohibitive. The *OCS Physics* curricula can be implemented at no cost.

After exposure to the *OCS Physics* curricula at the Joliet Junior College (IL), Lee College (TX), and National Science Foundation sponsored *Curriculum Development Workshop for Two-Year College Physics Faculty*[9] at Lee College in late March, 1992, the OCS curricula was adopted for a pilot section of MCC's College Physics sequence during the 1992 summer session. The College Physics sequence at MCC is a two-semester, pre-calculus-based course primarily for life science majors. The course consists of three one-hour lectures, now referred to as active learning sessions, and a three-hour laboratory each week. Although a five week summer course did not hold the promise of an ideal environment in which to pilot test a new curricula, it was decided that the curricula held such promise that it should be locally tested immediately.

The student response was overwhelmingly favorable. Numerous students, who had professed a fear of physics on a pre-course questionnaire, stated that they now regretted having waited until after their sophomore year to take physics, and moreover regretted having taken it as a summer course in such a compressed form. Several students commented that their career choices may have been different had they taken physics as freshmen. Equally as important, student gains on the Force Concept Inventory (FCI), an assessment tool gauging qualitative understanding of Newtonian concepts typically administered pre- and post-instruction, ranked with the largest reported in the literature for any choice of curricula (See Table I).

Table I: Student Gains on the Force Concept Inventory

Traditional Instruction Methods	N	Pre Instruction	Post Instruction	% Gain
University Physics				
Arizona State University (1984)	478	51	64	13
Arizona State University (1991)	139	52 (19)	63 (18)	11
Harvard University (1990)	--	68	77	9
Monroe C.C. (1992)	*28*	--	*62 (16)*	--
College Physics				
Arizona State University (1984)	82	37 (14)	53 (14)	16
New Mexico State University (1987)	48	38 (13)	44 (17)	6
High School Physics				
Arizona General (1991)	612	27 (11)	48 (16)	21
Arizona Honors (1991)	118	33 (13)	56 (19)	23
Arizona AP (1991)	33	41 (16)	57 (18)	16
Research-Based Instruction Methods				
University Physics				
Dickinson College (1987)	--	50	79	29
Dickinson College (1988)	--	41	57	16
Dickinson College (1989)	--	48	74	26
Harvard University (1991)	--	68	85	17
College Physics				
New Mexico State University (1987)	50	39 (12)	59 (14)	20
Arizona State University (1989)	116	34 (14)	63 (18)	29
Arizona State University (1990)	44	36	68	32
Monroe C.C. (Summer 1992)	*46*	*33 (15)*	*70 (12)*	*37*
Monroe C.C. (Fall 1992)	*57*	*30 (13)*	*73 (9)*	*43*
High School Physics				
Chicago General (1991)	18	28 (14)	64 (20)	36
Arizona Honors (1991)	63	28	66	38
Chicago Honors (1991)	30	42 (18)	78 (15)	36

All scores are percent correct, with standard deviation in parenthesis.
N is number of students taking exam post-instruction.
All data from ref. 2, 5, and 6, except Dickinson College data from *Curriculum Development Workshop for Two-Year College Physics Faculty*, and Harvard University data from Eric Mazur's *Peer Instruction Workshop.*

In the fall 1992 semester, the College Physics sequence was completely converted to an OCS-based curricula. Certain modifications, deemed more suited to the local student population, were implemented. However, the spirit of the OCS approach remained intact.

In addition to several alterations to the *OCS Physics* curricula, the majority of standard physics laboratory exercises were recognized as incompatible with the interactive, qualitative approach of the active learning sessions and removed from the laboratory curricula. In their place, a laboratory curricula was developed which integrated locally edited units of *Workshop Physics* not heavily reliant upon microcomputers, for example, many of the electricity and magnetism units in which an oscilloscope can provide a suitable microcomputer substitute, some sections of the *Tools for Scientific Thinking* which were converted into interactive demonstrations utilizing the department's lone Macintosh computer, and several standard laboratory experiments which had been redesigned to better mesh with the constructivist tone of the laboratory curricula.

Again, student response was overwhelmingly favorable, as reflected in the highest retention rate into College Physics II in many years and the consistently high marks given the course in a post-course evaluation poll, and gains on the FCI ranked with the largest reported in the literature for any curricula (See Table I).

As a test of the effectiveness of the OCS-based approach, in comparison to the standard methods of instruction used in the department, students in the department's Engineering Science Program were administered the FCI. These students had completed a traditionally structured calculus-based physics course. Although the FCI was not administered to this sample pre-instruction, and hence the gain on the exam is unknown, their post-instruction scores are consistent with those reported in the literature for similar groups of traditionally instructed, post-calculus-based physics course students, and substantially below those received by students in the College Physics sequence (See Table I). In light of these results, efforts have begun to implement at least some suggestions from the physics-education research community into the calculus-based sequence.

The drive to reform the introductory physics course is driven by physics-education research, and should not be misinterpreted as being driven by technology. Although technology may play a role in implementing many of the schemes developed by the research community, it is an implementation tool which can be prohibitively expensive, especially to members of the two-year college community. However, by implementing a curricula based on *OCS Physics,* and constructivist laboratory experiences, substantial improvements in student conceptual understanding, qualitative reasoning ability, and problem solving can be obtained, comparable to those obtained with more technology-intensive curricula, without the need for capital investment.

References

1. *The State of US Science and Engineering*, National Science Board, February 1990, 2.
2. D. Hestenes, M. Wells and G. Swackhammer, "Force Concept Inventory", *The Physics Teacher* **30,** 141-158 (1992).
3. F. Reif and J. I. Heller, "Knowledge Structure and Problem Solving in Physics", *Educ. Psychol.* **17**, 102-127 (1982).
4. A. B. Arons, "Students Patterns of Thinking and Reasoning", *The Physics Teacher* **21**, 576-581 (1983); **22**, 21-26 (1984); and **22**, 88-93 (1984).
5. I. A. Halloun and D. Hestenes, "The Initial Knowledge State of College Physics Students", *Am. J. Phys.* **53**, 1043-1055 (1985); and "Common Sense Concepts About Motion", *Am. J. Phys.* **53**, 1056-1065 (1985).
6. A. Van Heuvelen, "Learning to Think Like a Physicist: A Review of Research-Based Instructional Strategies", *Am. J. Phys.* **59**, 891-897 (1991);"Overview, Case Study Physics", *Am. J. Phys.* **59**, 898-906 (1991); and *ALPS Kit* and *OCS Student Study Guide*, Box 4013, University Park Branch, Las Cruces, NM 88003, 1991.
7. P. Laws, "Calculus-Based Physics Without Lectures", *The Physics Teacher* **44**, 24-31 (1991); "Workshop Physics: Replacing Lectures With Real Experience", *Proc. Conf. Computers in Phys. Instruction*, E. Redish and J. Risley, eds. (Addison-Wesley, MA, 1989); and *Workshop Physics*, Vernier Software, Portland, OR 97225, 1991.
8. R. K. Thornton and D. R. Sokoloff, "Learning Motion Concepts Using Real-Time Microcomputer-Based Laboratory Tools", *Am. J. Phys.* **58**, 858-867 (1990); and *Tools for Scientific Thinking*, Vernier Software, Portland, OR 97225, 1991.
9. Project Directors: Curtis J. Hieggelke, Joliet Junior College, Joliet, IL 60436 Tom O'Kuma, Lee College, Baytown, TX 77520

PROBLEM-BASED GROUP LEARNING IN AN HONORS GENERAL PHYSICS COURSE

Barbara J. Duch/David G. Onn/Cynthia A. Cuddy
Department of Physics & Astronomy, University of Delaware

Abstract

A small enrollment (24) honors section of a two-semester General Physics course for pre-professional (medical, dental, physical therapy, etc.) and biology majors was randomly divided into groups of four for solving problems and designing experiments. This section attended the twice weekly class of a large enrollment (150-200) course which also used group approaches (Onn et al., these proceedings). All problems and experiments were devised to relate physics principles to medicine, the human body and biology. Field experiences at the University of Delaware Sports Science Laboratory and local physics-related medical facilities were incorporated to demonstrate real world connections between the more abstract physics principles students were learning, and the ways these principles will be used in their chosen professions. Student responses to this approach will be presented and compared to regular sections of the same course.

Introduction

Many pre-professional students dread taking the two-semester General Physics course because they are not confident of their math and problem solving skills; while simultaneously, they feel pressured to get a high grade for professional school admission. These students also fail to see the connections between abstract physics principles and the knowledge that they need to succeed in their chosen careers. This course was designed to show students that physics was vital to their understanding of physiology, medicine, the human body, rehabilitation and other health fields.

Students worked in groups and learned to teach each other; effectively communicating what they knew — and what they didn't. They learned to depend on each other in order to successfully solve complex problems, and design and carry out open-ended experiments.

Background and Objectives

Research has shown that students' achievement is enhanced when they work together in a cooperative learning environment (Johnson, et al.,1991; Bonwell and Eisen, 1991). Use of cooperative groups fosters the development of learning communities in the classroom which reduces the high competitiveness and isolation of typical science courses (Tobias, 1990 and 1992; Project Kaleidoscope, 1991). Students who acquire scientific knowledge in the context in which it will be used are more likely to retain what they learned, and apply that knowledge appropriately (Albanese and Mitchell, 1993; Boud and Felletti, 1991).

In light of this research on learning, our objectives for developing this honors General Physics course were as follows:

- Provide "real world" applications of physics principles to focus and motivate premedical and other future health professionals.

- Emphasize open-ended experiments which the students themselves design.
- Require students to explain physics principles in their own words (in writing and verbally) without equations, symbols and formulae.
- Develop personal interactions and involvement among students through the use of permanent cooperative groups.
- Encourage students to "learn to learn" through a major research project related to physics and their career interests.

From our observations and thorough interviews with students by an independent professional, we believe that these objectives have been met.

Features of this Course

The honors students in this honors course attended the twice weekly meeting of the regular course sections where the principles of general physics were discussed (Onn et al., these proceedings). They also met once a week for a three hour class, instead of the conventional one hour recitation and two hour laboratory attended by regular students. During this three hour class students were assigned randomly to a permanent group of four. The three hour time slot allowed flexibility in scheduling the various activities described later.

Experiments were designed to be conducted in groups of four. Each individual in the group had specific responsibilities each week, with roles of responsibilities rotating weekly. Some problem sets were assigned to the groups while some were done individually. The tests were identical to those given the regular class with the exception of one medically related group problem.

Medical and health related applications

All problems and experiments related the basic physics principles to biology, medicine and the human body. For example, when the regular class learned about forces and torques, the honors students were challenged to discover how those concepts could be applied to determine how to minimize force on an injured hip. They found that if the center of mass was shifted more directly over the injured hip the forces were minimized.

When the regular class was studying equations of motion in two dimensions and resolving vectors into their components, the honors students were predicting the path of a basketball being shot by a college player. And when the player's femur was broken during the play, they analyzed the resultant pulling force on the leg immobilized using Russell traction.

Open-ended experiments

The experimental investigations conducted by the students were open-ended and avoided "cookbook recipe" results. Students were told to design an experiment, analyze the uncertainty of the data to be gathered and plan ways to minimize that uncertainty. For instance, when the regular class was learning about moments of inertia, the honors students were asked to estimate their own body's moment of inertia with their hands at their side. And then predict whether it would be the same, greater, or less if their arms were extended. The students consequently designed a way of measuring their own moments of inertia (in both positions) using turntables, pulleys and weights. They compared their results to predicted results and explained reasons for any differences.

Field trip experiences

Students visited the University of Delaware Sports Science Laboratory several times during the year. They observed how computer analysis of the muscle movement in children with cerebral palsy is used to advise doctors of the likely results of suggested surgical procedures. They also personally experienced stress-testing with EKGs, and lung volume tests. The University of Delaware Physical Therapy Department faculty demonstrated many techniques used in rehabilitation which were connected to the physical concepts being studied at the time. They also visited the Medical Center of Delaware in relation to medically related research projects discussed in the next section.

Research projects

In the spring semester, student groups chose a topic to research related to their career interests and physics. The research was conducted in and out of class. Students consulted with an expert (doctors and radiologists) for their research project. Students made oral presentations to the class and also handed in a research paper at the end of the semester. The topics chosen by the students were as follows: MRI as a Diagnostic Tool, Uses of Ultrasound In Pregnancies, Ultrasound in Physical Rehabilitation, Electrophysiology of the Heart, Surgical Corrections of Myopia, Current and Future Uses of Virtual Reality.

Results

Extensive evaluations were undertaken and are still being analyzed. However preliminary results are very positive.

Attitudes

Overall, students rated the course a 1.4 where 1 indicated excellent and 5 indicated poor. Students responded to other aspects of this course on final course ratings forms as shown below. The number of students responding was 20 per semester.

Scale: 1 = strongly agree .... 5 = strongly disagree

		mean	st. dev.
1.	I can apply general physics principles to problems in biology and medicine.	1.7	0.6
2.	I am comfortable working in groups.	1.3	0.6
3.	I am confident that I can analyze a physics problem.	2.0	0.8
4.	I would take another class designed like this one.	1.4	0.6

Rate the Effect on Your Learning of Physics:

Scale: 1 = very beneficial .... 5 = very detrimental

		mean	st. dev.
1.	Application of physics principles to biology and medicine	1.2	0.4
2.	Unstructured experiments	1.7	0.6
3.	Field trips	1.4	0.6
4.	Research project	1.5	0.5
5.	Group work	1.3	0.4

When students were interviewed by an independent consultant they all responded that working in groups aided their learning. Some typical comments from students on learning in groups:

- Group interaction gives a place to receive new ideas.
- The groups definitely help — not only if you don't know the answer, but if you have to explain it to others — you *really* have to understand it.
- In groups, we forced each other to *explain* the answers.

Responding to the question: What did you learn that was new and meaningful to you?

- Physics is really, really important in medical issues. I never knew it explained so many things.
- I learned how much physiology depends on and can be explained by physics. Also, physics is everywhere!
- I hadn't realized how much physics actually affected the human body or how interesting it could be.
- I learned the beginnings of how to apply the knowledge that I learn in the classroom to real life experiences. Too often, equations are learned, but two months later it is forgotten. This class left me with knowledge that hopefully I will retain forever.

Achievement

In the fall semester, final grades were compared between students in the honors section and a control group of students with matched grade point averages. The results are shown on the diagrams below. An in-depth analysis of the student performance on matching test questions will be conducted.

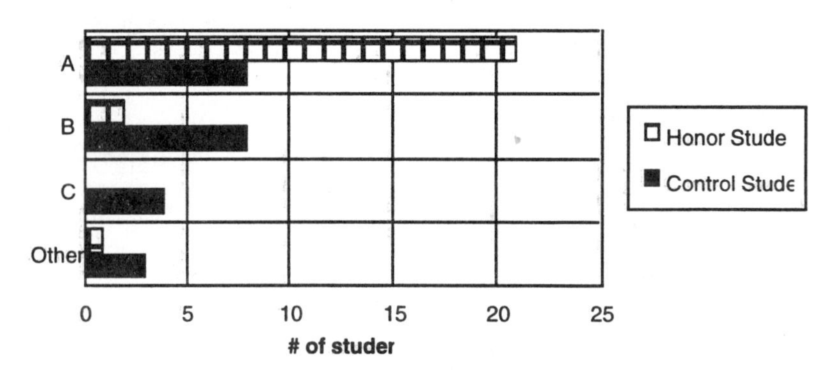

Summary

Preliminary results indicate that active group learning and connections to real world applications help students learn physics and apply that knowledge appropriately. Next year, the honors section of General Physics will have an entirely separate class, in which lecture, problem solving and experiments will be interwoven in the six hour per week meeting time. More group work and practical problems are being planned and a research project will be incorporated into the fall semester.

References

1. Albanese, M. A. and Mitchell, S. (1993) Problem-based learning: a review of literature on its outcomes and implementation issues. *Academic Medicine* (**68**, 52-81).
2. Bonwell, C. C. and Eison, J. A. (1991) *Active Learning*. 1991 ERIC-ASHE Higher Education Report No. 1 (George Washington University, Washington, D.C.).
3. Boud, D. and Feletti, G., eds. (1991) *The challenge of problem-based learning in education for the professions.* Herdsa, Sidney Australia.
4. Johnson, D. W., Johnson, R. T. and Smith, K. A. (1991) *Cooperative learning: Increasing college faculty instructional activity*. 1991 ERIC-ASHE Higher Education Report No. 4 (George Washington University, Washington, D.C.).
5. Onn, D. G., Cuddy, C., and Duch, B. J. (to be published) Reinforced Cluster Active Learning (RECALL) in a Large Enrollment Science Majors Class. *Proceedings of the Conference on the Introductory Physics Course*. Rensselaer Polytechnic Institute, Troy, NY, 1993.
6. Onn D. G., Duch, B. J., and Cuddy, C. (to be published).
7. Project Kaleidoscope (1991) *What works: Building natural science communities*, Volume One, Stamats Communications, Inc., Washington, D.C.
8. Tobias, S. (1990) *They're not dumb, they're different.* Research Corporation, Tucson, Arizona.
9. Tobias, S. (1992) *Revitalizing Undergraduate Science*

THE INTRODUCTORY PHYSICS STABLE EQULIBRIUM: A BALANCE BETWEEN ENGLISH AND CALCULUS

Toufic M. Hakim, Department of Physics
Jacksonville University

Abstract

New initiatives based on the less-is-more philosophy and the guided inquiry approach represent a long awaited substitute to the traditional and dull lecture style of teaching introductory physics. They may, however, be going too far away from the classroom setting too quickly, making their implementation more difficult and less popular than desired. Some transition time is needed during which a progressive program—more similar to the old style than not and with sufficient changes to address its common problems—could be adopted. A multi-level approach, aiming at establishing a good balance between the concepts and the mathematics, may serve as such a program.

Introduction

Studies[1] have shown that students in traditional programs of Introductory physics are not generally successful in: (a) coherently discussing physics concepts; (b) clearly connecting the theorems and definitions of calculus to the development of physical principles; and (c) systematically analyzing a physics problem by formulating a solution approach, carrying out the computation till the end, and interpreting the results. Numerous initiatives,[2] computer- or laboratory-based, have attacked these problems head-on. They are designed to match sociological, psychological and educational profiles in the current student market, but it may be too early to judge their success. Some of these changes are radical and move too far from the traditional way of teaching. This paper presents a transitory, easy-to-implement approach that targets every one of the problems above, while representing a gradual evolution from the old to the new school of thought. This approach is plausible in class sizes limited to 20 students, but versions of it could be implemented in larger settings, and it is most easily adaptable in departments where teaching loads are not prohibitively heavy. It is founded on the belief that conversation in a simplified language must precede any mathematical formulation and that calculus must be presented in parallel with the physics and not separately from it.

A Multi-Level Approach

The approach aims at fully integrating physics and calculus, but some of its levels could be used in an algebra-based course as well. It has as a building block a conversation-heavy syllabus, and incorporates problem-based group learning strategies and concepts-before-constructs tactics. As a final stage—to be tested soon—it develops into team teaching.

Conversation-heavy Syllabus

The syllabus is heavy on writing and discussion ['*You mean writing F net is equal to 'em a and tau net equal to I alpha?*']. It creates meaningful opportunities for talking about physics coherently and concisely, and provides a natural forum for confronting students' preconceptions and misconceptions.[3] Its main features are (a) essays and oral presentations, (b) letters and interviews, and (c) multiple-choice question designs.

a. The syllabus contains writing assignments regarding analyses of sections of the material (with the minimum necessary of mathematical formulation), summaries, student explanations of models and definitions, and answers to selected questions similar to those regularly found in contemporary textbooks. The essays, numbering no more than four 1-page and two 5-page write-ups, and the 2-minute talks are designed to pinpoint weaknesses and assess progress in the students' modeling of the physical principles, their level of understanding and their use of terminology. All essays are peer-critiqued and carefully reviewed and commented on by the instructor. The student critiques open up the floor for conversation—useful tactic considering that students can in some cases teach one other more effectively than the teacher can.

b. Students are also required on several occasions to write their own interpretations of phenomena, models or mathematical formulations in the form of letters to their grandmothers (or an older person) and they are encouraged (bonus points) to have them preceded by an interview, which they are expected to evaluate, in which they seek their grandmothers' interpretation and they offer theirs—which may not always be the generally accepted one. Many of them engage their friends in this fairly popular activity.

c. Students are once a week asked to create one conceptual multiple-choice question, and present an analysis for every suggested choice and justify the reason for the right one.

Caveats and benefits

While a conversation-heavy syllabus produces a large amount of paperwork (or diskettes) and proves to linger on some concepts, it adds significantly to the comfort and excitement level of the students vis-à-vis the class, and offers a good feedback-control mechanism. It helps develop pertinent verbiage, places physics within reach, and establishes the logical steps from observation to interpretation.

Problem-based Group Learning Strategy

Problem solving is at the essence of understanding physics. As a result, the approach is problem-based, rather than theory-based. ['*You mean the whole course is going to be a bunch of problems. No memorization. O Isaac!*] In order to achieve high efficiency and to reach out to most students, the approach used includes (a) step problem solving, and (b) group problem solving.

a. The step problem solving strategy consists of devising and assigning a list of problems incrementally increasing in difficulty, both in set-up and calculus, for every major concept. The goal is to teach the skill by repetition, not unlike how grammar is taught in middle school. The students work in pairs, supervised by an advanced student, to solve all problems on the list except for the last one—the most difficult—which is worked out individually and is the only one to be graded.

b. Three big problems are assigned during the semester to be solved by groups of 4's or 5's. These problems will have to be solved in steps, learned previously, and are evaluated based on the method of approach rather than the answer. All these problems deal with real conditions and situations that could be tested in a laboratory, rather than with esoteric questions.

Caveats and benefits

The problem solving session is popular in many departments, so this approach is fairly common, but when the group component is added and the list prepared, time consumption becomes a concern and creativity essential. The difficulty is in the group dynamics and the dilemma of fair evaluation, but that can be overwhelmed by the benefits. This strategy builds student loyalty and ownership value, and helps establish a student support system. From a physics point of view, the group projects will emphasize the modeling concept and its relevance to the real world, and the step approach is essential in developing problem solving techniques.

Concepts-before-constructs Tactics

Since it cannot be assumed that the students understand the most basic concepts of calculus,[4] it is important to introduce such concepts and talk about them in plain English, as applicable to physics. For example one could talk about pulling or pushing, long before discussing the concept of force. One could describe inertia by talking about people's unwillingness to change and how they must be disturbed in order to accept change. All derivatives could be introduced as a catch-up game: the velocity always tries to catch up with acceleration (and sometimes it can't), the linear momentum tries to catch up with the one force representing all the real forces. Or even as marginal change, as is done in economics ['*You mean like the derivative of student attitude toward physics on the first day of class is negative.*'] Explaining systems involving a time axis, one can talk about the real world being the shadow of the curve on the other axis (Our life is a shadow)—Mrs. O, the original lady, is chosen for instance to be the observer: one gives her a clock, a measuring device and a sense of direction and as many eyes as needed to observe, and she will take care of the rest. When setting up definite integrals or line integrals, one writes them long-hand like large sums and draws the pieces of the puzzle, before using them casually.

One has to also be careful when talking about laws; a reality flag is to be waived and the laws must be called such only in the right context. The students must be guided toward a solid realization that what they are learning are models that they can construct. These models are based on observations and have parameters of validity, they have to be tested and what we call laws are patterns elevated to higher levels of generality and respect, but only under certain conditions. The students should be guided to identify the need for and the way to formulation, and to assess models by criteria of validity and accuracy....And whose fault will it be if that stuff is not communicated to them? Granted, communication alone is not highly esteemed in cognitive development, but it must be there.

Even in calculus-based physics, students need to hear those expressions, understand them well before they see the construct, otherwise they will be too slow to make the connection. These points could be easily addressed if a team teaching approach is adopted.

Team Teaching

Through an NSF-ILI grant (Computer-Assisted Laboratory for Physics and Calculus, DUE-9252197), a full integration of computer-based physics and calculus will be implemented next year at Jacksonville University.[5] In that program, both instructors will jointly meet with the students—who have enrolled in both the calculus and the physics courses—for six hours a week + lab. The students will not be told which component is calculus and which is physics. All exams and homework assignments are integrated, the syllabi merged into one, and variations of the above strategies adopted.

Expected caveats and benefits

Results of this approach will be communicated at the end of the year but it is expected that the time to teach the two courses jointly will be much larger than twice the time to teach each course separately, and it is a program heavy on planning. But it is hoped that the value of the whole will be bigger than the sum of the values of its parts. Each component will be enhanced greatly, several perspectives added, more opportunities will be available for one-on-one interaction, and it will be a great learning experience for the teachers.

Requirements

For any of these ideas to be successfully implemented, there is a short list of requirements that must be met: (a) ample time must be spent focusing on the student needs and following-up on his/her progress, (b) meticulous organization is a must (detailed planning, logistics, good time management), and (c) choice of topics and level of depth must be customized.

General Teaching Strategies and Evaluation

Students have expressed great interest in the methodology, but detailed assessment is yet incomplete. Opinions differ on the group approach, but the emphasis on conversation and plain-English has been quite popular. Of course, several other strategies have been used in conjunction with them, namely the Harvard-minute, the 5-minute recap and the once-a-week multiple-choice party. During the last minute of every session, the students write down in three sentences what they learned during the sessions, what they liked and what idea(s) they did not grasp. In the first 5 minutes of the next session, they hear a response to a selection of the ideas they had judged vague and important ideas from the previous session explicitly reviewed. And once a week, 5 tricky multiple-choice questions are analyzed in detail during class time. Electronic mail will soon replace some of the above paper procedures.

Conclusion

The multi-level approach described above sits somewhere between the two main philosophical approaches to the teaching of general physics: (a) the traditional textbook-guided and lecture-heavy style of presenting the equivalent of 15 Halliday and Resnick-like chapters per semester; and (b) the preparation-intensive, computer-based, and discovery-oriented strategy of covering the minimum necessary amount of topics. The former, information-intensive, does not give the student sufficient time for an in-depth understanding of concepts and for the development of problem solving skills. The latter, designed with the customer in mind, focuses on those skills, and the literature is rich with its success stories. There have been recently numerous well-funded initiatives, following studies in pedagogy and cognitive theory, which have suggested improvements in the order, amount and type of material, the role and class of laboratory experimentation, and the usage of multimedia and computer analyses. It is quite exciting to see the popular interest in bringing a new face to Introductory Physics—maybe it has become a necessity. It is also professionally rewarding to have a wide choice of programs, but one risks to go in exactly the opposite direction toward the other extreme—as is the usually the case in education—unless some of these changes are incorporated gradually and selectively, are carefully evaluated, and remain focused on the student-customer and not on the adopted innovation. It is precisely those considerations which make the multi-level approach, described above, a good stepping stone on the way to adopting the new worthwhile innovations in introductory physics.

References

1. A. B. Aarons, *Am. J. Phys.* **50**, 13 (1982) and L.C. McDermott, *Physics Today* **37**, 24 (1984).
2. R. K. Thornton, *APS News: Physics News in 1992,* 2, **48** (1993) and references within; as well as J. Priest and J. Snider, *The Physics Teacher* **25**, 303 (1987); and E.H. Carlson, *Am. J. Phys.* **54**, 972 (1986).
3. L. C. McDermott, *Am. J. P.* **59**, 301 (1991) and A. Van Heuvelen, ibid. 891 (1991).
4. R. Frederick, *Cognitive Science* **11,** 395 (1987) and A. Van Heuvelen, *Am. J. Phys.* **59**, 898 (1991).
5. Other similar efforts can be found in: E.A. Murphy and R.S. Knox, Proceedings of the Conference on Introductory Physics, RPI, May 1993 and J.R. Hundhausen and F.R. Yeatts, *Laboratory Explorations in Calculus with Applications to Physics,* Harper Collins, New York, NY, 1993.

What Modern Physics? Two Threads in a Fabric

Charles H. Holbrow, Department of Physics
Colgate University

Introduction

How much modern physics should be in the introductory course and when? Some indication of the range of possible answers is given elsewhere in these proceedings.[1] Here I will take a narrow focus and try to show you how we weave quantum ideas into the first term of Colgate University's calculus-level introductory physics. I think you will agree that they qualify as "modern"— whatever that means. But I hope also to persuade you that these ideas provide a rich context for teaching things that we all agree should be part of a student's introduction to physics.

This paper describes innovations of content not of format. Over the past eight years we have also experimented with different styles of teaching the course. As many of the speakers at this conference have suggested,[2 3 4] the traditional lecture format is deficient in many respects. We have decreased the amount of lecture and increased the amount of small-class interaction, used more interactive lectures, and experimented with take-home laboratory exercises, computer use, and extended, context rich homework exercises worked singly or in groups.[5] We call these "focus problems"; they are similar to case studies.[6] An overview of the course and its goals, organization, and format are described by Amato, et al.[7]

The Atom as Star

In 1985 the Colgate Department of Physics and Astronomy decided to make the atom the central figure in the story line of its introductory course.[8 9 10 11] Since then, the first term has taught basic kinetic theory of gases, enough electricity and magnetism to understand the evidence for the atomic basis of electrolysis and the existence of the electron, the nature of waves, the wave-particle duality of matter, the rudiments of the quantum theory of light, and the atom as configurations with well defined discrete, quantized energy states. The story line requires the introduction of basic ideas of quantum mechanics; we also include some special relativity both because it is fun and for showing how experiment elucidates some aspects of the atomic nucleus. The principal conceptual tools are the conservation of energy and wave interference and diffraction. Aside from a brief review of elementary mechanics we draw on prerequisite high school physics for awareness of length, time, mass, velocity, kinetic energy and momentum. When force is invoked it is usually as time-rate-of-change of momentum.

Why Choose the Atom?

We wanted our introductory physics course to introduce students to the physics that physicists actually do. Listening to physicists talk about what they do, you quickly realize that this is very different from the content of introductory physics courses. In seminars and colloquia and hallways and laboratories physicists apply ideas of wave-particle duality and quantum properties to understanding the interactions and behavior of matter over an enormous range of energy scales. The atom, the electron, the nucleus, and the nucleons are major players in the little dramas of physics that we stage for each other.

If the purpose of introductory physics is to introduce students to physics, it seems a good idea to introduce students to these *dramatis personae* as soon as possible. Then they can understand sooner what interests their physics faculty, what weekly seminars are about, what *Scientific American* is presenting, and what is in the science section of their local newspaper.

There is also the Third Law of Introductory Physics: About 1/3 of the students enrolled in the first term of introductory physics will never — if they possibly can avoid it — take any more physics. If that is the case, we ask: "Of all the physics that we might teach in one term of `calculus-level' physics, what do we think it most important for that 1/3 to learn about?" We decided that for us who would teach it and for them who would learn it, the physics by which we have come to believe in and understand the atom would admirably introduce students to physics as it is practiced and have the best chance of leaving them with a residuum of useful understanding.

Indeed, we found later that reputable authority held similar views:

> If, in some cataclysm, all of scientific knowledge were to be destroyed, and only one sentence passed on to the next generations of creatures, what statement would contain the most information in the fewest words? I believe it is the *atomic hypothesis* ... that *all things are made of atoms — little particles that move around in perpetual motion, attracting each other when they are a little distance apart, but repelling upon being squeezed into one another.* In that one sentence, you will see, there is an *enormous* amount of information about the world, if just a little imagination and thinking are applied.[12]

From the Classical to the Quantum Atom

The first term starts with the classical hard-sphere atom and ends with an entity with energy levels and wave-like electron clouds. We start with a featureless point-sized object and finish with an electron cloud a few tenths of a nanometer in diameter surrounding a compact, dense assemblage of nucleons several femtometers in diameter, a system that can take on only certain definite configurations with well defined, i.e. quantized, energies. The course progresses from an object that obeys classical mechanics to one that can be properly interpreted only when we understand that matter and light exhibit a wave-particle duality at the heart of which lies Heisenberg's uncertainty principle.

Let me try to reveal the fabric of the course by describing just two of its threads. One thread is the nature of waves; how we use interference to identify them; how we use interference to measure them; and how we use them to measure things we cannot see. The other thread starts with simple statistics of hard-sphere atoms, and uses classical kinetic theory to introduce the ideas of average and standard deviation. The two threads intertwine when we discuss how the electron double-slit interference experiment implies that in some sense the electron goes through both slits at the same time. Wave-particle duality means that a single electron spreads through space and in some way can be in more than one place at the same time. Similarly, it can have more than one momentum at the same time. The uncertainty principle connects the standard deviation of the distribution of locations in space to the standard deviation of the distribution of momentums.

Waves

Interference from a slit is the central idea of our treatment of waves. You can go a long way with $d \sin \Delta = m \lambda$. Lectures describe how single-slit diffraction and double-slit interference arise in the simple cases. In laboratory students observe and measure such interference phenomena which they produce with He-Ne laser light. They examine a progression of structures of increasing numbers of slits which show them how the diffraction grating works. We want them to have an operational acquaintance with the diffraction grating to help them understand what a spectrum is and what atomic spectra tell us about atoms.

A discussion of interference establishes the extremely important property: If it shows interference, it is a wave. Then the observation of diffraction of electrons, neutrons and atoms experimentally establishes the wave nature of these particles and the validity of the deBroglie wavelength. We confront the problems of wave-particle duality.

Waves serve to probe the structure of matter. Students easily see that from an interference pattern made by a wave with a known wavelength you can infer information about the grating's structure. For simple cases they can infer the spacing between the evenly spaced slits of a grating and the width of the slits. We then use wave patterns to find the sizes of atoms, nuclei, and nuclear components and to show experimental evidence for structure inside these objects.

Mean and Standard Deviation

Atoms are first used in the course to explain the gas laws. Boyle's experiments and data are presented as examples of elegant empirical science. Then using Bernoulli's kinetic theory we show how even an atom as simple as the hard sphere has remarkable explanatory capability and leads to new relationships such as that between the kinetic energy and temperature of a gas.

This is also when we first introduce the concepts of mean and standard deviation of a distribution. The center of our discussion is a distribution of velocities such as might be obtained by measurement. Students learn to calculate the mean of this distribution and also of the distribution of the squares of the velocities. Then they learn to use the variance and standard deviation as measures of the spread of the distribution.

Students use these ideas repeatedly during the term. They frequently use them to analyze and characterize the uncertainty of their laboratory measurements. Towards the end of the first term they measure emissions from radioactive substances and verify that in this case the distribution is Gaussian with a variance equal to the mean. By the end of the term students can know that

$$\Delta x = \sqrt{\overline{x^2} - \overline{x}^2}$$

and what this means.

The Uncertainty Principle

The advantage of spinning these two threads, waves and statistics, side-by-side becomes apparent when we get to Heisenberg's Uncertainty Principle. The argument is not completely made that the uncertainty principle resolves wave particle duality, but the formulation of the principle

$$\Delta x \, \Delta p \sim \hbar$$

makes much more sense than is usual for its first presentation.

There are two particular benefits. First, it is easier for students to see that the principle is referring to a spread of simultaneous occupied locations. The quantity

$$\Delta x$$

is the standard deviation of that distribution. The connection to the conclusion from the electron double-slit interference that the electron in some sense passed through both slits is clear.

Second, the extraction of quantitative information from the uncertainty principle is much more easily justified. We are particularly interested in showing that forces in the nucleus must be different and greater than electro-magnetic forces. This requires establishing the order-of-magnitude of kinetic energy of confined particles which is customarily done by invoking the uncertainty principle in a manner often quite mysterious. Now if you know that,

$$\Delta p = \sqrt{\overline{p^2} - \overline{p}^2}$$

it is easy to understand that for a confined particle

$$\overline{p} = 0$$

and so

$$\Delta p = \sqrt{\overline{p^2}} \sim \hbar/\Delta x$$

is the r.m.s. momentum from which the kinetic energy can be estimated.

Concluding Comments

The story told here has been well told many times. It is just the chronicle of one aspect of the evolution of our understanding from the classical, basically Newtonian, view to something quite different, the quantum view. Our principal innovation has been to shape it so it can be usefully told in the first term of introductory physics.

The story of waves and uncertainty described above also has a number of important pedagogic features. There is spirality: the basic concepts — waves, interference, mean, standard deviation — come around several different times in quite different contexts. The story shows how physicists make an argument: properties of waves and statistics are used to draw inferences in ways that show students how physicists reason and what persuades them. There are many opportunities for students to use order-of-magnitude estimation and numerical reasoning, skills essential for their progress in physics and valuable for almost anything else they undertake to do.

I have chosen such a narrow focus in hopes of showing you concretely how it is possible to have it both ways: you can include modern physics while teaching concepts, skills, and a point of view that we all agree should be at the core of introductory physics. However, I fear that by concentrating on what is only a part of our course I give you a partial and misleading picture of introductory physics at Colgate. For a more synoptic view of Colgate's course, see reference 13 (in these proceedings) .[13]

References

1. Gordon Aubrecht, Charles H. Holbrow, and John S. Rigden. *"What Modern Physics? -An Instant Survey,"* (these proceedings).
2. Priscilla Laws (these proceedings).
3. Lillian McDermott (these proceedings).
4. Eric Mazur (these proceedings).
5. Patricia Heller, Ronald Keith and Scott Anderson, "Teaching problem solving through cooperative grouping. Part 1: Group versus individual problem solving," *Am. J. Phys.* **60** 627-636 (1992); Patricia Heller and Mark Hollabaugh, "Teaching problem solving through cooperative grouping. Part 2: Designing problems and structuring groups," *Am. J. Phys.* **60** 637-644 (1992).
6. Alan Van Heuvelen, "Overview, Case Study Physics," *Am. J. Phys.* **59** 898-907 (1991).
7. J. C. Amato, E. Galvez, H. Helm, C. H. Holbrow, D. F. Holcomb, J. N. Lloyd, and V. N. Mansfield (these proceedings).
8. C. H. Holbrow, J. N. Lloyd and J. C. Amato, "Putting modern physics first," *AAPT Announcer,* **17** 110 (#4 1987) and *Bull. Am. Phys. Soc.* **33** 81 (1988).
9. J. N. Lloyd, J. C. Amato, C. H. Holbrow and R. Williams, "Modern physics experiments in the introductory laboratory," *AAPT Announcer* **17** 110 (#4 1987).
10. C. H. Holbrow and J. C. Amato, "Modern introductory physics: an ascent through 14 orders of magnitude," *AAPT Announcer* **21** 75 (December 1991).
11. J. C. Amato and C. H. Holbrow, "Traditional teaching methods meet modern introductory physics: *AAPT Announcer* **21** 75 (December 1991).
12. Richard P. Reynman, Robert B. Leighton and Matthew Sands, *The Feynman Lectures on Physics*, Vol. 1, p. 1-2. Addison-Wesley, Reading, MA 1963.
13. J. C. Amato, E. Galvez, H. Helm, C. . Holbrow, D. F. Holcomb, J.N. Lloyd, and V.N. Mansfield, (these proceedings).

On Teacher's Misconceptions About Capacitors in Series

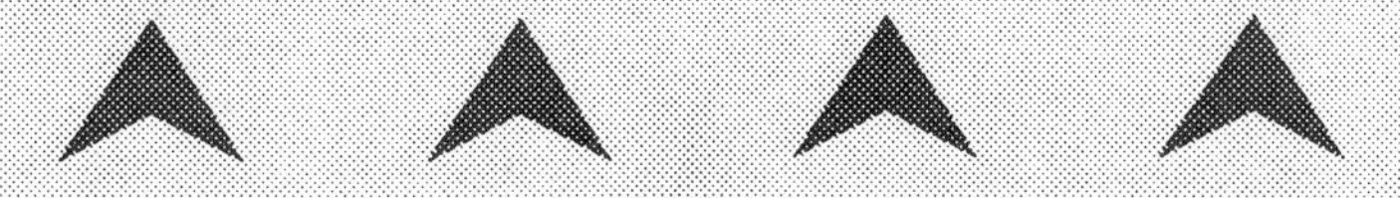

Ludwik Kowalski, Department of Physics-Geoscience
Montclair State College

According to physics textbooks, steady state charges on two capacitors connected in series with an electric battery are always equal. But direct measurements often contradict this prediction. A simple setup illustrating the inequality of charges is demonstrated at the poster session of this conference. It consists of two 0.1 μF capacitors made from different dielectric materials, paper and oil. The capacitors are charged to equilibrium (from a dc source) and then rapidly discharged through a galvanometer. First each capacitor is charged and discharged separately. Under this condition the momentary deviations of the galvanometeric needle are essentially the same, as they should be when charges are equal. Short delays of up to ten seconds, between the end of charging and discharging, have minimal effects on readings. Next the same two capacitors, and the source, are connected in series for two or more minutes to reach a steady state. They are then disconnected from the circuit and discharged through the galvanometer, one after another, in less than five seconds. The deviations turn out to be very different this time. The galvanometer shows that the electric charge on the paper capacitor is at least ten times smaller than the charge on the oil capacitor.

It is not possible to reconcile this observation with the electrostatic model of our textbooks according to which equilibrium charges on capacitors in series should always be equal. A theoretical analysis of this dilemma[1] demonstrates that distributions of voltages and charges at the beginning of a charging process are, in general, different from the distributions which prevail when (and if) a steady state of dynamic equilibrium is reached by the system. The capacitors, after being charged to voltages predicted by the electrostatic model, start discharging at a rate which depends on conductivities of the dielectric materials. Unfortunately, these nuances are usually hidden from students when the subject is first covered in the context of dc circuits. This often leads to serious misconceptions about capacitors in series.[2] The situation would improve if the topic were first discussed in the context of variable voltages or if we simply ignored the transients and concentrated on final voltages and charges. With the second approach the constant current through the capacitors can be calculated from Ohm's law

$$I = V/(R_1+R_2)$$

where R_1 and R_2 are leakage resistances of capacitors. The applied voltage, V, is the sum of potential differences across the capacitors, V_1 and V_2, where:

$$V_1= I*R_1 \quad \text{and} \quad V_2=I*R_2$$

The corresponding equilibrium charges on the capacitance are thus equal to

$$Q_1 = C_1*V_1=C_1*R_1*I \quad \text{and} \quad Q_2=C_2*V_2=C_2*R_2*I$$

Note that the ratio of constant charges is equal to the ratio of characteristic time constants, $(R_1*C_1)/(R_2*C_2)$, and that it does not depend on the magnitude of electric current. This conclusion is in perfect agreement with the complete analysis of capacitors in series.[1] The ratio of characteristic time constants can be shown,[3] to be equal to $(\rho_1 * k_1)/(\rho_2 * k_2)$ where ρ and k refer to the resistivities and dielectric constants of materials from which the capacitors are made. This explains why the equilibrium charges on capacitors made from identical materials happen to be equal. Only capacitors based on different dielectric materials will demonstrate the inadequacy of the electrostatic model.

Many students and teachers believe that the discrepancies between the predictions of the electrostatic model and reality result from the "leaky nature" of real capacitors and that such discrepancies "would disappear if resistances were much larger". This is clearly not the case; the ratio of constant charges does not approach unity when the current becomes vanishingly small. Only the time for reaching the equilibrium increases when the values of R_1 and R_2 are increased by a large factor. Capacitors in series can not be idealized in the same way in which we idealize lenses (very thin), surfaces (very smooth) or ropes (unstretchable and very light). A mathematical model of trajectories which ignores air resistance is approximately correct because its predictions converge toward reality when there is less and less air while predictions of the electrostatic model of capacitors in series do not converge toward true steady state conditions in the same way.

Perhaps future textbooks will use capacitors in series to demonstrate how an arbitrary idealization became a source of a serious misconception for more than a century. Practitioners knew about the limitation of the electrostatic model but textbook writers ignored the limitations and used the model without making a distinction between the initial and final states of the circuit. Unfortunately, the problems are still presented in the misleading way. Students think that solutions of problems refer to steady state conditions while textbook answers clearly refer to initial voltages. Many teachers also believe that textbook formulas

$$V_1 = V*C_2/(C_1+C_2) \quad \text{and} \quad V_2 = V*C_1/(C_1+C_2)$$

refer to steady potential differences. The misconception about capacitors in series would have been corrected long ago if students and teachers used the laboratories to address the "why do we believe" questions.[4] Physical reality does not have to agree with our ideas, but in physics our ideas must agree with reality. It would be nice, and highly desirable, if solutions of some textbook problems were tested against real experimental data rather than against the answers provided by authors.

It is important to emphasize that the criticism of the electrostatic model does not apply to the ac circuits in which periods are much shorter than R_1*C_1 and R_2*C_2. Voltages in such circuits change too rapidly for the steady states to be established and ohmic currents are small in comparison with the displacement currents. What can be said about the so-called "equivalence" formula, $1/C=1/C_1+1/C_2$, in the light of this discussion? The formula is usually derived within the framework of the electrostatic model and therefore the limits of its validity should also be clearly specified in our textbooks. The formula is appropriate for the ac circuits but it is not necessarily appropriate for the dc circuits. The paradoxical situation in which the equivalence formula is derived for the dc circuits electrostatic model (without any comments about the limits of its validity) should be corrected.

The meaning of equivalence is clearly more complicated when both dc and ac are to be considered. In the state of steady equilibrium the charge on the positive plate of the first capacitor is in general different from the negative charge on the second capacitor. Positive and negative charges on a single capacitor, on the other hand, are always equal. A set of two capacitors, made from two different dielectric materials, is usually not completely equivalent to a single capacitor. But the circuits can be equivalent in terms of the total potential energy stored, in term of equilibration times, or in terms of ohmic resistances.

A very simple setup, shown at this exhibit, is used to demonstrate a student-oriented activity designed to validate the correct theory of capacitors in series. The basic components needed (9-volt battery, two electrolytic capacitors of 470 μF, and two resistors, R_1=180 and R_2=20 kiloohms) were purchased, from Radio Shack, for less than four dollars. The resistors were connected in parallel with capacitors to simulate leakages through two different dielectric materials. The RC setups were then connected in series with each other, with the battery and with a switch. The time for equilibrating that circuit, after closing the switch, is about two minutes. A voltmeter across the 20 kiloohm resistor shows that at first the voltage rises to 4.5 volts, as it should, then starts going down until a constant value of about 0.9 volts is established. This value is simply the ratio of R2/(R1+R2) multiplied by the potential difference of the battery.

Ask students to read the voltmeter every ten seconds and to plot the readings versus time. Then ask them to compare the curve with the complete theory.[1] Explain to them that a steady electric current, no matter how small, is always essential in dynamic equilibrium. The electrostatic model of capacitors in series is incorrect because it ignores the current. Most electrical engineers know that constant voltages on capacitors in series depend on leakage resistances and not on capacitances. Physics textbooks, on the other hand, state that such voltages are distributed according to capacitances. Give your students a chance to decide who is right and who is wrong.

References

1. A. P. French, *The Physics Teacher,* March, 1993.
2. L. Kowalski, *The Physics Teacher,* May 1988.
3. L. Kowalski and M. L. West, Submitted to the *Physics Teacher* in April, 1992.
4. A.B. Arons, *The Physics Teacher,* May, 1993.

New Approaches to Motivating Students to Learn Physics and to Master Concepts

V. Gordon Lind, Department of Physics
Utah State University

Standard Approach to Teaching and Grading

The standard approach that I have used in the past and that is usually used by most of my colleagues teaching the introductory calculus based physics is as follows:

- Material to be covered is selected for the quarter or semester.
- A time table is worked out and a syllabus prepared.
- A list of problems and questions from the end of each chapter is prepared and assigned as homework.
- Tests are given over the material during a one or sometimes two hour time slot with most problems on the test being original. A few concepts type questions may be included along with problem solving.
- Test performance is usually according to a bell shaped curve. Grade cut-offs are determined and the students have their grades on each test determined with most students getting C's.
- Performance in the labs and on the assigned homework count 10-20% each and are included in the final grade.

Philosophy of New Approach

Expect more, specify expectations, be more cooperative with the student such as retakes on exams, flexible testing schedules, etc. and get a higher level of performance out of the students with more positive reaction to the course.

The students are anxious to succeed in their physics mastery and are willing to work hard to show that they can do it. If we make the A grade achievable by letting the student know what he/she must do to achieve it, i.e. if we make the grade within the reach of not only the more brilliant students but also those who are willing to put in the extra effort if they only knew what was required to get the top grade, then more students will achieve and learn physics to a higher degree of proficiency.

Students cannot claim mastery of physics until they have a mastery of **concepts** as well as having developed the skills and abilities to solve the problems. Therefore, the teaching of physics needs to emphasize both aspects and similarly to evaluate the student, both concepts and problem solving abilities must be tested.

Steps in the New Approach

Most of what occurs in the new approach is similar or only slightly different from the standard or more conventional approach. The basic steps are:

- Select and prepare material to be covered.

- Prepare a timetable and syllabus including examination times.

- Select problem assignments for A-level, B-level and C-level performance with substantial differences among what is required for each level especially in terms of mastery of the material.

- Allow students enough time on the tests to demonstrate what he/she knows an allow retakes (at least one) if needed. A testing center would be very helpful.

- Require satisfactory performance in lab.

- Require that all students do satisfactory homework on the problems expected of at least the C-level performer. The homework is not graded, just checked for completeness. The exams are the motivation for doing the homework and doing it correctly. The odd problems are used since they have the answers in the back of the book and the student can tell whether or not his/her solutions are correct.

- Give examinations on the problems assigned for the given level with only the input data changed so that the student can be familiar with the problem if he/she wants to because he/she has studied and worked it before, but the problem must be worked with new information. Therefore, the student must understand how to work the problem.

- Also give a concepts exam prior to the problem exam. Require high performance (80-90%) on the concept exam before allowing the student to take the problem exam, so the concepts must be mastered.

- Allow two or more retakes for all the exams during the quarter. The retakes should be totally new exams of the same level. The better of the two scores is used to compute their grades so a student is not punished for trying to do better. (A few students abused this retake possibility by not adequately preparing to take the exam the first time it was offered. In terms of fairness, however this didn't matter.)

Results Achieved by the New Approach

- Over 90% of the students opted for the A-level test even though the number of problems they were expected to be familiar with was very large (it included all the odd numbered problems at the end of the chapter but not the problems from the Additional Problems Section) and was 40% more than required for the B-level. No students chose the C-level (mastery of problems that end in 1 or 5 such as 21, 25, 31, 35, etc.). Whatever level, the student had to get 90% or better to pass at that level.

- Attendance at recitations (problem solving classes) was high.

- Attempts to cheat by the students was minimal.

- Students were frequently observed studying physics together and debating the best and most reasonable and easy way of working physics problems (much more so than other methods of motivation used).

- A real plus was the student-teacher relationship. We worked together as members of the same team. The problems and the mastery of concepts were our "mutual enemies".

- Students were frequently observed to be discussing physics concepts with one another and with more advanced students and other professors.

- The students had a positive attitude toward physics, commenting frequently on how interesting it was, how much they enjoyed it, how much it helped them in other classes and some switched majors to physics.

- As time went on and their mastery of physics improved, the students found that their physics course required less and less time to prepare for exams. Some began analyzing everything about them in terms of the physics they learned and attributed this to the fact that they had worked so many problems on so many topics it just came natural.

- Most students (70%) passed their exam with a score of 90% or better on the first attempt. This was the level of performance required of them to get an A or A- grade. If they took the B-level exam, they still had to get 90% or better and the same for the C-level. 45% of the students actually got 100% or perfect score on the problem exams and nearly 100% on the concept exam. On one exam 95% of the students got an A on the exam.

- The pace was high and the amount of work was great and the students considered physics required more time than other classes, but rather than reduce their performance to C-level they would almost invariably choose to drop the course instead (Only a few students chose to drop). It was clear that they were after an A or B grade and anything else was unacceptable to them. By dropping and picking up a physics class the following quarter from some other instructor the put themselves into a situation of most students getting C's and just as much work, but the relaxed requirement of not trying to keep pace and achieve mastery, somehow, was more appealing. The withdrawal rate was only slightly higher than normal (~12-16% instead of 10-14%).

- Attitude and positive feelings about physics and the particular course they took was very high. Three separate polls separated by 2 months all gave similar results, e.g. 75% said they were motivated to learn more physics and did more physics than material learned in other classes and they enjoyed it.

- The grade distribution was very high (see charts). At the end of the first quarter, the number of students with grades of B or better was twice the number of those below B. At the end of the second quarter, that ratio changed to 3/1 and it appears the ratio is similar and maybe a little higher for the third quarter which is just now being completed.

Reactions of the Teaching Assistants

Initially the TA's were opposed to the new system. They gave many arguments as to why it wouldn't work and the difficulties involved in administering it etc. By the end of the first quarter, however, the TA's were excited because they had had such a positive experience with the students. Because the students liked physics and had positive feelings and attitudes, the TA's caught their enthusiasm. They were also pleased to observe their students working so hard on physics, being interested in what they were doing and actually mastering the subject better than they had been accustomed to. I had a TA from China, one from India and one from the United States. Even, the Chinese TA commented on the students learning physics better than in his country with two years of study. He is an outstanding student himself and was very outstanding in China. The Indian TA commented "overall this system is good. The students enjoyed it. It was fair and I strongly feel that this system should be continued with some improvements".

Instructors Assessment and Perspective

I had three main goals when starting this class last fall. They were: (1) to use the modern electronic facilities in the classroom to a high degree as well as performing several demonstrations; (2) to design a system of motivating students and evaluating them so they could master physics and demonstrate their mastery on tests and to make the top grades reachable by almost any student who was willing to work hard enough; (3) to emphasize the teaching and testing of physics concepts, not just problems solving ability to a much higher degree than I had ever done before. All of these goals worked hand-in-hand together. I am very pleased with the outcome of the experiments tried and the degree to which my goals were achieved. I feel however, that I have many additional things to try to better accomplish goals (1) and (3) and there are also many improvements to goal (2), but frankly I was very surprised at the success I had at goal (2). The friendliness of the students and their attitude towards me, their class and physics as a subject is very rewarding. I want to continue the experiment with an improved list of problems selected for each level of grade to be earned. I would also like a testing center, and, thereby give more flexibility into testing times. The little bit of flexibility I did offer the students was greatly appreciated and allowed the students to take the test when they felt most ready for it, at least to some degree. I also want to introduce more variations into the problems and to get the students thinking about all the different variations possible for each problem, and thus increase their skills and range of adaptability. More interactive teaching and less lecturing is also a goal for next year. I am working on various ideas to accomplish that.

Course Grades for
Physics 221 & Physics 222
(Fall & Winter Qtrs.)

Textbook: Serway (Updated)

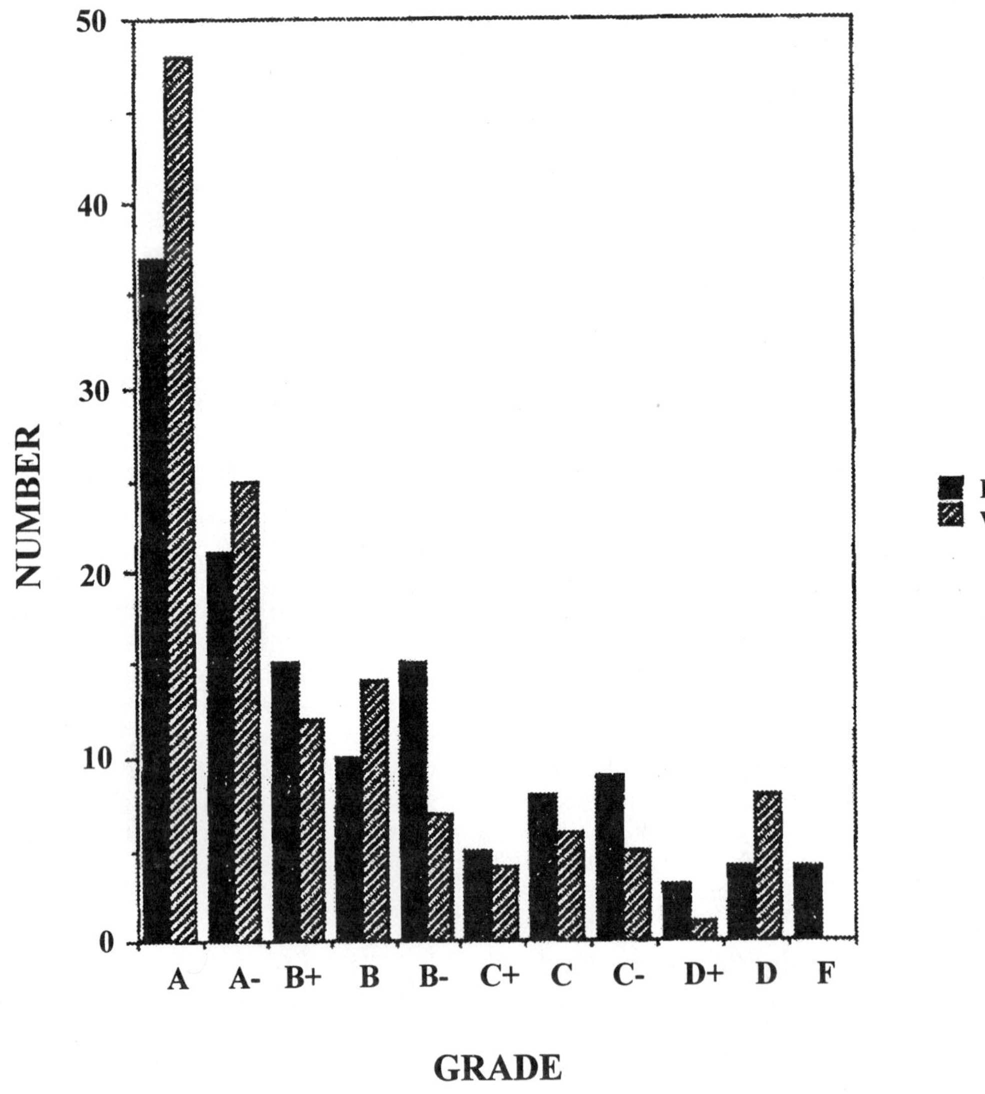

CONCEPTUAL EXERCISES IN ENERGY AND MOMENTUM

Jonathan Mitschele, Department of Physics
Saint Joseph's College

Introduction

Standard textbook problems emphasize numerical calculations, rather than qualitative analysis. This past year I used the curriculum materials on force and motion developed by Ron Thornton and David Sokoloff (*Motion and Force* and *Force and Motion*, available from Vernier Software[1]), which emphasize an experiential and conceptual approach to understanding, and was pleased with the results.

With that positive experience in mind, when my students had difficulties with concepts of work and energy, I developed a set of qualitative problems, which I used for drill. In class discussion, I made connections between the rather abstract models used in the problems and student experience driving cars — accelerating, braking, coasting, going up and down hills, and so on. Though they may seem elementary, I think these problems are comprehensive and I know that they challenge students to think carefully. My students found working through the problem set helpful; after they had completed the exercises, a large majority of the class had a firm grasp of the concepts involved. Qualitative problems can be very helpful to students, and so I offer these as examples of how one might construct similar problem sets to illuminate other concepts in physics.[2]

I should note that I had PASCO dynamics sets,[3] motion detectors, and MacMotion software[4] available for experimental verification; however, I believe these problems would serve as well in a more traditional lecture setting.

I have included in this paper the introduction to the problem set and the problems; when I give this assignment to students I provide a worksheet for their answers which is in the same form as the table used for the answers to the examples in the introduction.

Introduction to Energy and Momentum Exercises

The systems to be analyzed consist of a single block moving on a level or inclined surface (Figures 1-20). Forces that may affect the motion of the block include gravity, friction, and the force of a compressed spring; there may also be a constant applied force **F** of unspecified origin.

There are two steps for each exercise, drawing a free-body diagram and completing one or more rows of the accompanying worksheet.

Step 1: For each exercise, draw a complete free-body diagram (***use a pencil!***) which includes explicitly a vector representing each *physical* force (as distinct from force component) that acts on the object. It will be helpful in what follows to establish when, if ever, $\mathbf{F}_{\mathbf{net}} > 0$, $\mathbf{F}_{\mathbf{net}} = 0$, or $\mathbf{F}_{\mathbf{net}} < 0$.

Step 2: For each of these systems one or more combinations of the conditions listed at the top of each worksheet is possible. Determine the distinct combinations possible for each system, and complete a row on the worksheet for each combination, checking off each condition that applies. Your free-body diagrams should be helpful!

In determining the possible values of Δv, W_{tot}, ΔK, ΔU, and ΔE for Exercises 1-20, you should assume that the object continues moving in its initial direction during the time Δt over which we are determining these values. In other words, you should assume that while the object may be slowing down over the period of time involved, it does not reverse direction. It will be convenient to choose a coordinate system that has its x-axis parallel to and its y-axis normal to the surface on which the object is moving; this will make W_{tot} and ΔU easier to visualize.

As examples, I have worked through Exercises 1 and 4. The results are shown below.

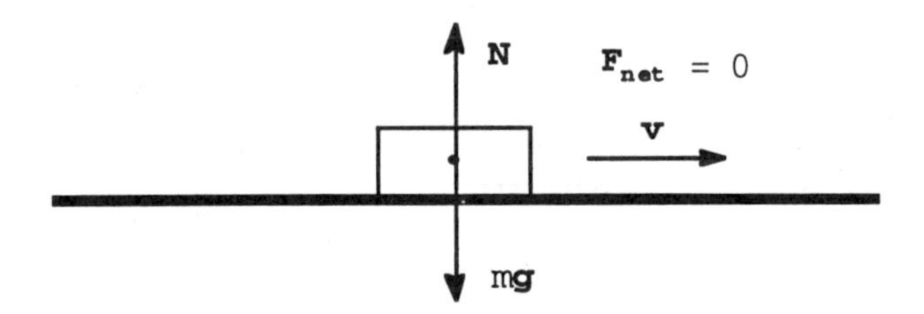

1. Object moving toward the right on horizontal frictionless surface with constant velocity **v**.

   Notes: Since the only forces acting on the object cancel each other, $\mathbf{F_{net}} = 0$. This leads to only one combination:

#	Isolated system	F_{net}		a			Δv			W_{tot}			ΔK			ΔU			ΔE			Δp
		=0	≠0	=0	>0	<0	=0	>0	<0	=0	>0	<0	=0	>0	<0	=0	>0	<0	=0	>0	<0	=0
1	√	√		√			√			√			√			√			√			√

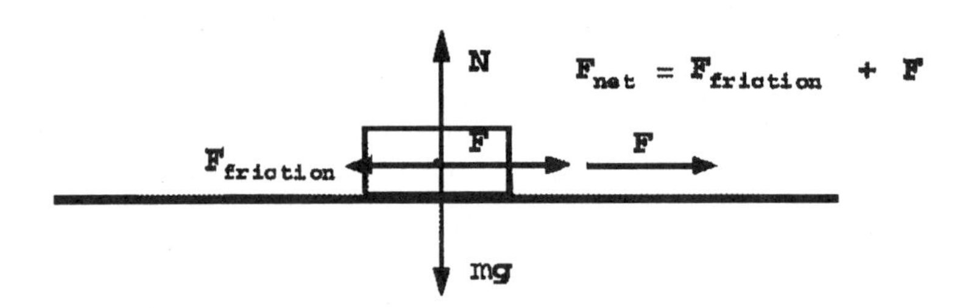

4. Object moving toward the right on horizontal surface (with friction) with constant force **F** acting in the direction shown.

   Notes: There are three possible values for $\mathbf{F_{net}}$, $\mathbf{F_{net}} = 0$ (4a), $\mathbf{F_{net}} > 0$ (4b), and $\mathbf{F_{net}} < 0$ (4c); in the free-body diagram above is for the case $\mathbf{F} > \mathbf{F_{friction}}$, which means $\mathbf{F_{net}} > 0$. These three possible values for $\mathbf{F_{net}}$ lead to the following three combinations:

#	Isolated system	F_{net}		a			Δv			W_{tot}			ΔK			ΔU			ΔE			Δp
		=0	≠0	=0	>0	<0	=0	>0	<0	=0	>0	<0	=0	>0	<0	=0	>0	<0	=0	>0	<0	=0
4a	√	√		√			√			√			√			√			√			√
4b			√		√			√			√			√			√			√		
4c			√			√			√			√			√			√			√	

Energy and Momentum Exercises

1. Object moving toward the right on horizontal frictionless surface with initial velocity **v**.

2. Object moving toward the right on horizontal surface (with friction) with initial velocity **v**.

3. Object moving toward the right on horizontal frictionless surface with constant force **F** acting in the direction shown.

4. Object moving toward the right on horizontal surface (with friction) with constant force **F** acting in the direction shown.

5. Object moving toward the right on horizontal frictionless surface with constant force **F** acting in the direction shown.

6. Object moving toward the right on horizontal surface (with friction) with constant force **F** acting in the direction shown.

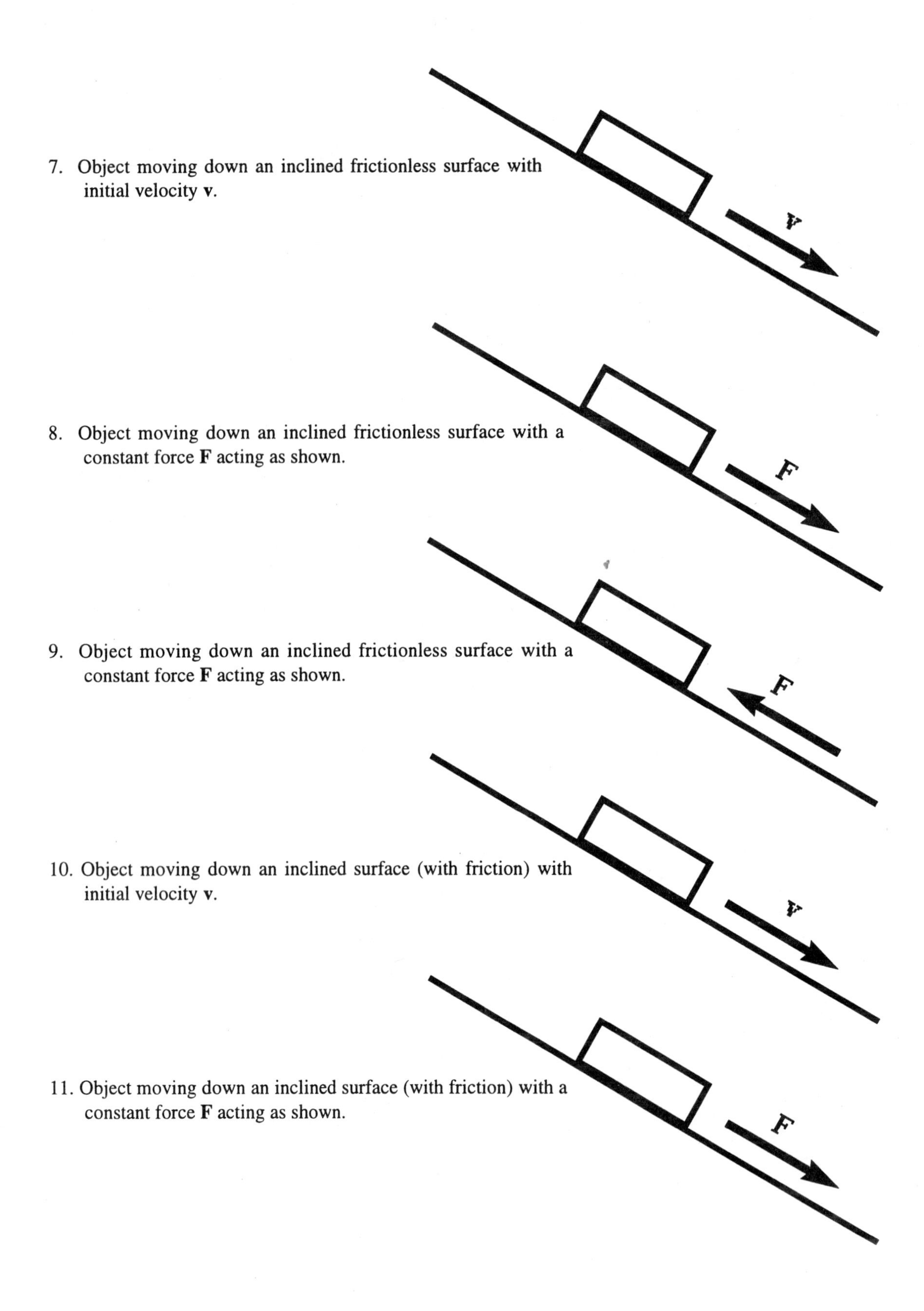

7. Object moving down an inclined frictionless surface with initial velocity **v**.

8. Object moving down an inclined frictionless surface with a constant force **F** acting as shown.

9. Object moving down an inclined frictionless surface with a constant force **F** acting as shown.

10. Object moving down an inclined surface (with friction) with initial velocity **v**.

11. Object moving down an inclined surface (with friction) with a constant force **F** acting as shown.

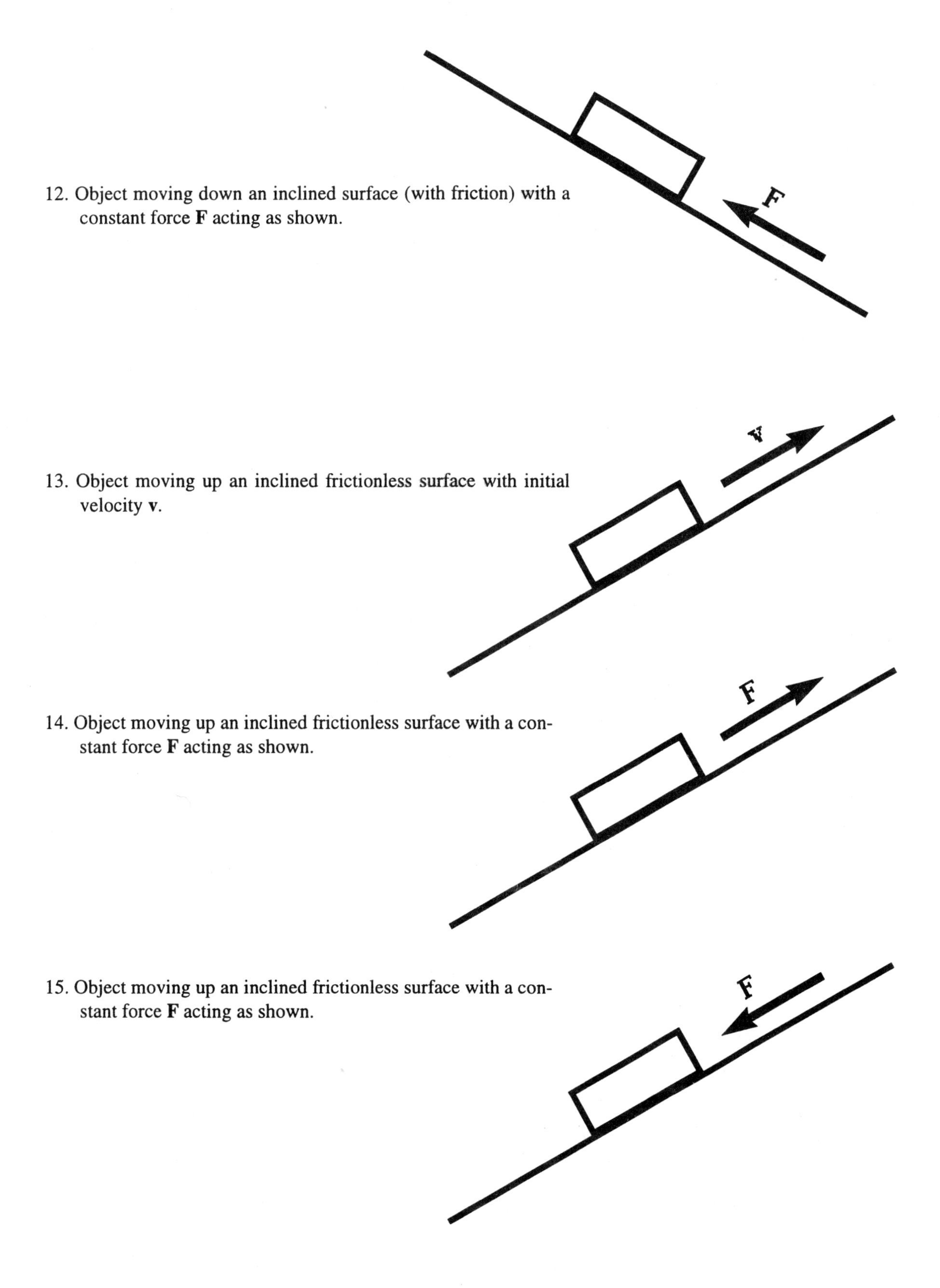

12. Object moving down an inclined surface (with friction) with a constant force **F** acting as shown.

13. Object moving up an inclined frictionless surface with initial velocity **v**.

14. Object moving up an inclined frictionless surface with a constant force **F** acting as shown.

15. Object moving up an inclined frictionless surface with a constant force **F** acting as shown.

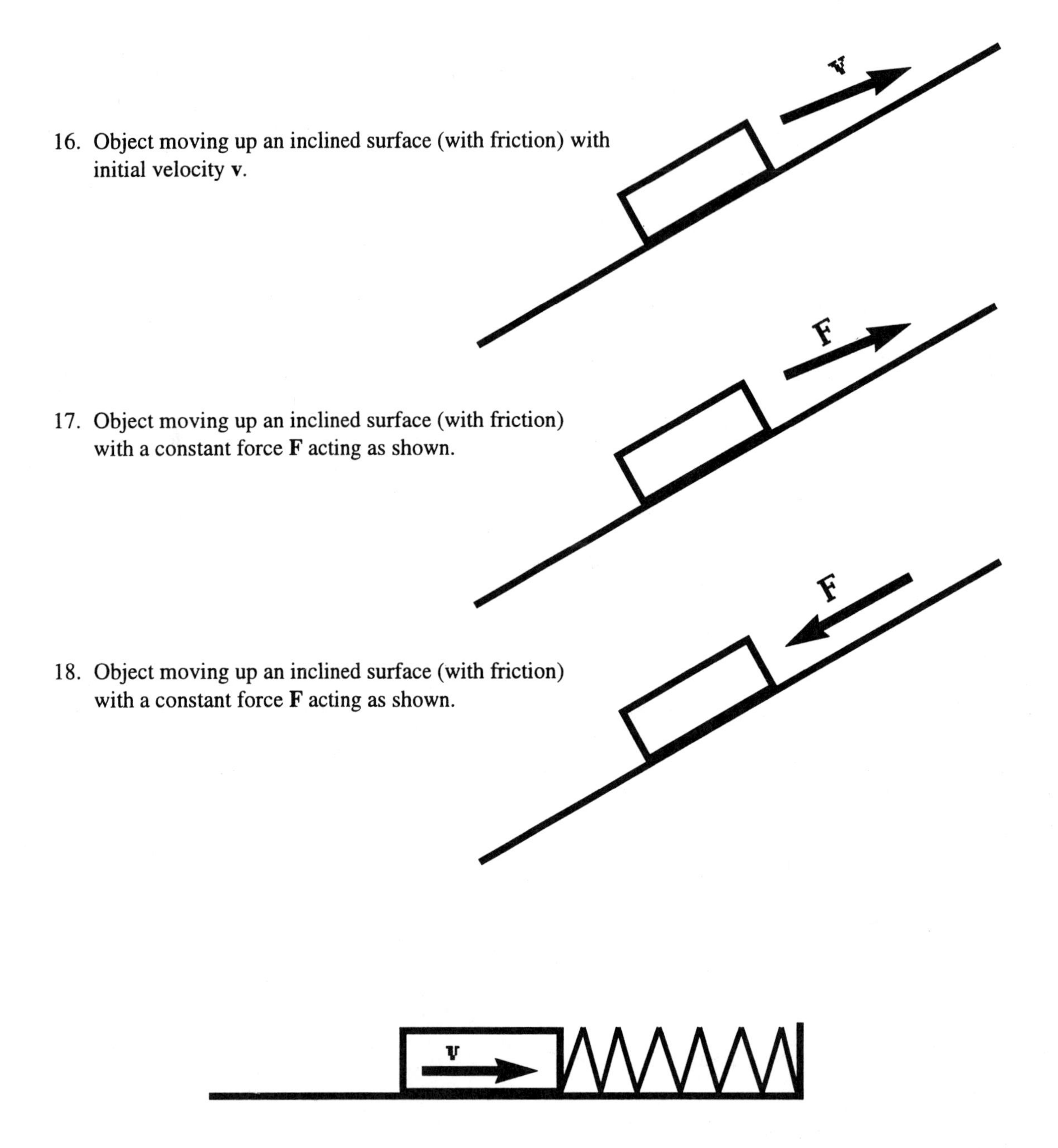

16. Object moving up an inclined surface (with friction) with initial velocity **v**.

17. Object moving up an inclined surface (with friction) with a constant force **F** acting as shown.

18. Object moving up an inclined surface (with friction) with a constant force **F** acting as shown.

19. Object moving with initial velocity **v** toward the right compressing a massless spring lying on a horizontal frictionless surface.

20. Object moving with initial velocity **v** toward the right compressing a massless spring lying on a horizontal surface (with friction).

References

1. Vernier Software, 2920 S.W. 89th Street, Portalnd, OR 97225; (503) 297-5317.
2. Please contact me if you would like a copy of the complete problem set with answer sheet.
3. PASCO scientific, 10101 Foothills Blvd., P.O. Box 619011, Roseville, CA 95678-8905; (800) 772-8700.
4. Morton detectors and MacMotion software are available from Vernier Software.

Thoughts on the High School/College Interface

Robert A. Morse
St. Albans School, Washington, DC

Introduction

These notes are a compendium of my thoughts on the subject of the High School/ College Interface based on the panel discussion, followed by answers to some questions recorded in the discussion. I base my answers on 25 years of high school teaching experience, half in public and half in private schools, experience as a Physics Teaching Resource Agent working with other high school teachers, and work on physics curriculum development and physics education research with Lillian McDermott, Mel Steinberg, John Layman, and recently with Priscilla Laws, David Sokoloff and Ron Thornton.

Discussion

The college and university Introductory Physics Course (IPC) has a significant effect on the high school physics course through several mechanisms.

For much of its history, the high school course has been modeled on the typical IPC, covering the same topics, in the same sequence, but with less sophistication in the mathematics. A few notable exceptions including the *PSSC Physics* and *Project Physics* courses, and to some extent Paul Hewitt's *Conceptual Physics* have taken some different approaches, but have not managed to change the entrenched model of the high school course. This is partly due to influence of the syllabi of the College Board Achievement test for the standard high school course and the AP tests for the Advanced Placement courses, and partly due to teachers of the high school course, whose perception of what an introductory course should contain and how it should be taught is based on their experience of the IPC. As a result, for many students the IPC in college repeats the high school experience, with more math and more students in the class. This is not a fully satisfactory situation. How might this interface be changed to improve the high school (and middle and elementary school) courses, the IPC for science majors and the physics experience for non-science majors?

Mechanisms

1. Colleges and high schools, through the College Board, have some power to drive the direction of high school teaching through the syllabus for the achievement test. It is important that the achievement test ask the right kind of questions. In addition, colleges could create more of a market for a broader course by recommending that high school students have courses in biology, chemistry and physics for admission.

2. Physics departments, in revamping the IP Courses have the opportunity to develop more meaningful courses for both majors and non-majors.

3. Teachers in the colleges, developers of the IPC's, and teachers in the high schools can work together to improve the sequence of experiences for all students, by taking advantage of each others expertise and experience.

Recommendations

1. On the basis of my own experience, and the considerable body of research on learning physics which you are hearing about in this conference, I suggest that the high school course for the broad spectrum of students should no longer attempt to "cover" the traditional syllabus. Instead, the course should aim at developing the conceptual base that we find missing in our students. Rather than give students the results of physics in algebraic equations, the equations should be an end point in constructing a robust qualitative conceptual understanding grounded in experiment and experience, and described in verbal, pictorial, and graphical terms before it is expressed in formal mathematical terms. The primary emphasis should be on mastery of techniques of model-building and sense making, of making and testing predictions–on learning to play what David Hestenes calls "the great game of science–modeling the real world" (1). In a one year course, students should cover fewer topics, but should have a much better knowledge of how they know what they know. This would ultimately provide a better base for the IPC, as students would come to the IPC with a better conceptual understanding of physics, and of the nature of knowledge in physics.

2. For a much narrower spectrum of students, we should continue to provide Advanced Placement courses in the high school which take similar approaches to the IPC. These courses can be useful both to students and to curriculum developers, as high school AP Physics teachers who are good observers of their students can collaborate with IPC writers in giving feedback on how their students work with the material. High school teachers can also help define the appropriate organization of physics knowledge for students. The IPC as currently codified presents difficulties for students in its use of language, in the sequence of ideas and in the way in which ideas are presented and developed. In the society of a small high school classroom with the freedom to mix class and lab work, teachers can know their students and how their students learn much better than a professor in a large lecture. Such teachers can serve as mediators in helping to detail how effective various approaches are with students. The success of a number of such collaborations between high school and college teachers, suggests that there can be more fruitful collaborations along these lines. An excellent example of a program which has done research in physics education, developed curriculum, educated underprepared teachers, and prepared teacher researchers and graduate students for research in physics education is the Physics Education Group at the University of Washington, run by Lillian McDermott. College and university physics departments would do well to follow the example set by McDermott's group which has had and is having a profound influence on physics education locally and nationally. (Both of the current high school teachers on this panel have been part of this group.) Examples of other collaborations include work by Hestenes, Wells and Swackhamer (2), by Steinberg (the CASTLE project (3)), and by Linn and Songer (4), to mention a few.

3. There should be alternatives to the IPC that address the needs of non-scientists, especially prospective elementary and middle school teachers, to know physics in a usable way, so that they may teach science in a useful way. An example is the Physics by Inquiry course (5) developed by Lillian McDermott's group. As many of the students in this category will not have had high school physics, courses developed along these lines are also profitable areas for collaboration in redefining the broad spectrum high school course.

4. In the long run, we should work to differentiate the high school physics experience for most students from the IPC. In the process, we should make the high school course accessible to a much larger number of students. The broad based high school course should lead into the college IPC. A narrower high school sequence should continue as in the AP course, while at the college level, a course similar to the broad based high school course should be available for students who need it. In the meantime, paying more attention to the contributions that people on both sides of the interface can make to physics education may improve physics teaching in both arenas.

Answers to Selected Audience Questions

1. *Are high school students today getting the same problem-solving experience we think we got?*

   This varies considerably depending on the teachers and curriculum. However, note the remarks by Lillian McDermott in her talk earlier that quantitative problem solving ability does not necessarily imply functional understanding. I spend less time on plug in formula manipulation, and more on problems that involve reasoning about the physics. In the process, I try to get students to think about how they solve problems and to distinguish between understanding the physics of the problem and reasoning about the behavior of the system before finding an appropriate and efficient path to solve it, then carrying out the mathematics and calculating and checking the result. In doing this I place an emphasis on *understanding the physics*. In the broad based course this is essential.

2. *Should we change the high school curriculum to cover a few items well?*

   See recommendation 1 above. I would have students learn to *do* physics extensively and carefully in a few areas, and learn *about* physics in a number of other areas.

3. *What is the minimal coverage the college people want from high school?*

   Let me turn this question around. I would like my students to leave high school with an experience of doing physics, of building a coherent conceptual model based on experience and experiment. I suggest that Newtonian mechanics should be a major focus and students should study one or two other areas in some depth, depending on the interest and expertise of the teacher.

4. *What to do for 95% who are not going to be physicists? What about the 80%? The ones who are not going to college?*

   In high school, I think the physics course should be broad based enough to reach a much larger fraction of students than the 20% or so who currently take high school physics, including those not going to college. At the college level, you may then distinguish between the courses for science and engineering majors and the courses for non technical majors. I suspect that there will continue to be overlap of courses between the high school and the colleges in physics for a considerable time.

5. *Why is the high school course a preparation for the college physics course?*

   As noted above, the IPC is the model of a physics course that most high school teachers and college teachers have in mind, what is expected on college entrance examinations, what most textbooks are written to teach, and thus what is taught.

6. *What is the impact of elementary and middle school on high school physics?*

   Good elementary and middle school programs can have a significant impact, when students arrive with a base of experiences which they have investigated and thought about. Presently, this is highly variable, depending primarily on the teachers students have had. This is an area where both high school and college courses which reach more students can have a long term feedback effect as future teachers take them and use ideas in their own teaching.

7. *What can college faculty do for underprepared high school teachers (locally)?*

   Get to know them, establish collegial links with schools and teachers, recruit them into appropriate courses, form alliances with experienced teachers (including Physics Teaching Resource Agents) to develop appropriate courses, look at models of long term out reach programs such as Lillian McDermott's program, join with them to study new developments in teaching introductory physics, etc.

8. *What is involved with making high school physics required?*

   If college admission requires high school physics, then more students will take it. That is probably too simple, but a number of colleges are now requiring three years of laboratory science for admission. I suspect that in many schools students accomplish this by taking one year of physical science followed by biology and chemistry. The physical science course might be a good place to introduce reforms in physics content and instruction which would affect a large number of students.

9. *Does college tuition drive Advanced Placement Physics enrollment?*

   I suspect it is more complicated than that. From my experience, student motivations are a mix of the intellectual challenge, prestige, the magic letters "AP" on the transcript, and the chance for placement. An informal look at my past students who did well on the exam shows only a few graduated early, but many used the placement to take more advanced courses.

References

1. D. Hestenes, "Modeling games in the Newtonian World," *Am. J. Phys.* **60** (8) 732-748, (1992)
2. D. Hestenes, M. Wells, and G. Swackhamer, "Force Concept Inventory," *Phys. Teacher* **30**, 141-158, (1992)
3. M. Steinberg, *Electricity Visualized: The CASTLE project,* PASCO Scientific, (1993)
4. M. Linn and N. Songer, "Teaching thermodynamics to middle school students: what are appropriate cognitive demands," *JRST,* **28**, 885-918, (1991)
5. L. McDermott, *Physics by Inquiry,* University of Washington, Seattle, WA

The Introductory Physics Experience at U.R.I.

A. C. Nunes/L. M. Kahn/S. S. Malik/S. J. Pickart
Department of Physics, University of Rhode Islande

Abstract:

To address the problem of slowly declining enrollments during the past two decades, and recognizing the mission of a state university to provide the best possible educational experience for students with a wide range of abilities, the URI physics department instituted a number of unilateral actions. These include: a new undergraduate program (Physics and Physical Oceanography); a new introductory sequence specifically for our majors; utilizing top undergraduates as paid teaching assistants; setting aside one room in the department which is *always* available and exclusively for student use; expansion of undergraduate research opportunities; and offering 'Physics Days' (a 2-3 day open house for the state's high school physics students) each year. These were in addition to already active SPS and Sigma Pi Sigma chapters, and senior research projects. Enrollments, numbers graduating, and departmental spirit have significantly improved.

Introduction

In 1987 The URI Physics department held a retreat to discuss the undergraduate program. This was in response to two decades of slowly declining enrollments in physics (See Figure 1.) (in the face of slowly rising enrollments both nationally and at the University of Rhode Island overall. . . see Figure 2.), and a general faculty dissatisfaction with the existing program. From this and subsequent discussions came a number of ideas and suggested actions which have been implemented by the department. We tried to address problems at four points; the High School level, introductory courses at the college level, breadth of programs, and the human experience.

High School Students

To try to stimulate interest for physics among high school students we began 'PHYSICS DAYS', an open house for all Rhode Island high school junior and senior science students, highlighting the department and recent advances in physics. These began in 1989. Undergraduates as well as graduate students and faculty participate. We host between 300 and 550 high school students each year.

Introductory Courses

As a State University, our students come with a wide range of backgrounds and talents. Our programs, in particular the introductory courses, must maintain the interest of the gifted students while providing as much help as possible for those of lesser aptitude. Many have to work to support themselves. We have attempted to take advantage of these conditions in part by using the introductory course as a selection device.

Undergraduate Teaching Assistants

Of those students passing with 'A's, some are asked to teach recitation and lab sections of the course in following semesters. They are trained and closely supervised (videoed, and critiqued by faculty during the semester), and paid for their time. All TAs spend two hours per week acting as tutors in the department's undergraduate room ('room 216').

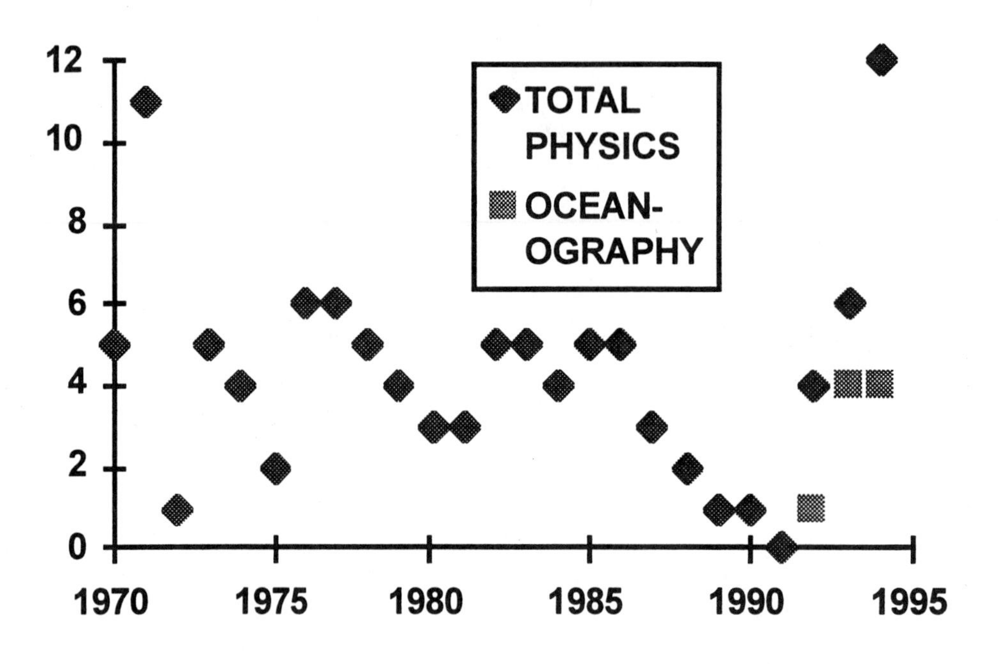

Figure 1. Physics Bachelors Degrees awarded at URI as a function of time. Numbers after 1992 are projected.

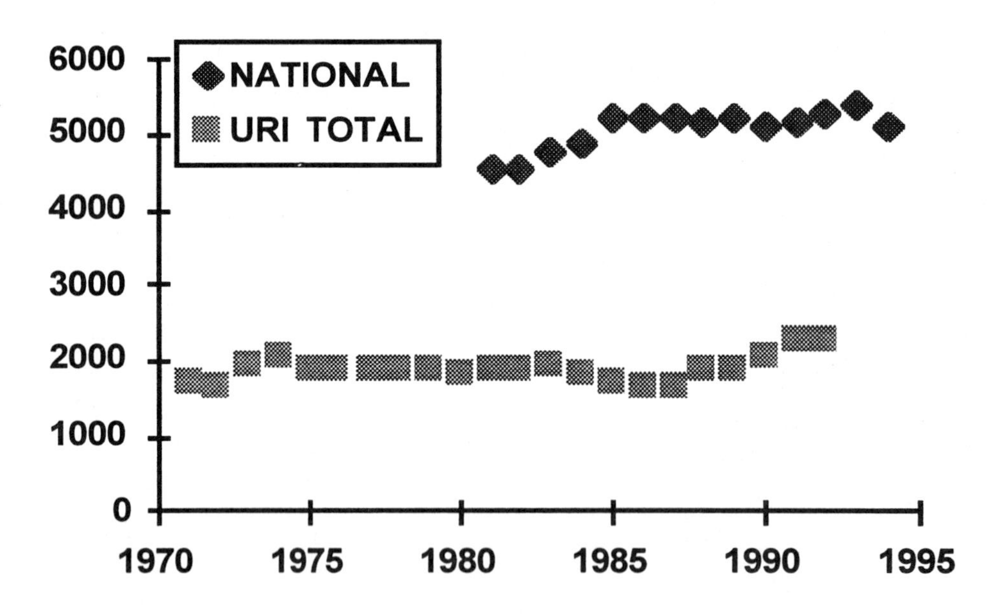

Figure 2. Physics Bachelors Degrees awarded nationwide (Source: AIP Publication no. R-151.29), and total URI Bachelors Degrees awarded.

Research

Top undergraduates may be asked at the end of their freshman year to participate in research during the summer with one of the faculty if funding is available. Upper level undergraduates are encouraged to participate in such projects and are often supported by the NSF Research Experiences for Undergraduates program.

Other Work

Students have the opportunity to work within the department as office, shop and technical help, and are paid for their time.

New Program

Fashions change, and it was felt that broadening our offerings might attract more students. This was done by collaborating with the URI Graduate School of Oceanography (GSO) to set up an undergraduate program in Physics and Physical Oceanography. This consists of the standard physics Bachelor of Science curriculum with one additional course in fluid mechanics (taught in the engineering department), one physical oceanography course at the first year graduate level, and the student's senior project done under the supervision of GSO personnel.

Human Experience (Room 216)

One room in the department was set aside exclusively for undergraduate use. It contains several computers, a VCR, benches and seats, and is secured with a combination lock. All physics undergraduates have the combination. During the day, a TAs are stationed there as a tutors. It has become a gathering place for our undergraduates where help with class work, spirited discussions, pizza parties and advice on any subject are available at nearly all times of the day or night.

Results And Assessment

During the past six years the spirit of our undergraduates has improved noticeably , and enrollments have increased. Projected numbers of bachelors graduating are significantly above past averages (see Fig. 1). We have undertaken a systematic study to more precisely determine the reasons for these changes by developing a questionnaire and administering it to all physics undergraduates. In addition, we are interviewing graduating seniors to try to pick up points not covered in the questionnaire. This is the first year we have done so. We hope to continue with this and use it to 'fine tune' the activities promoting the improvements in our department.

The questionnaire was administered for the first time this May. It is too long to reproduce here. You may have a copy by writing to us. This two page form included a number of both "multiple choice and essay type questions' asking: 1) What was it that caused the student to choose URI as his/her university; 2) Why did they choose to enter physics; 3) What enticed them to stay with us; and 4) What they like and don't like about the department. We received 18 responses out of a total undergraduate enrollment of 49. These included most of our Juniors and Seniors, and about a third of the Freshmen and Sophomores. One of our six Seniors were interviewed.

One very important source of information which we don't adequately tap is the students who decide to leave physics as undergraduates. Because this is most likely to occur with Freshmen and Sophomores who are not well known to the faculty, it is difficult to identify them before they are gone. We hope that the longitudinal nature of this study, pursued over several years will give us insight.

Although it is premature to present a detailed assessment at this time, it is interesting to note student response to our first attempt. The questionnaire itself, though the product of discussions both within and outside the department (including B. LaSere-Erickson and G. Erickson of our Instructional Development Office and J. Schaffran of our College of Human Science and Services, who also administered the interviews), received a good deal of constructive criticism from both students and office staff. It is gratifying to note that all of these criticisms were offered in a supportive spirit. All involved seemed genuinely desirous of helping. As a result, next year's questionnaire will be modified somewhat.

We have not evaluated all answers, but report here one preliminary analysis. One question asked the students to rate in order of importance to them a number of factors affecting their experience in the department. These included: cost (tuition), interactions with faculty, fellow students, graduate students, challenging courses, easy courses, research opportunities, teaching opportunities, room 216, paid work in the department, the Society of Physics Students, and program offerings. Simply putting a check mark on a master sheet corresponding to each student's answer left the sheet black. In other words, what is very important to one student is completely unimportant to another. This appears to be generally true across all levels, though statistics are poorer in such a break out. This seems to emphasize the diversity of our students' needs.

Conclusion

We are beginning a long term process to evaluate the importance of various aspects of our undergraduate program to our students. The introductory course is significant in this as its the student's initial exposure to the subject, and also to the kinds of people who make that subject their lives. It is their entry into the physics community, and it appears to be important to enlist them in collaborating in every aspect of department life as soon as possible. In this first look, it appears that the quality of life (including interpersonal relationships and the desire to be a part of the department 'team'), is at least as important to our students as the subject matter itself.

Is this a meaningful assessment?; Are we 'Stalking the Second Tier' (Sheila Tobias, They're Not Dumb, They're Different - Stalking the Second Tier, 1990, Research Corporation, Tucson, Arizona, 1990)?; Or, are the indicators (student's helpful attitudes, improved departmental spirit, increased enrollments) merely statistical fluctuations? We hope to be able to report more definitively after following this for a few years.

TEACHING TEACHING ASSISTANTS AT UT-AUSTIN

L. C. Shepley, Department of Physics
University of Texas at Austin

Introduction

This paper describes a course on teaching, taught by physics faculty to physics graduate students. This course covers teaching and lecturing techniques, critiques based on video taping, case studies, and observations of effective teachers. It is the main component of the UTAustin program for training graduate student Teaching Assistants (TAs). In the Physics Department at UTAustin, TAs teach laboratories and recitation sessions, grade, and tutor. The training program is guided by our Center for Teaching Effectiveness (CTE), and seeks to maintain the quality of undergraduate education, give graduate students communication skills, and provide TAs with a network of colleagues and mentors. The CTE resources given to teachers of this course, the advantages and disadvantages of the course, and its suitability for other institutions are also discussed.

Teaching Assistants Training Program

The original motivation for the TA training program was a mandate by the Texas State Legislature in order to promote better undergraduate education. Over the past decade, the course has proved its worth, and similar courses have proved valuable in most of the other UTAustin departments. In the Physics Department, the course, Phy 398T, is a one-semester course (3 means three semester-hours; 9 means graduate level; 8T is an identifier). It is taught by a member of the Physics faculty. I have monopolized it for a couple of years to improve both it and my teaching; it is not at all a one-person course.

In the UTAustin Physics Department, TA duties usually include teaching one to three sections (20 students each) of a laboratory (including grading and servicing lab equipment), tutoring, and holding office hours. Some TAs, particularly International TAs (ITAs), are graders; an ITA must pass an English certification process before being allowed to teach. Some TAs lead recitation sections, with special training (which I will not describe here) on teaching study techniques.

The Physics Department mandates that each TA take Phy 398T within one year of the initial appointment. UTAustin requires this course for any graduate student who is promoted to Assistant Instructor (AI); only AIs may be instructors of record for a course. A TA works under faculty supervision, though in practice has great freedom of action. In addition to Phy 398T, TAs are given a short pre-semester orientation and manuals. Each week the TAs for a lab hold meetings; one graduate student is appointed Head TA to serve as a mentor.

The CTE operates under the Vice President and Dean of Graduate Studies and the Executive Vice President and Provost. It supports the faculty who teach 398T courses by running a summer seminar and special workshops, providing a 398T handbook (300 pages) [Lewis 1992], and funding some equipment or travel. The CTE also directly supports TAs with reading material, special workshops, and a videotaping/critiquing service (I choose to provide this service myself).

The CTE also runs the ITA program. Each ITA must be evaluated for English proficiency. Those who are certified may teach, but those who are not may not. (Conditionally passed ITAs with student contact take a required ITA course.) The CTE runs a three-day orientation before the beginning of the fall semester for ITAs, covering techniques of teaching in an American university. Individual assistance for ITAs is also available through the Center.

In this article, I will concentrate on Phy 398T, as it is not a common component of TA training programs [Lambert/Tice 1993].

Phy 398T meets for two hours on Monday afternoons and one hour on Wednesday afternoons (since students attend the departmental colloquium). Enrollment is between 25 and 30, each semester. The course is given on a Credit/No Credit basis and is not stressful for the students.

There are four main goals for the course

1. *The students make friends who will help each other in teaching and research and offer constructive criticism.* These networks are important both for morale and for the learning of physics. Teaching is a new experience for most of the TAs, and the sharing of problems is important support.

2. *The students formulate a teaching philosophy and learn teaching techniques.* I do not try to create great teachers, but in Phy 398T students are exposed to good teaching and learn to recognize it. They learn the requirements for conducting a class fairly and honestly, and practice giving lectures, holding discussions, and other techniques. They are shown why it is easier to teach well than to teach poorly.

3. *The students learn techniques for teaching labs.* I concentrate on labs since most of our students teach them, they are different from other classroom experiences, and lab safety must be emphasized.

4. *The students learn how to give a seminar.* This is a major part of the course for two reasons: First, we have a requirement that a student give a public seminar followed by an oral exam (in lieu of a written general exam). Second, physicists are often judged on their work as given in seminars. Even students who are not interested in teaching careers are motivated by these reasons.

The main course requirements

Participation: No more than three absences are allowed. Each student must make minor presentations and a major one, which is videotaped and critiqued. Each student is encouraged to create a teaching portfolio [Seldin 1993]. (Few students fail to meet these requirements.) I assign non-mandatory homework, which the vast majority of students do complete. These requirements are described below:

Minor presentations are five to ten minutes in length and often include reporting on colloquia or seminars. A checklist is provided, and the reporters review both the style and content. Other presentations include interviews with experienced faculty, discussions of articles or book chapters dealing with teaching methods, or practice using specific techniques, such as questioning or leading a discussion. Often two or more students make minor presentations together. In each case the class criticizes the presentation, always in a helpful and friendly manner.

Major presentations are ten minute seminars on physics or pedagogy. They are strictly timed, in part so students can gain time management experience. I require that the content of the talk and all visual aids be approved in advance. My most frequent suggestion is to reduce the amount of material and the number and density of the content on their overhead transparencies.

Each student is videotaped during the major presentation, and three students make major presentations in one class period. After the presentations the class is given a quiz consisting of one question for each talk, supplied by the speakers and printed in advance. The class also makes written comments and suggestions (not anonymously). The three students are critiqued together, and all components, video, quiz results, comments, and suggestions, are valuable. This critique usually lasts about half an hour.

For homework, students compose a first day handout, namely a one or two page course description meant to tell the members of a class, on the first day of class, what the course is, what its prerequisites are, and how they are to be evaluated. For example, this past semester students concentrated on our lab for pre-medical students. Over an eight week period, my students discover the lab goals, what the students are like, the text and required materials, the detailed schedule, the grading policy, the policy on cheating, and so forth. Finally, my students were required, in groups of three, to complete the handout and print it using a word processor. The handout is then part of the students' portfolios. Homework is not onerous nor does it intrude on other courses or research.

I use a variety of teaching techniques in Phy 398T

I lecture only rarely. I do call upon guest lecturers, but not often. Discussion is the main technique, and I call on everyone at least once each period. I often break the class into groups of three to five; to aid these discussion groups I provide a one-page list of questions for which each group member provides a signed, written answer. The answers then defend their positions during a general class discussion.

Often these discussions involve a case study: A classroom or other situation is described in story form (people are given names, for example). The questions ask about the problems involved and their ramifications and how the problems should be resolved.

An example is the case of two lab partners. They have been told that their conclusions must be written separately, but are to discuss the lab and take data together. In the study, the students turn in almost identically worded conclusions. Questions may include: Does the similarity of the conclusions represent cheating or misunderstanding? What if anything should be done to punish the students? How are their rights to be protected? Might the students complain about an overly harsh TA?

Cases are drawn from actual experiences, my own or colleagues' and students'. The topics may be straightforward cases or may be provocative, concerning cheating, racial or gender discrimination, or favoritism. Responses from the discussion groups sometimes agree, sometimes strongly disagree, and the classroom discussions can become quite excited.

Discussions may also involve articles or chapters on teaching techniques. Sources include our department's TA Manual [Shepley/Kotz 1992], the CTE Handbook [Lewis 1992], brochures put out by our Graduate School [Reed 1992, Slate 1992], and selected texts [Arons 1990, McKeachie 1978, for example].

Phy 398T is evaluated each semester by the students, anonymously. Sometimes a criticism will be that it is not rigorous enough, but the relaxed nature of the course is preferred by most students. In these evaluations, students are asked specifically about the goals of the course (given above, they are included on the first-day handout): Are they appropriate, or should they be modified? Were the goals met, and if not how can my teaching be improved? My students are not shy about suggestions, and the course has evolved to be both useful and attractive to the students.

Why this course may be suitable for other institutions

A course on teaching techniques, given by a physicist to physics students, one semester long, is in many ways more effective than even an elaborate orientation or a course taught by a non-physicist. In Phy 398T, the students are given continual encouragement over a long period, during which they form networks and learn how to use resources. Phy 398T solidifies the information obtained from orientation and manuals, and it should not replace these other training methods. Since it is taught by physics faculty, problems specific to physics are discussed, and students learn of faculty involvement with good teaching.

Various concerns about the course include whether it is hard for a physicist to teach, since it clearly is different from a technical graduate course. Preparation and follow-up time can be rather extensive at first, but it need not be onerous. I do not find it difficult since good support is available (both from the CTE and from other physics faculty). I learn as I go, and it is best to plan on teaching it for several semesters to foster this learning process.

I should also emphasize that Phy 398T is taken seriously by the department and The University and is counted the same as any other course in determining a teaching load.

It may be the case that some faculty will resent even the existence of a course like Phy 398T. In the course, students do criticize the teaching and lecturing styles of physics faculty members and visitors, but I am careful that they criticize both their fellows and all others in a helpful and friendly way. Most people easily see that the benefits outweigh any perceived threat, because the students become better TAs and are much more at ease when explaining their research work and formulating their own ideas. Furthermore, I am a source of teaching tips to some of my colleagues, who recognize that a teacher of Phy 398T will, of necessity, develop good pedagogical style and an extensive repertory of teaching techniques.

When the course was initialized over a decade ago, it was basically a simple discussion session in order to assure students that very little work was required. That was a mistake, for TAs tended to resent a requirement that they take an insubstantial course. As the course developed substance the students' morale has improved: Most TAs enjoy teaching, and all appreciate the experience in giving seminars. Still, Phy 398T does not intrude upon other courses or research, and that is a factor in its acceptance by students and faculty.

To sum up, Phy 398T is a useful way to help TAs become better teachers and better physicists. It has succeeded here at UTAustin, where it is supported by the administration, our Center for Teaching Effectiveness, and the Physics faculty. I certainly recommend that this type of TA training program, based on university support and a departmental course, be considered by other institutions.

Acknowledgments

I wish to thank the Center for Teaching Effectiveness, especially Marilla Svinicki and Karron Lewis, and Austin Gleeson for support and guidance in teaching Phy 398T. Support for my travel to the Conference was provided by the Physics Department and the Center for Teaching Effectiveness, The University of Texas at Austin, and a grant from the Texas Advanced Research Program (TARP-085-Matzner).

References

1. A. B. Arons (1990): *A Guide to Introductory Physics Teaching* (Wiley, 1990)
2. L. M. Lambert and S.L. Tice (1993) (Editors): *Preparing Graduate Students to Teach* (American Association for Higher Education)
3. K. G. Lewis (1992) (Editor): *Teaching Pedagogy to Teaching Assistants—A Handbook for 398T Instructors* (3rd edition) (Center for Teaching Effectiveness, The University of Texas at Austin)
4. W. J. McKeachie (1978): *Teaching Tips—A Guidebook for the Beginning College Teacher* (7th edition) (Heath)
5. L. Reed (1992): "TAs/AIs Talk About Teaching" (Center for Teaching Effectiveness, The University of Texas at Austin)
6. P. Seldin (1993): *Successful Use of Teaching Portfolios* (Anker)
7. L. C. Shepley and N. Kotz (1992): "Manual for Teaching Assistants, Assistant Instructors, and Graders" (Physics Department, The University of Texas at Austin)
8. A. Slate (1992): "Handbook for TAs and AIs" (Office of Graduate Studies, The University of Texas at Austin)

USE OF SPREADSHEETS IN THE LABORATORY COURSE AT A 4-YEAR LIBERAL ARTS COLLEGE

George F. Spagna, Jr., Department of Physics
Randolph-Macon College

Abstract

The introductory calculus-based physics sequence was modified to meet newly revised College requirements for "computer-intensive" courses. Students made use of the *Quattro Pro 4.0* spreadsheet environment as the preferred tool for calculation and graphical representation of phenomena. In addition to traditional problems assigned from the text, "spreadsheet problems" which required the students to manipulate various parameters to explore a wider range of solutions were developed. This approach was applied in both semesters, and included problems ranging from kinematics and dynamics through electricity and magnetism, circuits, and optics.

Example problems are presented. Student reaction and the instructor's evaluation of the effectiveness of this mode of instruction are discussed.

Background

The faculty of Randolph-Macon College have recently replaced a general education requirement for "computer literacy" with a requirement for "computer proficiency". The earlier requirement originated at a time when entering students were unlikely to have prior experience with computers, and included both "hands-on" exposure to applications (word-processing, spreadsheets, statistical packages, etc.) and an exploration of social and ethical issues surrounding the growing use of computers. Since most entering students now seem to have already some minimal exposure to computer applications, it was determined that a more realistic educational objective was to expose them to computers as intellectual tools within specific disciplines. Students must demonstrate minimal proficiency with word-processing by completing a standard performance test, and are then required to complete at least one "computer intensive" course as part of their undergraduate program. While it is not required that this course be within the student's major program, it is expected that many students will elect to complete this requirement within their major department.

Implementation

The introductory calculus-based course (Physics 111-112, "Physics for Scientists") was offered as a computer intensive course during the current academic year. Students were introduced to basic spreadsheet applications using *Quattro Pro 4.0* (the College has a site license agreement with Borland). Access was readily available in the main computer labs, and several students had their own computers and personal copies of the software. (We had also planned to make the software available on the network in the College library, but that network turned out to be incompatible with this application.) They were expected to use the spreadsheet as their primary calculating tool, for constructing numerical models of physical phenomena, and for graphing relations among variables within assigned problems. The spreadsheet environment was not intended to take the place of necessary algebra and analytic manipulation of basic relations. In several problems the students were expected to obtain a result analytically and then compare that solution with a numerical calculation. Implicit in this modeling scheme is the idea that students can readily change parameters of a given problem to explore the range of possible solutions with greater facility than simply calculating solutions to more problems.

Classroom instruction on using the spreadsheet was integrated into the discussion of physics material. For example, when the notion of velocity as the derivative of displacement was introduced we also showed students how to calculate derivatives numerically. Similarly we introduced numerical integration when discussing displacement as the integral of velocity. In addition, several handouts were distributed to help the students become more familiar with the features of the spreadsheet, and to reinforce the techniques presented in class.

First semester problems were largely assigned from the textbook. These were enriched by asking the students to manipulate the given conditions of the stated problem in order to further explore the range of solutions, since many were hesitant to "play" with the solutions on their own. Second semester problems were not taken from the text, and involved more modeling than simply calculation. The normal scheme was to assign problems on a Wednesday, making them due the following Monday, thus allowing time for students to seek additional help as needed. Typical problems are presented in the Appendix.
The spreadsheet was also integrated into the second semester of the laboratory. In addition to using *Quattro* to analyze and display data, the students were given access to a "template" which performed a weighted least squares fit and determined the reduced chi-squared value for each functional form tested. In the first semester of the laboratory this function was carried out initially by hand, and then using a program resident on the campus mainframe. The transition to a spreadsheet environment was done to make the program more "user friendly" and so that students would be using a single software package rather than several.

Challenges

The population of this course comes from two groups. Students who have an interest in physics as a major field take it as their introduction to the discipline. We prefer that it be taken in the first year, but we find a growing number of students who explore the idea of majoring in physics as sophomores. This course is also required of students majoring in chemistry, who typically take it as juniors or seniors. Students in the classroom can be expected to have some real interest in science - this is not "physics for prisoners".

Despite the availability of extra help as needed, several students remained reluctant users of the tools we made available. More than one student openly expressed willingness to accept a lower grade in the first term, rather than put forth the effort to develop real skill with the spreadsheet software. What began with the intent of giving students greater freedom to explore phenomena without requiring the drudgery of repetitive problem solutions became a battle to get some students to work with the computers at all. Some students were willing to learn new skills, but were then reluctant to apply them beyond the narrow scope in which they were originally applied. Numerical differentiation and integration were applied in several areas, yet some students had to "relearn" the techniques for each problem. Growing exposure to computers in high school has not completely eliminated "computer phobia" among our entering students.

In spite of student resistance the number of applications of spreadsheet software is likely to grow rather than shrink. Some of the resistance is institutional, in the sense that there is not yet a large cadre of upperclassmen with this experience. Some of the resistance is probably structural, since the intellectual model of liberal education tends to emphasize breadth of exploration at the introductory level rather than depth. If the introductory course can be viewed as a "survey of physics", then the spreadsheet problems may give the impression of requiring too narrow a focus.

Conclusions

This was our first iteration towards developing an appropriate level of computer usage within the introductory course. Even those students who initially resisted using the computer were able by the end of the course to quickly program the spreadsheet and generate graphs displaying their results. Although student unease may well have distracted attention from learning physics to learning *Quattro*, fuller integration of computers rather than less seems the appropriate response. This is partly because of the mandate to develop and present computer intensive courses, and partly because of our expectation that computer applications are likely to increase rather than decrease. Additional work stations will be available with an upgrade of the network in the Library computer center in the next academic year, which should make access even easier. As more students pass through the course, it is expected that a support group of "computer proficient" junior and senior level students will become part of the learning environment.

Appendix: sample problems

Physics 112
1993 February 17

Spreadsheet problem: Coulomb's Law
due: Monday, 1993 February 22

Turn in your solution on diskette in a *Quattro* file named **coulomb.wq1** . Do not turn in printed output!

Use named cells for the Coulomb force constant k and for *e* the charge on an electron. Use named cells for each of the charges below:

An α particle ($q_\alpha = +2\ e$) is located at the origin. A proton ($q_p = +\ e$) is located at x = 1 μm.

Calculate and plot as separate series the force on an electron due to each of these particles as a function of x as the electron is moved along the x axis from x = - 5 μm to x = + 5 μm. Be careful to account for the *direction* of the force, with positive forces defined here as pointing along the positive x axis.

Calculate and plot as a third series the *net* force on the electron. Put all three series on the same graph, with appropriate legend identifying the different series.

Note: be careful that none of the x cells is specifically at the positions of either the proton or the α particle, or you will generate error messages.

Physics 112
1993 April 7

Spreadsheet problem: calculating the magnetic field near a wire
due: Monday, 1993 April 12

As before, turn in your solution on diskette, in a file named **Bfield.wq1.** Do not turn in printed output!

A straight wire segment 50 cm in length is centered at the origin along the x axis. The wire carries a current of 2.5 A. Use the Biot-Savart Law to integrate numerically the magnitude of the magnetic field at 10 cm intervals along lines parallel to the wire from x = -25 cm to x = +25 cm for R = 5 cm, 10 cm and 25 cm. Calculate the magnetic fields for these same points assuming that Ampere's Law could be correctly applied.

Plot these six field calculations as a function of x.

Continuous Quality Improvement of the Learning of Physics by Engineering Students

James D. White, Department of Physics
The Pennsylvania State University

Abstract

A Continuous Quality Improvement (CQI) team consisting of faculty members and a student from the College of Engineering and the Department of Physics has been established at The Pennsylvania State University. The charge for this team is to investigate "the learning of physics by engineering students." The CQI approach is based on the popular business style, Total Quality Management (TQM). The basic tenets are to analyze business practices from the ground up with contributions from all points of view, and to implement lasting change only after fully testing the abilities of that change to improve productivity. This management style reduces the need to "re-work" unsuccessful solutions after they have been fully implemented. The team has conducted surveys of engineering and physics faculty, students currently taking the introductory classes, senior engineering students, and students who have dropped out of the engineering sequence entirely. Based on these data, the team has chosen areas for improvement and is establishing experiments to test possible improvements. This presentation will outline the steps that the team has taken and will be implementing next term.

Commonly, whenever a new professor begins teaching a course, that professor either adopts previous instructors' styles and formats or "scraps" them and designs the course from scratch. The former approach often resulted in the course becoming stagnant, whereas the latter entailed an inefficient "re-working" of the entire course design and format with little benefit from previous mistakes. This pair of approaches has produced very few net improvements in introductory courses. Echoing Sheila Tobias's invited talk, it is clear that the physics community must attempt to step outside itself and adopt management techniques that encourage not only reform, but lasting incremental improvement. This paper is about an attempt being made at Penn State to embrace a management technique based on this premise.

For a number of years, businesses across the nation have adopted a Total Quality Management (TQM) approach to increase productivity and minimize waste. This technique has been implemented in the past few years at Penn State, under the pseudonym *Continuous Quality Improvement* (CQI). CQI teams boast successes in areas ranging from food services to career counseling projects. Inspired by these successes, the deans of the College of Science and the College of Engineering have sponsored their own experiment into the CQI method by establishing a CQI team to look into "the learning of physics by engineering students." The Team consists of faculty members and a student from both the College of Engineering and the Department of Physics. Although the terminology and some of the CQI methodology have been somewhat altered for this academically-oriented project, the team has made significant progress in its exploration.

Continuous Quality Improvement (CQI) is a process of group problem-solving. Social scientists have outlined CQI as a long list of steps that should be strictly followed. Our team has distilled it down into a much shorter list.:

1. A team should be formed by a sponsor, a manager with the authority and desire to implement changes. This ensures that the person who holding the purse strings is committed to improvement and change before a lot of time is spent figuring out what change should be implemented.

2. The team must focus on a business practice or problem that needs improvement. The team studies all aspects of the activity, including what expectations the "consumers" and the "suppliers" might have. This allows team members, individuals directly involved with the business enterprise, to analyze their practices from the bottom up.

3. The team designs and conducts experiments to assess the value of proposed changes. These experiments must have definite measures of outcome, ensuring that data are taken and used to help guide business practices.

4. Upon achieving successful outcomes, lasting change is implemented. By basing policy changes upon solid data, CQI gives such change legitimacy in the eyes of the workers, a boon to management-personnel relations.

The sponsors at Penn State, the deans of the College of Science and Engineering, carefully chose team members representing many of the differing viewpoints in the Penn State physics and engineering communities. In addition, a TQM facilitator from the president's office joined the group to aid in the process. During the first year, the team requested that we be joined by the director of the Instructional Development Program, as we became aware that we have very little experience with the type of data collection and analysis that we hope to perform. The team members are as follows:

Larry Burton — Department Head, Electrical & Computer Engineering
Joe Conway — Professor, Engineering Science & Mechanics
Mike Dooris — Sr. Planning Analyst, Research & Planning (Facilitator)
Diane Enerson — Director, Instructional Development Program
Reinhard Graetzer — Associate Professor, Physics
Howard Grotch — Department Head, Physics (Team Leader)
Anil Kulkarni — Associate Professor, Mechanical Engineering
Paul Sokol — Associate Professor, Physics
James White — Graduate Student, Physics
Jim Whitmore — Professor, Physics

For the first year after the team was formed, we analyzed how the teaching of physics to engineering students was being conducted We discussed and mapped out the teaching and learning processes. Currently, the introductory course meets each week for two 50 minute lectures and two 50 minute recitations devoted to problem solving. The lectures are given by a professor in a 250-seat auditorium. The recitations, with a maximum of 40 students, are lead by graduate teaching assistants. The data collection employed four major information-gathering techniques. The first entailed surveying nearly everyone even remotely involved with the introductory class or the physics program. Data were collected from students in the introductory class, from senior engineering students who had finished the physics sequence, from students who transferred out of engineering, and from all of the physics and engineering faculty.

In the style of S. Tobias's *They're Not Dumb, They're Different —Stalking the Second Tier*[1] our second data collection technique involved having Mike Dooris, our facilitator, visit the introductory class for a term. He participated in the class as a student and kept a journal of his experiences. The third technique was to analyze data that have already been taken for other reasons at Penn State. For example, the team made a statistical comparison of math placement-exam scores to ultimate physics grades. Lastly, we administered the "Force Concept Inventory Test"[2] to students before and after the introductory class; These data will serve as a baseline for assessing future improvements in conceptual understanding.

The remainder of this paper is devoted to outlining the team's major conclusions from the preliminary data and discussing some areas that the team will focus on initially in its experimentation. Through the student surveys, it was verified that students spend far too little time working on physics outside of class (an average of four hours per week). Forty four percent of the students complete half or less of the assigned homework before attending lectures. Experiments in course administration (grading, homework, quizzes, attendance policies) were conducted over the past term to identify practices that increase the students' time on task. For example, study time for classes where only quizzes and exams were graded were compared to classes where quizzes, exams, and one randomly-chosen problem from each homework set (assigned twice a week) was graded. Very simple, common-sense solutions have had documented success. The department is also attempting to improve students' study habits by increasing the counseling of students and distributing a *Study Strategies Guide* developed by Diana Enerson.

Personal contact between students and instructors needs to be encouraged. Currently, 50% of students are taking advantage of neither office hours nor the Learning Resource Center (a drop-in tutoring room), and only about 10% use either of these resources more than three times a term. This problem is being approached by improving the teaching assistants' preparation and by experimenting with alternative course structures to increase the interaction among students and between students and instructors. Rather than simply working through problems on the board, teaching assistants were encouraged to cultivate group activities by breaking their classes into small working groups, and encouraging their students to help each other complete a specified task.

From the results of the Force Concept Inventory Test, it is clear that we are not doing any better than anyone else in our conceptual mechanics instruction. A major experiment to take place this fall, with the financial support of the Dean of the College of Science, is the addition of a lab to the introductory mechanics class, based on the activities developed in "Workshop Physics." A number of sections will be run concurrently, some integrating the lab and others using the standard two lecture/two recitation per week format without any lab work. Other approaches under consideration for improving students' conceptual understanding include incorporating computerized instructional testing and tutoring software. The software would concentrate on the conceptual traps students fall into, as identified by research into the learning of physics.[3] In conjunction with this, we will be seeking funding to develop the "Learning Resource Center." This center has already been established; however, it still lacks the necessary resources to warrant calling it anything but a tutoring room.

Initial statistical studies of data from the last two years have shown that there is a very good correlation between low scores on a number of the math placement-exam questions (given to all Penn State students prior to their freshman year) and ultimate physics failure. This fall, an experimental, one-credit remedial/refresher course will be offered; entrance will be optional and will be based upon scores on the math placement exam. The main question to be addressed in this study, with its control groups, is: "Can the university bridge the gap between high school and the demanding introductory physics class for students whose study habits and math skill are initially weak? Or is it enough to suggest to students that they just wait to start physics until they get some college class experience, and another university math course?"

The Penn State experiment into this management technique is still in its infancy and, as the name implies, CQI is a continuous process. The CQI method is nothing more than common sense; however, in a large-university setting, common sense does not always prevail over inertia. The formulation of this CQI team at Penn State has allowed individuals who want to improve the undergraduate program to influence more than just their own course. The team has also given the department head some ammunition during these tight economic times for acquiring new funds for the department's teaching program.

References

1. Tobias, S., *They're Not Dumb, They're Different—Stalking the Second Tier*, Research Corporation, Tucson, AR, 1990.
2. Hestenes, D., M. Wells, and G. Swackhamer, "Force Concept Inventory," *TPT* 30 141 (1992).
3. For example: McDermott, L. C., "What we teach and what is learned: Closing the gap," *Am. J. Phys.* 59 (4), 301 (1991), McDermott, L. C., "Research and computer-based instruction: Opportunity for interaction," *Am. J. Phys.* 58 (5), 452(1990), McDermott, L. C., "Research on conceptual understanding in mechanics," *Physics Today* 37, 24 (July 1984).

STRANGENESS AND CHARM IN THE INTRODUCTORY PHYSICS COURSE

(THE ROLE OF CONTEMPORARY PHYSICS IN THE INTRODUCTORY COURSE)

T. P. (Ted) Zaleskiewicz, University of Pittsburgh-Greensburg

Fred Priebe, Palmyra Area School District, PA

Introduction

One of the recurrent themes throughout the Conference on the Introductory Physics Course (CIPC — or the *Resnick* Conference) was:

> "What role should contemporary physics topics play in the content of the introductory general physics course?"

In particular, this question was addressed in the Panel Discussion, "What Modern Physics?" and at the Contributed Poster, "Strangeness and Charm in the Introductory Physics Course". A wide range of opinions on the role of contemporary physics in the introductory course was revealed during these sessions. As members of the Contemporary Physics Education Project (CPEP), the authors have concerned themselves for several years now with this question — and more importantly — appropriate answers to this question.

CPEP at the CIPC

CPEP is a non-profit organization devoted to the improvement of contemporary physics education. CPEP considers its primary target population to be introductory physics instructors and their students. As representatives of CPEP our goals in attending the CIPC were two-fold; to discover changes in teaching approaches and physics content at the introductory level of instruction and to share new teaching materials developed by CPEP with those attending the conference.

The CIPC has provided important input to the process by which CPEP can evaluate its role in modifying both the content and instructional strategies used in the introductory physics courses. The CPEP membership hopes to work hand-in-hand with the CIPC organizers and participants
in order in bring about these changes.

In what follows, we will present our personal experiences in developing teaching materials about contemporary physics and incorporating those materials into our introductory courses. We will focus on the contemporary physics topic; the Standard Model of Fundamental Particles and Interactions.

What is the Standard Model?

The Standard Model is a theory that describes the most elementary or fundamental particles in the universe and the interactions that occur among these particles. The fundamental building blocks of the universe are the six quarks and the six leptons (and their anti-particles). The quarks combine in groups of three to form baryons(protons, neutrons, etc.) or in quark - antiquark pairs to form mesons (pions, kaons, etc). The fundamental interactions (forces) in the universe are described in terms of a set of elementary bosons called exchange particles. In general practice, gravity is not included in the Standard Model theory.

Why Teach the Standard Model in the Introductory Course?

The Standard Model is a product of the '60s and '70s that matured and became widely accepted in the 1980s. It is a revolution that changed our view of the structure of matter — just as the Copernican revolution changed our view of the solar system. Certainly as important as knowing that an atom has a nucleus surrounded by an electron cloud — the quark structure of the universe should be a little piece of the 'philosophy of life' that each of us carries around. What better place to impart this knowledge than in the first physics course that a student encounters?

The Standard Model is now largely accepted by the physics community. Although most agree that it is not the *whole truth*, there is a general consensus that it will play a large role in any *complete picture* of the structure of the universe. Particle physics has outgrown its chaotic image of earlier times and now offers a systematic interpretation of the structure of matter that can be readily presented in the introductory course.

Particle physics is still an evolving field however. Students find excitement in reading (or more realistically watching TV) about the new discoveries that are still quite common. As of this writing, there is an intense search underway for the top quark, there is much exciting new data on the dark matter question and on the origin of the universe. This truly contemporary character of particle physics, carries an important message for the introductory student — "Doing physics is an evolving process — often with social, political, and cultural facets".

Finally, new teaching materials are now available to assist instructors in presenting particle physics to their students and in *achieving a comfort level* with those materials themselves. New editions of most introductory physics texts now contain a chapter (or more) which covers particle physics topics. The Contemporary Physics Education Project (CPEP) has developed the *Standard Model of Fundamental Particles and Interactions* CHART. (see Fig. xx). This Chart has been praised by educators and particle physicists worldwide as an excellent aid to presenting the main features of the Standard Model. CPEP has also developed ancillary materials to support use of the CHART in the classroom (see below). In terms of *instructor support*, CPEP has offered a series of Workshops on "Teaching with the CHART". These Workshops have been presented at the national meetings of the American Association of Physics Teachers and in a variety of settings at the local and national level.

Strategies for Teaching the Standard Model

Instructors contemplating teaching the Standard Model in their introductory course almost uniformly ask, "How can I fit it in?" Our response is to use the strategy endorsed in several sessions at this conference — Less is More — but be sure the 'Less' includes particle physics!

Most introductory physics textbooks have become encyclopedias — the authors never intending for them to be taught cover to cover. Instructors should pick and choose topics to design a coherent and logically structured course. There is no reward doing a poor job of teaching *everything.*

In certain sessions at the CIPC, teaching physics as a *process* was emphasized. Encouraging students to think logically and to develop problem solving skills should be the real goal of the introductory course. To some extent it does not matter what vehicle (i.e., physics topic) is used to teach these skills.

If, as an instructor, you agree with the above reasoning you probably have more latitude in selecting course topics then you at first thought. The authors have discussed this problem of *what to leave out* with a wide cross section of instructors. Their responses suggest omitting or giving greatly reduced coverage to any or all of the following traditional topics; a-c circuits, thermodynamics, statics, geometrical optics, and fluid dynamics as a way to *make room* for contemporary physics topics. The next decision is where (at what point in the course) to teach the Standard Model material. One technique is called *Resnick Sprinkling.* Legend has it that this idea was first suggested publicly by Robert Resnick during the Conference on the Teaching of Modern Physics

(CTMP) held at Fermilab in 1986 (ref. 1). Instead of starting or ending the course with a unit on particle physics — you simply *sprinkle* the concepts throughout the course at the appropriate places.

When you discuss *forces* in dynamics, take a few moments to stress that there are only three fundamental interactions in nature and that (except for gravity) most of the *every-day* forces we experience are really veiled expressions of the electroweak interaction. During the E and M segment of the course, discuss the operation of particle accelerators - - how electric fields increase the energy of a charged particle (e.g. proton or electron) and how magnetic fields focus and steer beams of charged particles. During the *collision theory* part of the course, collide electrons or protons instead of billiard balls. Many of these interludes can be motivated if you have a copy of the *Fundamental Particles and Interactions* wallchart in the lecture room and/or if each student has a notebook size copy of the CHART.

Our own experiences in teaching this material are that students respond favorably to learning about particle physics — independent of the teaching style employed. One of us uses the tried and true lecture - recitation mode in teaching college engineering students while the other makes extensive use of cooperative learning techniques with his high school students. In both situations, students report that they find their study of the particle physics material interesting and rewarding.

A Brief History of the CHART and CPEP

The initial suggestion for a **wallchart** summarizing the properties of the fundamental particles and interactions was presented by Fred Priebe at the CTMP in 1986 (ref. 2). Priebe was motivated by the lack of an *up-to-date* wallchart (similar to the chemists' periodic table) which would present the important basic ideas of particle physics. He was also encouraged by the interest and enthusiasm shown by his students in designing proto-type wallcharts as a class project.

At follow-up meetings in 1987, the ad hoc Fundamental Particles and Interactions Chart Committee (FPICC) came into being and began serious work on a field-test version of the CHART. Membership on the original FPICC included particle physicists and high school and college educators — thus assuring a balance of theory and practice.

This field-test version of the *Fundamental Particles and Interactions* wallchart appeared in the December 1988 issue of *The Physics Teacher* magazine accompanied by a lengthy explanatory article (ref. 3). Hundreds of enthusiastic responses were received from educators and physicists in all 50 states and over 50 countries. Based on this abundant input the current version of the CHART was formulated and released in 1990. In the meantime, the FPICC had broadened its goals and evolved into the non-profit Contemporary Physics Education Project, Inc. (CPEP) — an organization devoted to the improvement of contemporary physics education.

A 'walk-around' the CHART

In this section we will take a 'walk-around' the CHART (see Fig.xx), discussing the physics content as well as some of the pedagogical features of the CHART (For a more detailed discussion of the CHART, please see ref. 3).

The symmetric lay-out of the CHART is a reminder of the importance of symmetry in particle physics; fermions on the left, bosons on the right, and properties of the interactions in the center. The color-coded backgrounds delineate related areas. For example, the electroweak components *weak* and *electromagnetic*, are rendered in *two* shades of yellow to emphasize their closely related nature. Furthermore, whenever the electromagnetic interaction appears on the CHART — central figure, boson table, properties of interactions table, etc. — it is always shown in bright yellow. In a similar manner, three colors are used in the fermion tables to delineate the different generations of quarks and leptons. This color coding is intended to be a pedagogical aid to help students see the over-all unity of the information presented on the CHART.

The *diagram of the structure within an atom* is the center piece of the CHART. It shows not only the nucleus and the surrounding electron cloud but also the quark structure of the nucleons. It should be emphasized that this is a diagram of an atom — not a picture! Students need to be told that the diagram is not to scale, electrons, quarks, etc. are not hard spheres, and that this is a static representation of a very dynamic system.

In the upper left corner of the CHART are the tables of fundamental *fermions* — quarks and leptons. An important point to make here is that quarks carry color charges and therefore experience the fundamental strong force whereas leptons do not. Symmetry is evident here in that there are three generations of leptons and also three generations of quarks (assuming the existence of the *top* quark).

In the upper right corner of the CHART are the tables of fundamental bosons. These bosons are the *force carriers* or *exchange particles* of the corresponding interactions. If you are 'into' *Resnick Sprinkling* you might take a few moments during the E and M section of your course to introduce the photon as the exchange particle of the E and M interaction. This should make is easier for the students to accept the role of the other exchange particles when they are presented.

The symmetry of the CHART is continued with a table of *sample fermionic hadrons* on the lower left and a table of *sample bosonic hadrons* on the lower right. The entries in these tables are clearly **not** elementary particles but are rather combinations of quarks (and/or anti-quarks). Three quarks can combine to form a baryon (fermionic hadron) while a quark-antiquark combination yields a meson (bosonic hadron).

Directly below the *central figure* is the table *properties of the interactions*. (The gravitational interaction is not generally considered to be part of the Standard Model and is included here only for comparison purposes.) The color coding emphasizes that there are only three fundamental interactions; gravitational, electroweak, and strong. An important point here is that the relative strength of the interactions depends on the interaction distance. At the smallest separation shown the weak and electromagnetic interactions are almost equal whereas at greater distances the electromagnetic is much larger than the weak.

Across the bottom of the CHART, three figures represent (from left to right) neutron decay, electron-positron annihilation, and decay of a charmed meson. Green shaded areas represent the gluon field, red lines the quark motion, and black lines the motion of leptons. Just as with the central figure, it should be stressed that these are static representations of dynamic events.

Ancillary Materials

In order to enhance the effectiveness of the CHART as a teaching aid, CPEP has developed an array of support materials designed for both students and instructors. These materials are available commercially through our distributor Science Kit (SK — ref. 4).

Our most recent product is an *Activity Packet* (SK# 71957-40) on the CHART and particle physics which was mailed to 16,000 high schools in the United States. The packet contains classroom activities and teacher 'help' sections. It was designed so that even teachers who are unfamiliar with particle physics topics would be able to *Resnick Sprinkle* their courses with cutting-edge material from contemporary physics.

Another prominent ancillary teaching aid is an interactive software program (SK# 70252-00, color; SK# 71957-02, monochrome) which uses animation and graphics to explain many of the details of the CHART. The program also has interactive segments on other facets of particle physics such as accelerators, detectors, and conservation laws. Although currently available in only Macintosh format, CPEP plans to issue an IBM version in the near future.

In addition, the selection of support materials includes two 'hands-on' activities — the *Mini-Rutherford Laboratory* (SK# Z64712-02) and the *Simulated Particle Detector* (SK# Z64712-01) — which will give your students a taste of the 'believing without seeing' nature of experimental particle physics. Materials for these activities may be purchased from Science Kit or can be prepared from instructions in the *Activity Packet.*

Finally, the CHART itself is rendered in three versions; the original large wallchart (SK# 71957-00) as well as in *poster* (SK# 71957-01) and *notebook* (SK# 71957-30) formats. The poster version has been seen in many faculty offices and has reportedly been used by students as dorm decoration.

Acknowledgments

In the seven (plus) years of its existence CPEP has received encouragement, advice, and financial support from a variety of individuals, institutions, and corporations. In particular, we would like to thank Roland Otto, Director of the Lawrence Berkeley Laboratory Center for Science and Engineering Education for his assistance in the early years of the project.

Financial support has been received from the U.S. Department of Energy, Lawrence Berkeley Laboratory, Stanford Linear Accelerator Center, CERN-Microcosm, IBM, Rockwell International, Martin Marietta Astronautics Group, American Association of Physics Teachers, and Burle Electron Tubes.

In addition, we would like to recognize the contributions of the members of the CPEP corporation and the financial support of their institutions as well as the assistance we have received from so many other members of the particle physics and educational communities.

Final Comment

During the CIPC, much time and many words were spent in trying to answer questions such as;

"What is contemporary (or modern) physics?"
"How much time should I devote to it?"
"What should I omit to cover it?"

A interesting response to questions like these came from a graduate of a three week Modern Physics Institute for high school teachers (ref. 5) offered a few years ago;

> " — by using *Resnick Sprinkling* in my courses everything gets so intertwined and interwoven that keeping track of old physics and modern physics is hopeless — **it's all just physics!**"

References

1. Gordon J. Aubrecht II, ed., *Quarks, Quasars, and Quandaries*; Proceedings of the Conference on the Teaching of Modern Physics (American Association of Physics Teachers, College Park, MD, 1987), p.257.
2. F. Priebe, in papers prepared for the Conference on Teaching of Modern Physics, Gordon J. Aubrecht II, ed., AAPT, College Park (1986).
3. William T. Achor et al., *Phys.Teach.* **26**, 556 (1988)
4. Science Kit, 777 East Park Drive, Tonawanda, NY 14150, Voice: (800) 828-7777, Fax: (716) 874-9572
5. *"Summer Institutes in Modern Physics for Secondary School Physics Teachers",* National Science Foundation TPE-8955227, Principal investigator; G. Samuel Lightner, Westminster College

Editor's Note

A colored "Fundamental Particles" chart accompanied this paper. For a copy of the "Standard Model of Fundamental Particles and Interactions" write: CPEP.MS 50-308, Lawrence Berkeley Laboratory, Berkeley, CA 94720

Part III

Contributed Poster Session II

FIRST YEAR STUDENTS' CONCEPTIONS ABOUT VECTOR KIMEMATICS*

Jose M. Aguirre, Langara College, Canada
Graham Rankin, Keantlen College, Canada

Introduction

Research has been carried out to identify students' preconceptions regarding several vector characteristics [1,2] (Aguirre & Erickson, 1984; Aguirre, 1988). This research was done before the students had been formally instructed in vector kinematics. Data was first collected in individual interviews; then later a questionnaire was used in classroom settings in conjunction with a group interview technique. In the latter, it was found that the preconceptions identified previously in few individuals were common among students of that age group (15-17 years).

This article reports on further work to find out what effect instruction has upon these preconceptions. Instruction, in our case, means the traditional 'expository' method of teaching, in which the instructors introduce the topic (vector kinematics) according to a prescribed program or textbook. The instructors for the data reported in this article did not use any special strategies that took into consideration the students' prior knowledge. In our study, several classes at the college level were asked to respond to a few questions (see Questionnaire) regarding several vector characteristics relating to orthogonal components in kinematics. These first year college students had recently completed a course in mechanics, which included two-dimensional kinematics.

The Characteristics of Orthogonality (Independence) of Vector Components in Kinematics.

In a previous study, Aguirre and Erickson[4] introduced a list of 'Implicit' vector characteristics. These were called 'implicit' because physics instructors usually consider them so obvious that they are not explicitly introduced during instruction, or if they are presented, not enough emphasis is put on them. Four of these vector characteristics are:

(VC-1) Composition of orthogonal velocities
(VC-2) Independence of direction of components
(VC-3) Independence of time of components
(VC-4) Independence of magnitude of components

The above were selected for this study (see Appendix A for brief description of each). We believe that an understanding of these characteristics is crucial in order to comprehend vector kinematics and other physics topics in which other vector quantities are introduced.

Description of Research

The experimental situation or task used in the data collection was a 'frictionless' inclined plane. The apparatus used consisted of an inclined plane (air-table) and a square plastic block that acted as a slow-motion projectile (see Figure 1 on page 2). The projectile was provided with a constant linear (translational) velocity across the

**A version of this paper has been published in Physics Education, Vol.24, 5*

incline by means of a spring-loaded plunger. The spring's kick was always applied along a horizontal line acting at the centre of mass of the block in order to avoid rotation, and this X-motion was demonstrated to the students before the air-table was tilted. An 'arrow' was drawn on the block's face to show the orientation of the block at all times. On the tilted air-table, the block would also be "acted upon" by a component of the force of gravity, and this Y-motion was also shown to the students. Thus the separate X-motion and Y-motion were demonstrated, but the students did not see any motion that involved a combination of these components until after they had answered the pertinent questions.

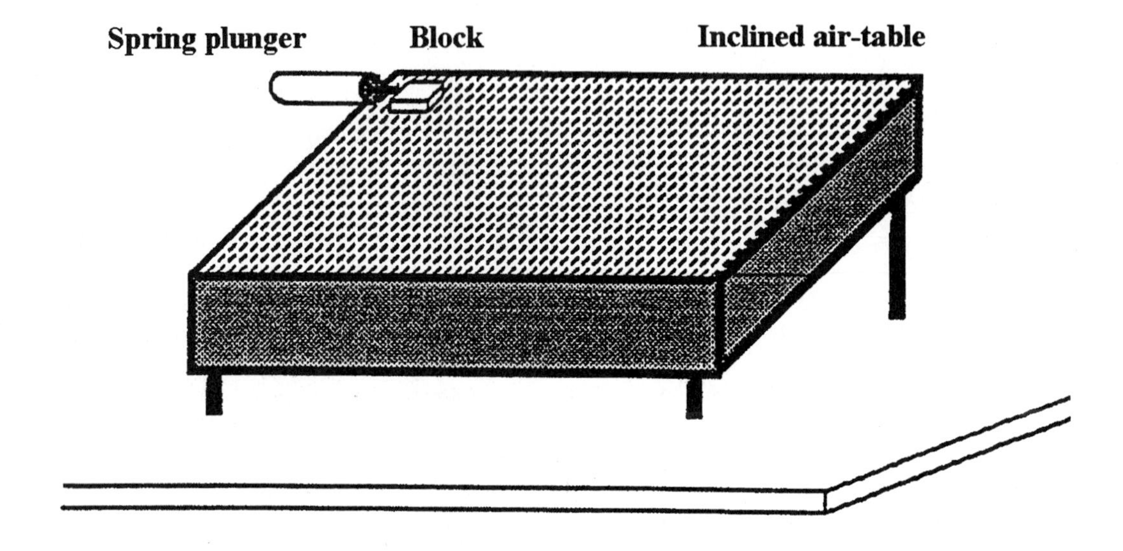

Figure 1. Task: Air-table as Frictionless inclined plane.

Student Sample

The sample consisted of 73 first year students from four different colleges in the Vancouver area (Canada); all students had taken at least one course in mechanics.

Administration of Task and Questionnaire

Each of the four classes was shown the task apparatus, its functioning was demonstrated, and each student was provided with a Questionnaire with the four questions related to the experimental situation (see Figure 1). During the questionnaire administration each question was read aloud and explained with reference to the apparatus. The alternative choices for questions 3 and 4 of the Questionnaire were selected from the responses given by pre-instruction students to similar questions. The sequence of events for the air-table task and questions follows; the actual questions are included in the Questionnaire.

Questionnaire About Motion

1. Draw the *path* that you think the block would follow in its motion on the inclined air-table:

2. On the path you have drawn above, select three well-spaced points along the path; at each point draw the block and the arrow on it to depict its orientation (that is, showing the way the block is pointing).

3. If the block takes 1 second to cross the incline in the horizontal "X" direction (if this were the only movement influencing its motion); and it also takes 1 second to cross the incline in the vertical "Y" direction (if this were the only movement influencing its motion), then how long would it take to go along the path you have just observed when both motions are simultaneously involved? Will it be:

   A) equal to 1 second

B) less than 1 second
C) greater than 1 second

4. If the average vertical speed (in the "Y" direction) is increased (by increasing the slope of the inclined plane) from 4 m/s to 5 m/s, then what will be the horizontal speed (in the "X" direction) of the block if a spring's kick similar to the first one is applied on it? Will it be:

   A) equal to 6 m/s
   B) less than 6 m/s
   C) greater than 6 m/s

Question 1: (Probing the students' conceptions on the composition of orthogonal velocities.) First, the air-table was positioned horizontally with the block in contact with the loaded spring, which was released while the air-table fan was functioning. Students observed the block moving straight across the table with roughly constant speed. Next, the air-table was positioned as an incline with the fan 'on', and the block was 'dropped' from the top edge. Students saw the block gaining speed down the incline. Finally, the block (arrow pointing in the horizontal direction) was placed in the upper lefthand corner of the incline (see Figure 1) in contact with the spring which was ready to be released, and the air-table fan was operating. Students were asked to think and silently predict the path that the block would follow after releasing the spring, which will act at the block's centre of mass. The fact that the 'spring's kick' acted at the centre of mass of the block was clearly indicated to the students. Students then answered question 1.

Question 2: (probing the students' conceptions on the independence of direction of the components). Students were asked to read and answer this question. After responding Question 2, students were shown the experimental situation as explained in the final section of the introduction for Question 1. The block was positioned in the upper lefthand of the incline and a kick was given by the spring. The effect was that the block followed a parabolic path leaving the incline at the lower righthand corner (as in Figure 2A).

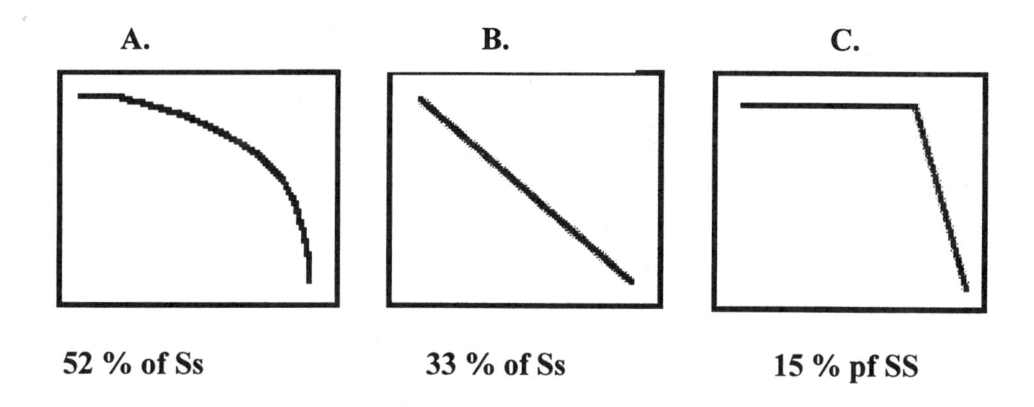

Figure 2. These curves represent the paths drawn by the students as responses for Question 1. Above each curve is the percentage of students (Ss) who drew them.

Question 3: (probing the students' conceptions on the independence of time for the component motions). Students were told that it would take the block 1 second to cross the air table if the table were in a horizontal position (pure X-motion or no effect of gravity), and it would also take 1 second to move down the incline without the action of the spring's kick (pure Y-motion). Then, students were asked to read and answer this question.

Question 4: (probing the students' conceptions on the independence of the magnitudes of the components). Students were told that the speed of the block produced by the spring's kick was 6 m/s if the air-table were horizontal (they were also reminded that that speed was roughly constant on the frictionless air-table) and that the average vertical speed down the incline (the inclination used for questions 1 to 3) due only to gravity, without the spring's kick, was 4 m/s (or that the final speed at the of bottom of the incline was 8 m/s); then, the students were asked to read and answer Question 4. As a caution note, the reader must realize that the values of the speeds given in this question were deliberately larger than the ones that could be actually calculated considering the air-table dimensions and the magnitude of the spring's kick.

Data Analysis and Discussion

Students' responses are summarized in Table 1. The results reveal that a significant percentage of students had not changed their preconceptions about vectorial kinematics despite instruction. The table shows that only about half of the students held views consistent with the physicists' approach (Conception A in Table 1) when dealing with orthogonal components. A discussion of the results for each question is presented below.

TABLE 1

Percentage of student responses for the different conceptions relating to the four vector characteristics as embedded in the four questions of the Questionnaire.

	VECTOR CHARACTERISTICS			
STUDENTS' CONCEPTIONS[1]	VC-1 (Q. 1)	VC-2 (Qt. 2)	VC-3 (Q. 3)	VC-4 (Q..4)
A	52%	34%	52%	52%
B	33%	48%	7%	32%
C	15%	6%	40%	11%
U[2]	0%	12%	1%	5%

n = 73

1. For conceptions see Figures 2 and 3, and Questionnaire.
2. Uncommitted students

Question 1: About half of the students (52%) predicted that the two velocities influencing the motion of the block will combine producing a kind of parabolic path (see Figure 2A; Conception A in Table 1). One third also predicted a combination of the two velocities but resulting in a straight line diagonal path (see Figure 2B; Conception B in Table 1). And 15 percent predicted a two-step sequential path (see Figure 2C; Conception C in Table 1). Apparently, these students thought that the velocity caused by the spring has to wear off before the velocity due to gravity takes over; that is, they have a non-combinatory view of the composition of velocities.

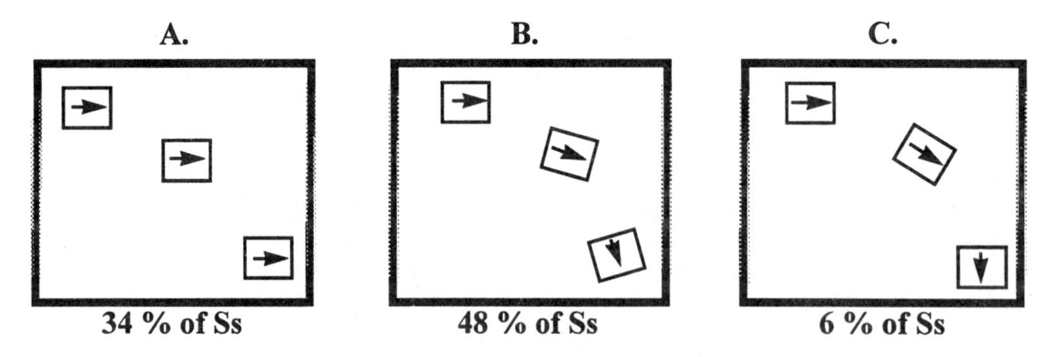

Figure 3. These figures drawn by the students represent the three consecutive positions of the block moving along the include and depicting its orientation. Above are the presentages of students (Ss) who drew them.

Question 2: Almost half of the students (48%) drew the block with its orientation — depicted by the arrow drawn on it — always tangential to the path at any point (see Figure 3B, Conception B in Table 1). This response was interpreted as either a confusion or the maintenance of a 'misconception'. Perhaps they confused the direction of the instantaneous resultant velocity with the direction of the horizontal component. It was assumed that students would use the arrow drawn on the block's face as a 'hint' to visualize the-orientation of the moving block, which at the start of the motion (and during the whole motion) in fact represents the direction of the constant horizontal component of the resultant velocity. Based on both the failure of using this 'hint' and on the earlier individual interview data,[5] in which students did use the orientation of the block to show the direction of the horizontal velocity, we can assume that these students did not really grasp the concept of independence of direction when composing simultaneous orthogonal velocities and maintained their intuitive view. One third of the students showed that the initial (horizontal) orientation of the block remains unchanged while it moves along the parabolic path (Figure 3A; Conception A in Table 1). Showing with this, apparently, their understanding of the independence of the directions of orthogonal components. A small percentage of students (6%) held the belief that the orientation of the block would change gradually from a horizontal to a vertical heading (Figure 3C; Conception C in Table 1). A significant percentage of students (12%) were uncommitted on this question.

Question 3: A high percentage of students (40%) selected alternative C in the questionnaire; this was interpreted to mean that they were maintaining the preconception that the elapsed time for the resultant motion is greater than either of the times for the separate orthogonal components. Most of these students used a "Pythagoras' Theorem" on time in searching for a numerical solution. Only half of the students (52%) have correctly grasped the time-independence characteristic (alternative A) when composing orthogonal component velocities . This is not very promising considering that this concept is crucial to comprehend composition of velocities. A few students (7%) also predicted that the time for the resultant motion would be less than either of the times for the separate orthogonal components (alternative B).

Question 4: Students were cautioned, before responding this question, that they had to think to what would happen to the horizontal speed (which is a result of the spring's kick) after increasing the slope of the incline. However, despite instruction and the cautionary remark only half of the students (52%) predicted that the horizontal component would not be affected by the increase in the vertical component (Alternative A). A large group of students (32%) predicted that the horizontal velocity would decrease due to the increase in the vertical component (Alternative B). The high percentage holding this conception was surprising considering that students saw and were told that the incline was a frictionless surface; it may be that they still considered some friction causing a decrease in the horizontal speed. A few students (11%) predicted that the horizontal velocity of the projectile would increase if the slope of the incline (air-table) were increased (Alternative C). A possible interpretation of this result is that these students might have confused the horizontal component with the resultant velocity. There were also a few uncommitted students (5%). Whatever the interpretation used to account for students selecting alternatives B and C, it is clear that these students have not grasped yet the concept of the independence of the magnitudes of the orthogonal components.

As a concluding remark for this discussion of results, it should be mentioned that while analyzing the data a comparison between the conceptions held by college students and those preconceptions identified for high school students was being conducted. As has already been shown some similarities were found, which were partially used in constructing Table 1.

Other studies have been done on other physics topics about the influence of instruction upon preconceptions. Clement[6] found that many physics students have a stable alternative view of the relationship between force and acceleration: motion implies a force. Other examples, among many others, of similar studies have been done by Viennot,[7] McCloskey,[8] Leboutet-Barrel,[9] Cohen et all,[10] Peters,[11] Halloun et al.[12] In most of these studies it was found that a relatively high percentage of college students maintain their intuitive conceptions despite instruction. Interpretations of these results were that in some cases students failed to understand the formal explanations because they appear to be counterintuitive; in other cases students could hold both the formal conception and their own. They use the former for school work and the latter for practical out-of-school situations.

Conclusion and Implications for Instruction

The purpose of this research was to study the first year college students' conceptions about some basic characteristics of vector kinematics; these students had already taken at least one course in mechanics. The analysis of the data (as summarized in Table 1) has shown that only about 50 percent of the student sample had grasped the formal vectorial treatment of composition of orthogonal component velocities. The other 50% of the students either maintained their preconceptions or were uncommitted, but Table 1 shows that the majority of this 50 percent apparently kept their intuitive beliefs. These results should be a serious concern for physics instructors; and efforts should be made to enhance the students' understanding of vector kinematics. Part of the problem could be that instructors either do not deal explicitly with some of the independence characteristics, or do not place major emphasis on them during instruction. If both an explicit discussion of the mplicIt independence characteristics and the consideration of the students' conceptions were implemented, it may result in a better comprehension of vector kinematics in the part of college students.

Appendix A

The implicit vector characteristics
(expressed in a conceptual mode)

A brief description of each of the four implicit vector characteristics of the study are presented below. These characteristics were defined with respect to high school pupils for translational movements occurring in a uniform gravitational field near the earth surface.

(1) Composition of orthogonal velocities: If two orthogonal velocity components (v_x and v_y) are simultaneously involved in the motion of a body, they will combine to produce a resultant motion with magnitude and direction different from those of the individual components.
(2) Independence of direction of components: If two orthogonal velocity components (v_x and v_y) are simultaneously involved in the motion of a body, the direction of the instantaneous velocity of each component remains constant.
(3) Independence of times for component motions: If two orthogonal velocity components (v_x and v_y) are simultaneously involved in the motion of a body, then the time taken for the resultant path is equal to the time taken for each of the 'x' and 'y' motions considered separately.
(4) Independence of magnitudes of components: If two orthogonal velocity components (v_x and v_y) are simultaneously involved in the motion of a body, then the magnitude of the instantaneous velocity associated with one of component (e.g., v_x) at a given instant is unaffected by the other component (v_y), even when v_y is changed in some way.

References

1. J. Aguirre and G. Erickson, *J. Res. Sci. Tea.* 21, 439 (1984)
2. J. Aguirre, *Phy. Tea.* 26, 212 (1988)
3. Reference 1.
4. Reference 1.
5. Reference 2
6. J. Clement, *Am. J. Phys.* 50, 439 (1982)
7. L. Viennot, *Eur. J. Sci. Educ.* 1, 205 (1979)
8. M. McCloskey, *Sci. Am.* 234, 122 (1983)
9. L. Leboutet-Barrel, *Phy. Educ.* 11, 462 (1976)
10. R. Cohen, B. Eylon, & U. Ganiel, *Am. J. Phys.* 51, 407 (1983)
11. P. Peters, *Am. J. Phys.* 50, 501 (1982)
12. I. Halloun & D. Hestenes, *Am. J. Phys.* 53, 1056 (1985)

ASSESSMENT OF UNIVERSITY PHYSICS IN PUERTO RICO

José L. Alonso, Moisés Orengo, Antonio Martinez, J. C. Cersosimo
University of Puerto Rico
& José Peñalbert, University of Turabo

Abstract

Introductory Physics in our universities is being evaluated in an effort to determine the causes of low levels of science literacy in our society as well as the dwindling numbers of students majoring in Physics. Our study was based on the administration of a pre and post course test using Hestenes Force Concept Inventory. The results clearly evidenced the lack of effectiveness of the course in transmitting and constructing fundamental concepts. This is consistent with the present situation of Introductory Physics in the United States. Our systematic assessment of the Introductory Physics includes: course, faculty, and student profiles. In addition, the teaching methodology is probed and the students (majors and non-majors) aptitude and attitude towards Physics is assess using questionnaires and polls. This study is the first phase of a curricular reform aimed to increase the number of students majoring in science and engineering. We expect that this assessment will provide a working tool for Physics professors to implement innovative and more effective teaching methods.

Introduction

A new awareness of the necessity of assessing teaching practices in higher education is sweeping the United States today. In particular, introductory courses in science and mathematics are being evaluated in an effort to determine the causes of low levels of science literacy in our society. Recent results indicate that the state of science and math education in Puerto Rico, gauged by the large attrition rates in gate keeper courses and the shrinking numbers of students majoring in these fields, is similar to the national trends.

As a first step to improve science teaching in the island, five[1] institutions have joined efforts to assess the teaching and learning of the University Physics course. The project is sponsored by the Puerto Rico Alliance for Minority Participation (PR-AMP).

Our scheme includes assessment of the teaching methodologies and teaching resources, faculty profiles, student profiles, and student learning. The complete sample includes approximately 1,300 students enrolled and 40 professors teaching the calculus-based Introductory Physics course. We have administered questionnaires to a considerable fraction of the human resources related to the course (students, faculty, TA's). We also administered pre and post tests based on David Hestenes 'Force Concept Inventory'. In section II we present Hestenes test results. The recommendations for course improvement are presented in sections III. In section IV the results of the course content survey are presented. In section V we discuss the future work.

Conceptual Test

A conceptual test, based on David Hestenes Force Concept Inventory was administered at the beginning (Pre-Test) and at the end of the semester (Post-Test) to college students registered on the first part of the course. In order to decrease the time constraint of administering the test during lecture hours, only 15 questions were selected from the Inventory. Nearly 1,200 students took the Pre-test (75 % of the study sample), while 520 students took the Post-test (32%). The mean score on the Pre and Post tests were 27 % and 35 %, respectively. Note that the expected score from a random selection of the multiple choice was 20 %. A typical test results for one institution is presented in Figure 1 (below). The mean scores on the pre and post tests for each institution are given in Figure 2 (below). Inspection of Figures 1 and 2 clearly indicate that the traditional teaching methodology used in the introductory physics course is not effective in developing thinking skills and concept development in the students. A result addressed several times in this conference.

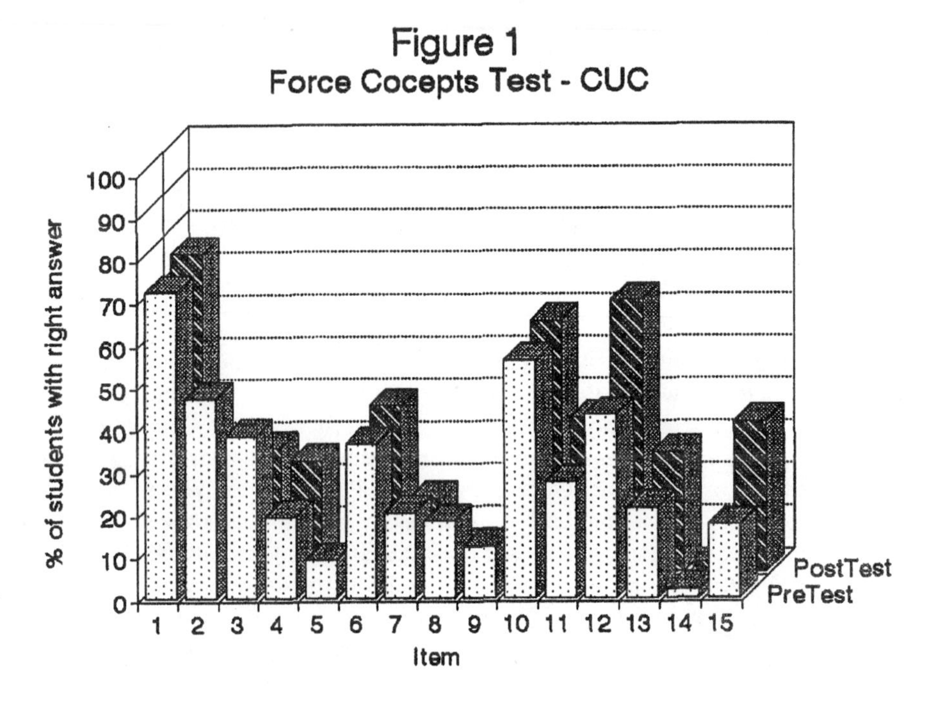

The Pre-Test was also administered to more than 100 High School students and to 17 High School physics teachers. The mean score for high school students was 23 %, while that for high school teachers was 38 %. The percentage of students that took high school physics for each institution is presented in Figure 3 (page 3).

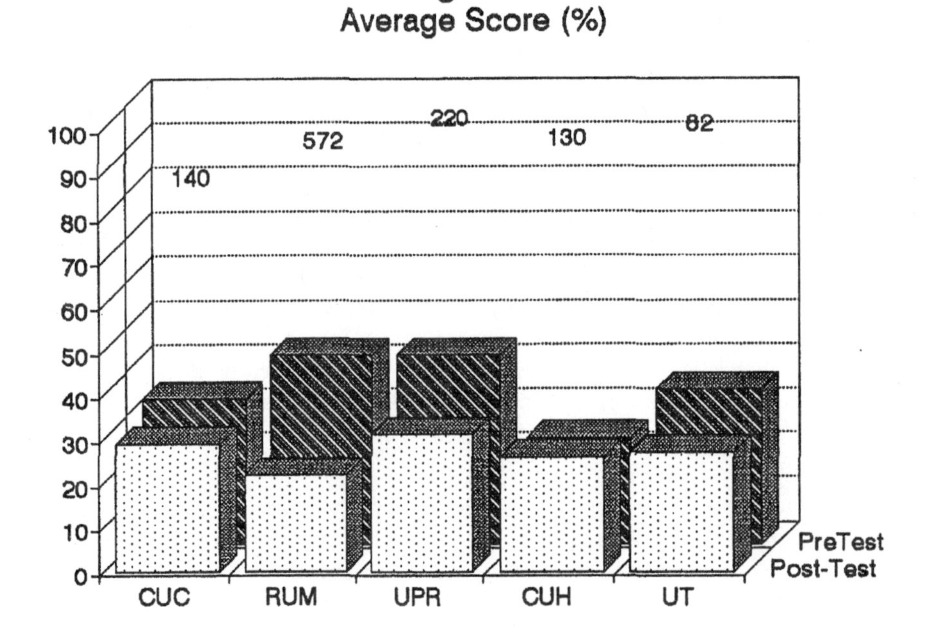

Figure 3
% Students with physics in High School

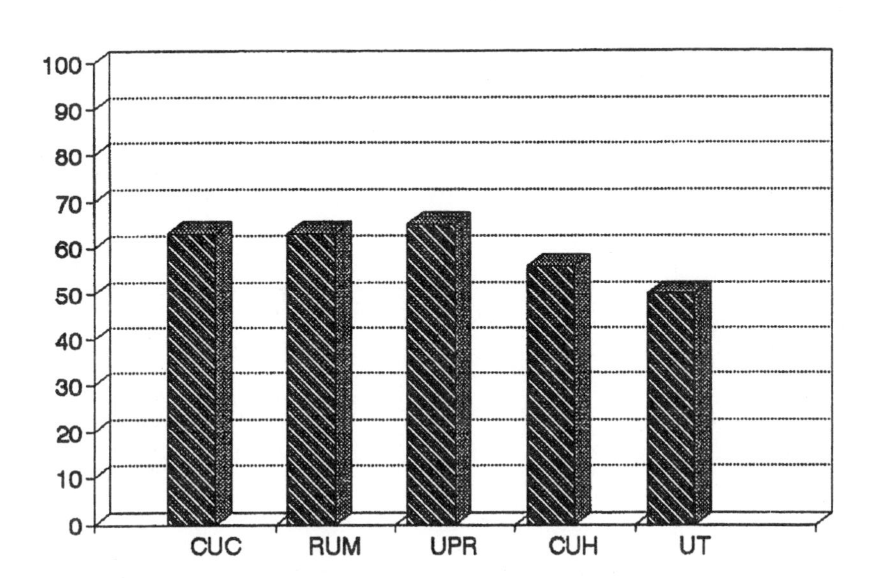

Recommendations for Course Improvement

In this section we present our recommendations for course improvement, based on our assessment study.

1. Lecture and laboratory periods should be well coordinated, preferably integrated.

2. Lecture should include more Hands-On demonstrations. An inventory of available demonstration and audio-visual materials (per topic) must be created.

3. Student participation in class must be increased by problem solving and Hands-On activities.

4. Implement the use of computers to simulate physical situations, acquire data, analyze data, present results and reports.

5. Recipe type laboratories should be substituted by inquiry oriented experiments. Train laboratory instructors in the use of teaching strategies.

6. Use technological tools like video to provide a standardized content.

7. Implement collaborative learning.

8. Produce laboratory manuals (in Spanish) emphasizing error propagation, significant figures, graphical representations and adapted to the equipment use in the lab.

Course Content

In a content survey among Physics, Chemistry, Biology and Engineering faculty, although there is no clear consensus on the minimum course core content, our preliminary findings indicate that the following topics should be either de-emphasized or eliminated.

- AC CIRCUITS
- RIGID BODIES
- KINETIC THEORY
- THERMODYNAMICS
- MAGNETISM IN MATTER

Future Work

1. We plan to expand our study to include the assessment of the second part of the course. In particular, we are interested in the design and administration of an instrument to establish student misconceptions in electricity and magnetism.

2. We will assess the effects of the implementation of the recommendations suggested by this work in the acquisition of concepts and in the development of effective problem solving skills.

References

1. Cayey University College-CUC, University of Puerto Rico (Mayaguez Campus)-RUM, University of Puerto Rico (Río Piedras Campus)-UPR, University of Turabo-UT. Humacao University College-CUH

What Modern Physics? An Instant Survey

Gordon J. Aubrecht, II, Department of Physics, Ohio State University
Charles H. Holbrow, Department of Physics, Colgate University
John S. Rigden, American Institute of Physics

Introduction

What modern physics should be taught in introductory physics courses? How much of it and when? What should it replace in the present introductory curriculum?

These are not simple questions. Professors of physics disagree about the answers; they even disagree about the questions. What does "Modern Physics" mean, they ask? Is it the physics of electrons, x rays, and radioactivity-phenomena discovered nearly a century ago? Is it Einstein's special relativity, now nearly ninety years old? Is it the quantum ideas of wave mechanics now approaching retirement age? What about the neutron, fission, fusion, lasers, general relativity, quantum field theories? Is the standard model, with its quarks, gluons, and quantum chromodynamics, "modern"?

Three Questions

Can physicists extract agreement about "modern physics" from the various interpretations of and answers to these questions? We tried to find out by posing them directly to our audience. We wanted them to get individually involved with the questions and have their ideas stirred up. We split the audience into six groups of about fifty people each, and gave each group a marker and a transparency with one of three different questions on it. Each group was told to organize itself quickly and generate answers to its question. In this way each question was to be answered by two different groups.

The questions were:

1. What is the most important modern physics now in the introductory course?
2. What is the most important modern physics that should be added to the introductory course?
3. What should be taken out of the introductory course?

We hoped that the responses to the first and second questions would provide a working definition of what "modern" meant as well as show the variety of topics that are covered by this umbrella word. We thought the answers to the first question would show to what extent "modern" physics was already in the introductory course, while the answer to the second question would indicate what pressure there was to add more. The third question was put in as a reality check and to find out what in the traditional curriculum people were willing to trade off for more "modern" physics.

Many Answers

After about ten minutes of dynamic and chaotic interaction, the groups handed in their transparencies, and we put them on two overhead projectors and discussed them. The contents of these transparencies, edited and reorganized, are given in Tables 1 to 3.

TABLE 1: What's In?

Large Idea	*Specific Topic*
Atomic structure of matter Idea of quantization	Bohr atom Davisson-Germer experiment Photoelectric effect experiment Bohr Model
Idea that there are limitations to classical laws of physics	
Wave-particle duality; the Uncertainty Principle	
Special Relativity	Maxwell's equations as examples of Lorentz invariance
Miscellany	Standard model Superconductivity
NO AGREEMENT	One group rejected the question on the grounds that they had no modern physics in their courses and did not want to have any; or they had some of these topics, but did not agree that they were "modern"

Divination

What can we reasonably infer from the responses to these questions? Informed by the vigorous discussion that accompanied the display of the transparencies, we can make four points. First, for many physicists, "modern" means non-Newtonian physics, consisting chiefly of quantum mechanical phenomena and ideas, with an occasional mention of Einstein's special theory of relativity. Some argued that our goal should be "contemporary" physics which is whatever contemporary physicists do, and may include use of contemporary tools such as the computer. Second, there is substantial resistance to including modern physics in the introductory course,

TABLE 2: What Should Be Added

Big Topic	*Specific Topic*
Probability & Uncertainty	Wavefunction Blackbody radiation Double slit experiment
Atomic View of Nature	Bohr Atom
Energy Quantization	Nuclear radiation Photoelectric effect Stern-Gehrlach experiment
Special Relativity	
Chaos	Newtonian motion is only predictable over a short time
Miscellany	High temperature superconductivity Blackbody radiation Quantum Coupled states Standard Model Feynman diagrams as a model of interactions
Other Answers	The answer depends on your story line The question makes no sense

expressed in a few cases with strong feeling. Some argued that there is much that is modern in the "classical" approach. Third, there is notable agreement about what to take out of the curriculum whether it is to make room for modern physics or to practice "less is more." Fourth, it is important to have a clear story line and it will determine to a major extent the topics and ideas a teacher uses in the introductory course.

TABLE 3: Take-Out Menu

Responses of Group 1	*Responses of Group 2*
Fluids	Fluid Dynamics Thermodynamics
Rigid Body Motion	Rotational Motion Moments of Inertia
Macrophysics	Geometrical Optics Special Relativity AC Circuits Magnetic Materials Friction

"Modern" is . . .

"Modern physics" loosely defines the physics discovered between 1895 and 1940, from x rays to nuclear fission. It also includes the ideas of special relativity and quantum mechanics used to interpret those discoveries. Because it omits the last fifty years of physics, many find the term a poor guide to what material should go into introductory physics.

To include the physics done since the 1930s, we might speak of "contemporary physics." This is not very satisfactory either, because, on the face of it, contemporary physics ought to be the physics going on right now in physics laboratories and offices. The June 14th issue of Physical Review Letters shows that this includes topics such as black holes, superstrings, braid theory, the mass of the tau neutrino, Auger electron spectroscopy, heavy-ion collisions, why large grains come to the top when you shake a can full of powder, plasma waves, defects in crystals, Fermi surfaces, plasmon instabilities, and femtosecond energy relaxation in polymers.

What about the rubric "twentieth century physics"? To ±5%, it encompasses everything, phenomena and ideas, since the discovery of x-rays. Its very comprehensiveness, however, shows that our problem is not what to call "new physics". Our problem is rather to decide what of the vast amount of new physics to teach to beginning students of physics. Can we show them the forest and not lose them among the trees? And just what is this forest we're showing them, anyway?

The Post-Newtonian Era

Twentieth century physics is post-Newtonian. Its greatest achievements build on or displace, Newtonian physics. Newtonian space and time have been replaced by four-dimensional space-time; the concept of force goes unused in quantum mechanics; the law of universal gravitation has become the geometry of curved space-time; action at a distance has been replaced by an electroweak field and a chromodynamic field, each with its characteristic quanta.

Even in domains where Newtonian approximations hold, the style and spirit of present day physics are in conflict with Newtonian traditions. Electronic computation is diminishing the importance of the calculus, Newton's creation, and mathematicians are making big changes in the calculus curriculum. The study of chaos has reminded us that Newton's system of the world is not a reliable clockwork.

Here are the elements of a serious dilemma. For over three hundred years, an important goal of introductory physics has been to make Newtonians out of Aristotelians. The textbooks, the iconography, the beliefs and practices of physics teachers all have been directed to this goal. Now a century of physics has eroded the foundation ideas of this course. What should be done about it?

We could do nothing at all. After all, Newtonian physics works well in the macroscopic world of our everyday experience. It is a reliable guide to building bridges, factories, machines, and much of the apparatus that sustains our civilization. For most people that may be enough; it may, indeed, be all that we can ask of men and women taking one university physics course. A person who understands the basic Newtonian ideas is certainly technologically literate in some important respects. And how
much technological literacy can we reasonably expect students to acquire from a single physics course?

We could embellish Newtonian physics with a selection of interesting phenomena and technology such as those listed in the second columns of Tables 1 and 2. We might teach Newtonian physics and issue warnings and examples of realms where it won't do. There are pitfalls of confusion in this approach. Students, especially students taking physics, want to know what "the truth" is. After they have worked hard to understand subtle arguments and concepts, they become restless and resentful when you tell them these arguments and concepts have serious limits on their validity. We could argue that it is good to remind students that even in the exact sciences truth is contingent, but it is clear that such an
argument is self-serving to the extent we use it to escape explaining what physicists really believe is true.

Some might argue that introductory physics should carry on the tradition of physics as natural philosophy, as a model of the world. The ideas of physics are a remarkable intellectual treasure of all of humanity. They are the structure of our understanding of the material universe. The ideas, several of which are listed in the first columns of Tables 1 and 2, are interesting and richly complex and interconnected. Shouldn't students, within the limits of practical pedagogy, be introduced to the most important of these ideas even if it risks downplaying Newtonian physics?

These three models characterize, some might say caricature-ize, a range of possible introductory physics courses. Our instant survey showed that there were approximately as many participants who had used no modern physics in a two-semester course (or equivalent) as those who had taught at least one of the topics listed. Although there a were a few eloquent advocates of something toward the natural philosophy end of the range, most of the conference participants were teaching a traditional Newtonian course supplemented with more or less non-Newtonian material. Of these, perhaps one-third aspired to include more non-Newtonian material in their courses. The discomfort and concern occasioned by the discussion about content may arise from the conflict and inconsistency that ensues when non-Newtonian physics is included in a course with a story line that is chiefly Newtonian.

What Comes Out

It is worth noting what teachers of introductory physics are willing to sacrifice from the present curriculum. There are two points and an irony to be noted here.

First, the topics listed in Table 3 are already severely diminished or gone from many introductory physics courses. Geometrical optics and AC circuits are relegated to laboratory exercises in many curricula. Friction shows up less and less. Rolling rigid bodies are in decline. Thermodynamics and fluid mechanics have been deleted from many physics courses. For many teachers of introductory physics this list does not create new room for new topics or approaches.

Second, it is remarkable that both groups chose fluid physics, classical thermodynamics (but not statistical physics), and rigid body motion for omission from introductory physics. This reflects the intense pressure to reduce the amount of material in the introductory physics course, but it also exhibits the dominance of single-particle physics in the introductory curriculum, a dominance that is itself "modern".

This is ironic. At least half of the topics noted in the Physical Review Letters that we have taken as representing contemporary physics involve fluid dynamics or thermodynamics or both. Two major conceptual structures in heavy current use are well on their way to being dropped from introductory physics.

Story Lines

Taking note of this irony does not mean we should all rush to put fluid mechanics and thermodynamics back in the introductory curriculum. Rather it reminds us that what we put into the curriculum depends upon the story line. And here there was considerable agreement: an introductory physics course must have a story line. Many attendees agreed that a strong central theme was crucially important to the success of the course.

The issue then is that of what story line to choose. The story line will determine where any individual course will be within the range of models as described above. For example, one proposed story line emphasized experimental evidence for quantization in Nature as distinct from continuum behavior. Quantization such as is seen in the Stern-Gerlach experiment is presented as a major distinction between classical and post-Newtonian physics. The centrality of experiment in physics is stressed by examining several crucial experiments.

Story lines that would connect physics to students' active experience in their immediate world were strongly urged by a number of conference participants. They expressed concern that abstractions such as Feynman diagrams were too remote from student experience to be usefully taught in an introductory course. To which was made the reply that abstraction of the real world from direct knowledge of events is an essential part of physics. Physicists do not believe only in what they can touch and experience directly; they use abstractions to model the world.

Conclusions

Reading the auguries of our instant survey, we may safely prophesy that the introductory physics course is changing. The sheer volume of new physics forces some change. The transformation of our society from mechanical to electronic, from manufacturing to information processing, forces more change. Change will accelerate as more and more contemporary physicists, comfortable with post-Newtonian physics, become teachers of the introductory physics courses. The most modern physics that we bring to our classes is the modern viewpoint on physics of the contemporary physics teacher.

These changes will gradually reflect the major shift in physics away from the Newtonian paradigm. Any time the paradigm changes there is bound to be serious intellectual discomfort. The diversity and contention of views we heard from participants to some extent reflects that discomfort. But they also reflected a widespread commitment and eagerness to develop and adapt introductory physics to be responsive to these changes. Like the Red Queen and Alice in Through the Looking Glass, all physics teachers must run very fast to keep in the same place. The remarkable presentations and discussions of this conference show, that like the Red Queen, many physics teachers are willing to run twice as fast as that in order to get somewhere else.

USING SPREADSHEETS IN THE UNIVERSITY PHYSICS COURSE*

Gordon J. Aubrecht, II, Department of Physics
Ohio State University

Abstract

I am using spreadsheets as self-tutoring modules in a new version of the engineering physics course being developed at The Ohio State University. The use of such spreadsheets arises naturally out of the purpose of the course. I demonstrate the advantages of employing spreadsheets, and explain the methods I am currently using to integrate the spreadsheets into the course.

Philosophy

The real world is not organized as a physics problem is. Often one knows much more than is relevant to solving a problem. Thus it is important for students to develop intuition. Students will have to learn how to winnow what's necessary from what's given. Many other times there is not sufficient information, and one must learn how to find it, or at least how to describe the world as it is. Therefore, one should pose problems that give students too little or too much information to be able to analyze them, so that they must determine what sorts of information are missing and propose methods for obtaining them.

Mathematical skills of future scientists and engineers will be different from those of a generation ago. Computers and calculators make rote or manipulative numerical skills obsolete. We need to be teaching our scientists and engineers of the future how to use the tools available. Thus, the course should also address numerical skills as well as the traditional arithmetic and algebraic skills.

Course problems should also emphasize the transferable skills we wish to develop. The student should be challenged to determine whether sufficient information is given to solve a given problem. The use of exact solutions to approximate problems as well as approximations to exact problems will be encouraged. Students should be encouraged to estimate answers as well as calculate them. Modeling skills must be tested in the problems. Graphing and functional relationships enter into all science, and so there must be appropriate problems designed to elicit the extent of student knowledge. All these considerations show that the current problems in texts are inadequate and the emphasis too one-sided. I have been thinking about the kinds of problems one should develop for some time.[1,2]

Scientists and engineers spend years to develop "physical intuition" to learn what variables are important. It is a difficult skill to develop, but the earlier we begin introducing students to the necessity of using good judgment, the better for their learning. Student skills can be built, but only if we give the students homework problems and examination questions that allow them to develop the necessary experience. The things we say as teachers are sometimes belied by the sorts of problem we choose to ask students to solve. If we wish to teach specific things, we have to ask questions that use those specific skills and knowledge.[2]

An emphasis on transferable skills, those useful to the student in many areas of physics as well as elsewhere in school and after graduation, could minimize the differences. Examples of such skills include graphing and

*Work supported in part by the U.S. Department of Education under FIPSE grant #P116B01237.

recognition of functional relationships, dealing with error and uncertainty, and use of scaling and symmetry arguments. The University Physics teacher assumes that students will be able to pick up these ideas on their own, whereas experience has shown that these skills must be taught explicitly.

How Spreadsheets Can Help

Much difficulty is caused by the varied mathematical sophistication of the University Physics students. Most take calculus concurrently, and some students struggle for understanding, while some students have already had calculus. Spreadsheets minimize such initial differences among students. Our spreadsheets on Macintosh computers are being used to introduce the students to problems that will help them gain insight into numerical estimation techniques and to modern ways of tackling problems. The computer spreadsheet we have chosen to use with the course (Excel) is supported on both Macintosh and MS/DOS platforms and so can be used in identical forms on the major microcomputers that are accessible to university students. Other computer tools such as M.U.P.P.E.T., CUPLE, Interactive Physics, and Physics Explorer can supplement the spreadsheets.[3,4,5]

At the beginning the book *Models of Reality* we use spreadsheets to help students learn about vectors. We have students change angles and lengths. These spreadsheets are made available to the students, and they already work without extra effort on the part of students. As later chapters treat different topics amenable to spreadsheet use, we continue to supply them with working spreadsheets. Gradually, we intend them to be able to create their own spreadsheets, so we give them practice in filling cells in, etc.

The spreadsheets help develop the idea that models exist, mathematical or pictoral models, that can allow visualization to succeed in the students' minds. The use of different parameters in the entries help students see the outcomes within one model. This can lead them to an understanding of the role of the model in varying descriptions. For example, a two-dimensional model can also show one-dimensional outcomes for certain choices of parameters. Students may change the value in a cell and find that there is no change in the result; this way they can learn about the value of determining whether the outcome numbers depend on *all* the inputs. Extraneous information can be found in spreadsheets, too.

Course Spreadsheets

We have developed and utilized several different categories of spreadsheets. These are called by different names (in order of increasing demands upon the student):

i) *Data*: These spreadsheets contain data gathered from MBL sensors and put into spreadsheet form so it may be manipulated (by students).
ii) *Examples*: These are spreadsheets constructed as part of a worked example in the book.
iii) *Text*: These are spreadsheets referred to in the text or shown in figures. Students are allowed to see the "guts" of the operations performed (or claimed to be performed). These are often less visual than Examples or Models.
iv) *Models*: These spreadsheets are complete and self-contained and allow students to change parameters and see the resulting change immediately.
v) *Templates*: These are spreadsheets with blanks to be filled in by students as part of an assigned problem. This guided development is intended to help them get good spreadsheet skills and to give them a model for their own construction of spreadsheets.
vi) *Student-constructed spreadsheets*: Spreadsheet problems in the book cause students to create their own spreadsheets. While these are intended to be entirely programmed by students, we hope that the experience with our previously-constructed spreadsheets (Text, Examples, and Models) and their experience filling in the blanks in Templates will give them a good basis for their own work.

So far, over a hundred spreadsheets of various types have been constructed. The reaction from students indicates that the materials seem to be reasonably comprehensible to them.

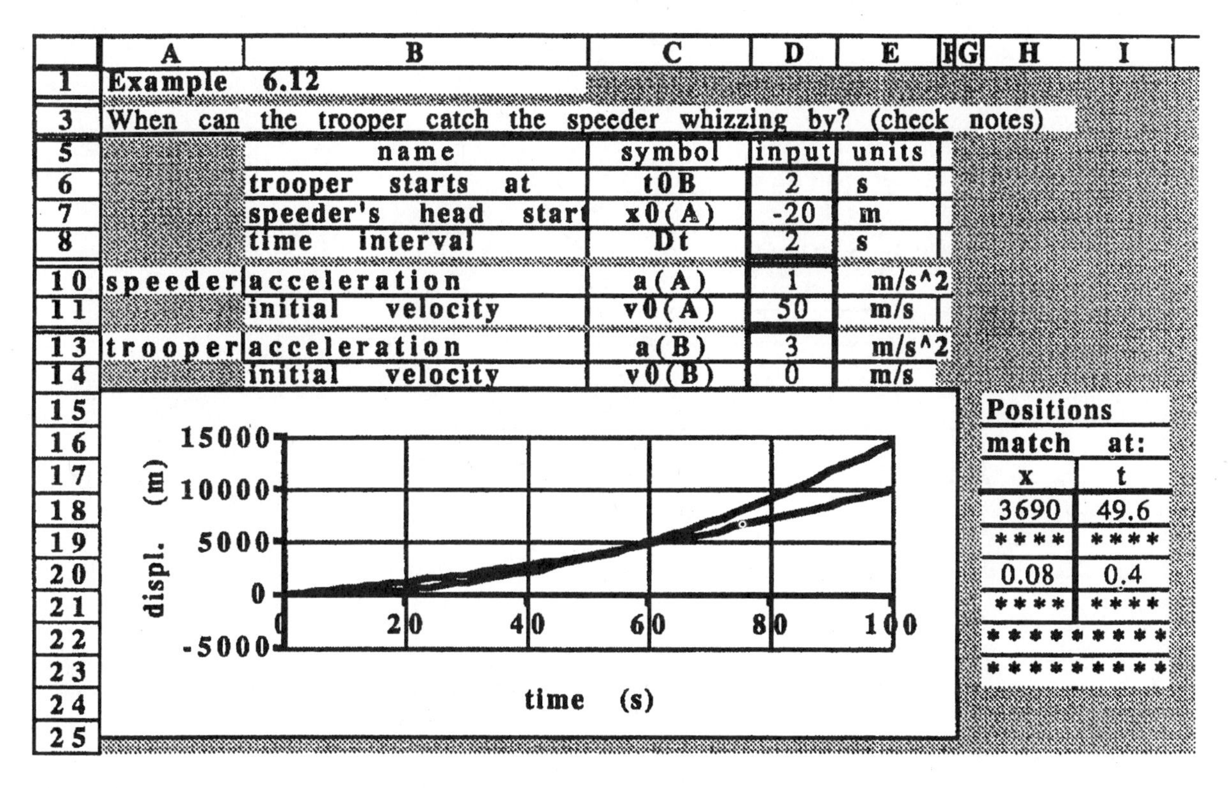

Example 6.12
When can the trooper catch the speeder whizzing by? (check notes)
name | symbol | input | units
trooper starts at | t0B | 2 | s
speeder's head start | x0(A) | -20 | m
time interval | Dt | 2 | s
speeder | acceleration | a(A) | 1 | m/s^2
initial velocity | v0(A) | 50 | m/s
trooper | acceleration | a(B) | 3 | m/s^2
initial velocity | v0(B) | 0 | m/s
15000
10000
5000
0
-5000
displ. (m)
0 20 40 60 80 100
time (s)
Positions match at:
x | t
3690 | 49.6
**** | ****
0.08 | 0.4
**** | ****
****
****

Figure 1. An example spreadsheet opening screen.

Figure 1 shows an example of the opening screen of an example-type spreadsheet. Figure 2 shows an example of the opening screen of an model-type spreadsheet. Figure 3 shows an example of the opening screen of an template-type spreadsheet. Note the similarities in the screens. We have tried to keep a common style to them. The calculations are accessed by scrolling down the screen.

	A	B	C	D	E	F
1	Model 13.4					
3	Do these two waves coming from left and right form a standing wave on this string?					
5	name	symbol	input	units		To see, push
6	frequency-left	fL	12	Hz		button "run" to
7	frequency-right	fR	10	Hz		start a time
8	amplitude-left	AL	1	cm		sequence.
9	amplitude-right	AR	1	cm		
10	time	t	1.090909091	s		run
11	wavespeed	v	20	m/s		

-------- incident left ------- incident right ——— superposition

Figure 2. A model spreadsheet opening screen.

	A	B	C	D	E	F
1	Template 20.1					
2						
3	One-dimensional gravitational field find the force on a test mass					
4	due to (the gravitational field of) a source mass, as a function					
5	of the position of the test mass.					
7	formula used	F = (G*m1*m2)/r^2				
9	constants:	G	6.67E-11	N m^2/kg^2		
10		m_1	1	kg		
11		m_2	1	kg		
13	r	F				
14	1.00E-10					
15	1.00E-09					
16	1.00E-08					
17	1.00E-07					
18	1.00E-06					

Figure 3. A template opening screen.

The difference between the example and the model are essentially nonexistent. One is a program used in a text example, while the other is used in end-of-chapter problems. The template has the invitation to spreadsheet construction. By this time, the second quarter has begun. Students have had experience changing parameters and seeing how the model world responds. They can begin to program, if they have not already done so. As part of the transition, the problems ask students to take a one-dimensional model spreadsheet and make it into a working two-dimensional model, or to extend a model in various ways. The intent is that they become more and more conversant with the process of construction of a spreadsheet as the year progresses.

Spreadsheets are first introduced in labs, where students find that they can save effort, and they begin to learn the essentials there. However, it is very hard to convince them of the utility of using the spreadsheets for calculations in the course outside of lab. We are currently thinking about ways to convince students that they should use the spreadsheet programs and play with the numbers in the spreadsheets during the course. Methods used so far include extra credit for turned-in spreadsheet problems, asking about spreadsheets on quizzes and exams, and having occasional recitation classes in the computer lab.

The development of student skills is being stressed. The spreadsheets constitute perhaps the most important example of this project's emphasis on skills.

Acknowledgment

I am happy to acknowledge the work of Ken Bolland and Mike Ziegler in this work, which has been a collaborative effort. All of us have written these spreadsheets.

References

1. G. Aubrecht and J. Aubrecht, "Constructing Objective Tests," *Am. J. Phys.* **51**, 613 (1983).
2. G. Aubrecht, "Testing—A two-way street," in G. Born, ed., Communicating Physics, Univ. of Duisburg Press, 1986. G. Aubrecht, "Is there a Connection between Testing and Teaching?," *J. Coll. Sci. Teach.* **20**, 152 (1991).
3. E. F. Redish, "The coming revolution in physics instruction," in L. A. Steen, ed., Calculus for a New Century, *National Academy Press,* 1988. See also the article in Academic Computing, Nov. 1988, page 18: "From here to the future: The impact of the computer on College Physics Teaching."
4. W. M. MacDonald, E. F. Redish, and J. M. Wilson, "The M.U.P.P.E.T. Manifesto," *Computers in Phys.* **2**(4), 23 (1988).
5. J. M. Wilson and E. F. Redish, "Using Computers in Teaching Physics," *Phys. Today* **42**(1), 34 (1989).

CUPLE — The Comprehensive Unified Physics Learning Environment, An Interactive Innovation for Effective Education

Eric Bluntzer, Center for Innovation in Undergraduate Education
Rensselaer Polytechnic Institute

Introduction

Modern cognitive theory has redefined standards of learning and cognition. Classical ideas and methods of instruction are outdated and not as effective as they were once thought to be.[1] The CUPLE educational system provides a unified base for learning that makes use of the latest research in cognitive theory[2] in order to create a situation where students can learn more effectively at a lower cost.

The problem with the classical classroom environment is a lack of mental activity by the student. This is to say that the student is not required to think for anything to occur. Material is presented to them and may occur around them. However the student is allowed to be passive. They do not have to participate, and in many cases they are encouraged not to. Some instructors have recognized this problem, and attempt to alleviate it by requiring classroom participation. This is effective, but only keeps a few students interacting with the topic at hand at any given time. Studies show that the skill level of the lecturer makes little difference in the learning that occurs in a lecture environment.[3] As far as conceptual learning is involved, an excellent lecturer does about as much for the learning process as a poor lecturer. There is no guarantee that the students will be paying attention. If the students are paying attention, the information will stay with them better if they manipulate it and have an opportunity to place it into their world view.[4] Even if they do place it in their world view, it is likely that they will adopt a version that matches their preconceptions more than it matches what is accurate.[5] Listening is not enough, the student must interact. Nowadays, interaction is provided by the lab section of a course. The present implementation of the laboratory causes there to be a separation between the material taught in lecture, and the recipes followed in lab. Though more effective than lectures, traditional laboratories cover a fraction of the undergraduate physics course, and have their own barriers to learning.[6]

Historically, interaction has been achieved by having a mentor work with a few students. Mass education has caused the student/teacher ratio to increase, requiring the students to rely on instructions to perform their experiments. The personalized feedback provided by a mentor was lost. Learning became a matter of lecture and lab. Modern technology has provided us with a means to restore active personalized instruction by making use of the computer. The computer is a tool which can provide a flexible interactive situation. The CUPLE system does this in a cost effective manner. With proper implementation, universities may dramatically increase the quality of their instruction while decreasing the cost of doing so.

CUPLE Overview

CUPLE consists of a set of tools that are given context in a multimedia environment. A computer shows the student sees a structure of topics. By using this structure, they can specify their topic of interest. This will display a variety of available modules for instruction. The student selects one that they feel is appropriate or they feel comfortable with. At any point they may return to use one of the others. A typical tool is a computer-assisted laboratory. This laboratory will introduce itself and any equipment that is involved. Laboratory procedure is available, along with videos of procedure at key points. Making use of this video, setup is demonstrated, and common pitfalls are addressed. A brief background is present with a links to in-depth tutorials for the student that does not feel comfortable with the relevant theory. The laboratory makes use of microcomputer hardware to gather data directly from the laboratory material. In the case of an oscillating spring, it detects the springs motion, and plots graphs of position, velocity, and acceleration. Mathematical operations can be performed directly upon these graphs to observe the meaning of differential functions, and the relevance it has to the data. A video sequence of the experiment can be captured. This can be used to coordinate a repayable image of the experiment with the gathered data, and see what is occurring at each point in time. The mathematics are demonstrated visually, and anomalous readings are more easily understood. This step is invaluable for the examination of experimental error. The student may also gather their own data directly from the video sequence. At each frame, the weight on the spring is selected with the mouse. This creates a series of points which records position and time. This hand gathered data can be exported to a spreadsheet for analysis and compared to the computer gathered data.

The Unified Environment

By making use of these various tools in a centralized environment, the concepts that are accented by each are unified. Physics theory that is introduced in a tutorial can be manipulated on many levels. The mathematics behind the theory can be expanded upon using WinPhys, Maple, or other modeling tools. By doing this, the step between theory, equation, and graph is made. The student is provided with real examples which they manipulate to see how the resulting graphs behave. This blending of interdisciplinary concepts goes further. Along with an interactive manipulation of mathematics, the behavior of the theory is demonstrated with video. Graphed data from the video shows how the example relates. Instead of requiring a student to take material from different classes home and making a discovery that they are related to each other and the world at large, everything is provided and unified.

Interaction

CUPLE provides more than a platform for embedding knowledge. It gives the capability for students to interact with that knowledge. Students must interact to learn. Interaction is provided on many levels.[7] The student may choose the tools to learn from that they prefer. They can change the conditions on animations, equations, and other demonstrations to see the result. They gather an manipulate data from a variety of sources in an attempt to induct general concepts of behavior from specific examples. Without this interaction, students are passive learners, which is to say no learners. CUPLE also tests students concepts. This is feedback in the form of questions that the student must answer correctly before they can proceed. Students must have feedback to test their ideas, verify them, and internalize them with confidence.[8]

System Cost

A typical CUPLE station serves two students. It includes at least a 386 processor with 4M RAM. IBM's PSL or Vernier's ULI are used to gather laboratory data. A VideoLogic or MMotion is needed for analog (passthrough) video. An ActionMedia II or Video For Windows compatible video board is required for digital video. Use analog or digital video, not both. The workstation must be running Microsoft Windows and Asymetrix Multimedia Toolbook. Each station will require basic laboratory hardware to use for running laboratory-style experiments. This may sound expensive, but when analyzed, it is cost-effective. The present style of teaching requires several hours a week of lecture and recitations. This is time that is not interactive, and has been shown to be education-ineffective. By restructuring the class structure, we can use this time interactively while decreasing the cost of instruction. If we consider a Physics course of 700 students, we can break them up in two manners.

Present method of instructrion

	students /section	hours /week	# sections	instructors /section	instructor-hours
lectures		2	2	1	4
recitation	28	2	25	1	50
lab	24	2	30	1	60
total hours		6			114

Instruction via the interactive classroom

	students /section	hours /week	# sections	instructors /section	instructor-hours
lectures	350	1	2	1	2
recitation					0
workshop	35	5	20	1	100
total hours		6			102

There is a lower instructor/student ratio in the proposed scenario, however the instructor plays more of a support role, and not required to be a one man show. Rather, the instructor monitors progress and addresses specific questions. The instructor becomes a teacher instead of a lecturer.

Customization

Presently CUPLE includes material covering the topics of mechanics, optics, some electromagnetics, and some thermodynamics. However CUPLE is not meant as a user-only product. It includes extensive support for making the creation of robust new modules that are both user and author friendly. Icons linking to CUPLE tools and utilities can be embedded using a menu. This gives instant access to reference materials such as a glossary, periodic table, calculator, library of examples that can be performed at home, three dimensional graphing tools, all of the modeling and video tools, and any piece of software that is windows compatible. Creating hyperlinks and control is trivial. An object-oriented database of your resources is maintained so you may search and alter the software you use without needing to change any of the tutorials that have been created. At Rensselaer we have a library of two hours (~1 Gigabyte) of assorted video clips that can be used in our modules, and accessed via PCs from a networked server or the campus mainframe. Embedding a video is a matter of selecting a title.

Groups have found authoring of CUPLE to be sufficiently convenient, that they have used it for creating educational material in chemistry, management and other fields. We suggest using undergraduates to develop modules. This way, material comes from an undergraduate viewpoint, and does not make inappropriate assumptions. The development can be guided and reviewed by faculty to ensure thoroughness and accuracy. By doing this the knowledge of a doctorate can be combined with the interface of a graphics designer and the power of a computer hacker at the cost of a work study or two. Having these people as resources are far from necessary, but most institutions have a variety of talents that can help with their development effort.

References

1. McDermott, Lillian C., "Research on Conceptual Understanding in Mechanics," *Physics Today*, July 1984
2. Redish, E.F., Wilson, J.M. and McDaniel, C.K. "The CUPLE Project: A Hyper- and Multimedia Approach to Restructuring Physics Education," *Proceedings of the MIT Conference on Hypermedia in Education,* Cambridge, MA: *MIT Press*, 1992.
3. Halloun, I.A. and Hestenes, D. "The Initial Knowledge State of College Physics Students," *Am. J of Phys.*, Vol.53 (1986), p.1043.
4. Laws, Priscilla W., "Workshop Physics: Replacing Lectures with Real Experience," *The Conference on Computers in Physics Instruction Proceedings,* 1990
5. Champagne, A., Klopfer, L., Anderson, J.H., "Factors Influencing the Learning of Classical Mechanics," *Am. J of Phys.*, 48 (1980) p. 1074
6. Bluntzer, E., "The Role of Digital Video in the Interactive Learning Environment", *The Physics Teacher*, submitted
7. Hwang, Y.F., "The Effectivness of Computer Simulation in Training Programmers for Computer Numeric Control Machining," *Dissertation Abstracts International,* Mar 1990, p.2810
8. Beichner, R.J., "The Effect of Simultaneous Motion Presentation and Graph Generation in a Kinematics Lab" *Dissertation Abstracts International,* Dec 1989, p.1617

A MULTI-DEVICE COMPUTERIZED CLASSROOM IMAGE PRESENTATION SYSTEM

Richard A. Crowe, Wm. D. Heacox, & John T. Gathright, Department of Physics
University of Hawaii at Hilo

Abstract
We have designed and incorporated a process control computer into a classroom image projection system for use in physics and astronomy lectures. The system enables the instructor to display desired images from one or two video disk players and a slide-to-video projector on a large-screen television easily visible to the entire class. The control process is efficient (single keystroke) and transparent to the students. The system is mounted on a movable cart, which is especially useful in moderate-sized classrooms where other fixed audio-visual equipment is usually lacking.

Hardware and Software
The UH Hilo image presentation system consists of five major components: a large-screen television monitor (Mitsubishi CK-3502R), two video diskplayers (Pioneer LD-V2000 and Pioneer LD-700), a slide-to-video converter(Elmo TRV-35G), and a control computer (IBM PC/AT clone) with appropriate software to access the video disks and the slide tray. The TV rides into the classroom on a cart designed to carry all the audio-visual equipment, which is especially useful in moderate-sized classrooms of 35-40 students where projection equipment is often lacking. Custom-control software was written in C and assembly language by JTG. The system was completed in March 1989 with the interfacing of the slide-to-video projector to the computer. A schematic of the image display system appears in Figure 1. At the logical center of the diagram, the EI & S 8652 relay board is a memory-mapped device that requires one 8-bit bus slot in an AT-class microcomputer, and switches one of eight input lines to a single output line. The configuration appearing in the schematic shows five of the relay positions being used to select one of three devices; three of the relay positions route an NTSC signal to the display monitor, while the other two are wired into the manual remote control for the Elmo TRV-35G, allowing the computer to activate the slide forward/backward mechanism. When the relay corresponding to the slide advance (or reverse) line is selected, under software control, that line is held low for a few tenths of a second, just as though the user had pressed the appropriate button on the hand-held unit. This causes the TV screen to go blank, but the duration of the operation is so short that it is perceived by viewers as part of the process of changing the slide. The computer sends commands to the laser disk player via a parallel port, using software supplied by Visual Database Systems. The Willow VGA-TV Monitor Driver Board produces a broadcast quality NTSC signal from VGA graphics software, enabling us to display computer graphics simulations on the large TV monitor.

Classroom Usage
Prior to conducting the class, the instructor may prepare a simple,-text-based "device control file'', which is translated by the software into commands appropriate for selecting the desired sequence of images from particular video devices. A sample page from a device control file appears in Figure 2. The first line of each set of image commands is a label by which the image can be called up directly using key words. The second line specifies the device (VD1 or VD2 for the two video disk players, ELMO for the slide projector). The third line commands the device to find the desired image. For example, "F7037S'' commands the video disk to Search for Frame #7037, "O24SP'' is a command to Search for Chapter 24 and Play, or "F'' commands the slide

projector to Forward one slide. The fourth line is an image title, and the next 12 lines function as text to prompt the instructor. Striking the "Enter" key commands the system to exhibit the next image in sequence. The instructor can also command the display devices directly from the keyboard, change the device being projected, and move around in the image file at will (forward or backward by jump or step). The process is completely transparent to the students, since accessing each image takes only about 1 second. Device control files can be stored on floppy disk for later use, which has proved to be incredibly convenient. It would be comparatively trivial to incorporate other image-storage devices, such as a CD-ROM player, into the system, which is also flexible enough to permit convenient control to interactive video disks such as the AAPT Physics Cinema Classics. One notable problem is that the quality of the slide projector is not quite adequate to reproduce fine, high-contrast material, such as in slides with small text. Thus, we have supplemented our computerized system with an Elmo EV-308 Presenter, which projects photographs and transparencies directly onto the TV screen using a small CCD camera. This has also been useful for class demonstrations in which the equipment is hard to see from a distance (such as for the Meissner effect). Student reaction to our system has been quite favorable, and the popularity of our introductory astronomy survey course has been boosted as a result (enrollment has doubled to over 100 students since the fall of 1989). One of us (RAC) has maintained student feedback statistics for his classes since March 1989 when the complete system became available. In total, 101 students have responded to "A/V materials and procedures were well-integrated logically with the rest of the course"; of these, 79% strongly agreed and 17% agreed with the statement, while the mean response was 4.7 on a 5-point scale. We believe that the system has allowed us to develop high-quality presentations that rival those of the best professional planetariums.

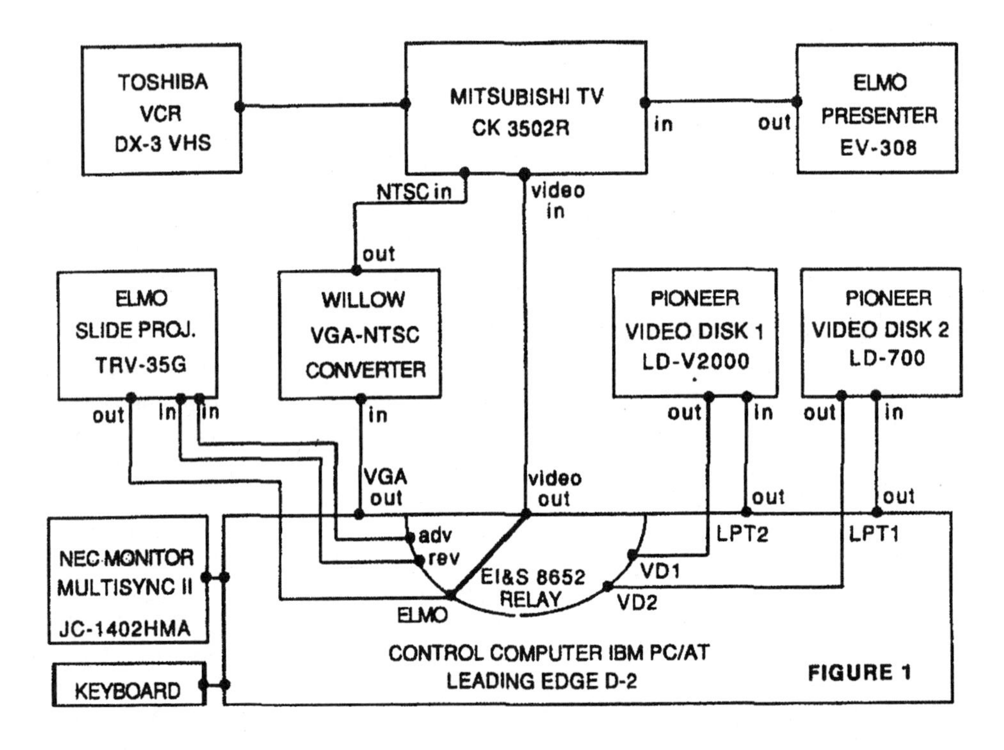

FIGURE 1

MECHANICS: FREE FALL
ELMO
F
FREE FALL MOTION ON THE EARTH

Galileo dropping objects from the Leaning Tower of Pisa:

after 1st second, Distance from top = 16 ft, Velocity = 32 ft/sec down
after 2nd second, Distance from top = 64 ft, Velocity = 64 ft/sec down

after 3rd second, Distance from top = 144 ft, Velocity = 96 ft/sec down

Exercise: What is the acceleration of the object in ft/sec/sec?

MECHANICS: LUNAR GRAVITY
VD2
O24SP
FREE FALL MOTION ON THE MOON

On the moon, the feather and the hammer hit the ground at the same time, if dropped from the same height at the same time. The acceleration due to gravity is not dependent on the mass of the object dropped, but -rather depends on the mass (and the radius) of the planet. In the case of the moon, the acceleration due to gravity is about 16.5% that of the Earth; if you weighed 170 pounds on Earth, you would weigh 28 pounds on the Moon.

Exercise: How far would an object fall from rest in 1 second on the Moon?

MECHANICS: ORBITAL MOTION
VD1
F7037S
FREE FALL MOTION IN ORBIT

If a projectile is fired from the surface of the Earth at a high enough speed, the Earth curves away from the projectile so that the projectile never falls to the ground; at this speed, the projectile falls around the Earth; it is then said to be in orbit. Newton then realized that in a similar way, the Moon falls around the Earth, and that gravity keeps the Moon moving in this path; if gravity was not exerting a force, the Moon would move off in a straight-line path, according to the law of inertia.

Exercise: What critical speed is needed to get the projectile in orbit?

WHAT'S F? WHAT'S M? WHAT'S A?:

A NON-CIRCULAR SDI-TST-LAB TREATMENT OF NEWTON'S SECOND LAW*

Richard R. Hake & Ray Wakeland, Department of Physics
Indiana University

We have devised a non-circular lab treatment of Newton's Second Law which actualizes a thought experiment by Arons and melds Socratic Dialogue Inducing (SDI) and Tools for Scientific Thinking (TST) techniques. A low-friction cart on a horizontal runway is attached to a horizontal segment of a string through a "force probe" transducer attached to the cart. The string passes over a pulley so that a vertical segment of the string suspends a weight W. For the *partially loaded* cart (Standard Body SB), W is released and the Force-Probe- Output(t) ≡ FPO(t) = constant, and acceleration a(t) = constant (as read by a sonar "motion detector") are plotted in real time on the computer screen for the accelerating SB. These two plots yield one FPO(a) point for SB. Students repeat this measurement for various W to obtain a *linear* plot of FPO(a) for SB passing (nearly) through (0,0). Students are Socratically led to understand that these measurements serve to *operationally* define a force scale in terms of the acceleration (operationally defined in terms of position vs time snapshots or sonar signals) of an SB. Using this scale, the students then take data which demonstrate that F(a) *is linear* for a fully loaded cart (slope greater than for SB) and for an empty cart (slope less than for SB). Students are then led to (1) *operationally* define, as a property of each body, the body's *inertial mass* as the slope of its F(a) line, (2) understand F = ma as a *non-circular* law of nature, (3) consider the relationship of inertial and gravitational mass.

> *One afternoon several years ago the writer was asked to proctor an examination in elementary physics to be administered to a large room full of army trainees. As he strolled the room waiting for the examination to begin he overheard many snatches of excited, apprehensive conversation - of which one significant piece has haunted him ever since: "Sure, I know* ***F*** *=* ***ma****, but what's* ***F****? what's m? what's* ***a****?"*
>
> Robert Weinstock, *Am. J. Phys.* **29**, 698 (1961)

Introduction

Common logical deficiencies and circularities in textbook presentations of Newton's Laws are the subject of an extensive literature[1] and have recently been discussed by Arons[2a] and by Hestenes.[3] Arons[2b] has suggested that introductory-course students might be led to a satisfactory understanding of the meaning of F and m in Newton's Second Law **F** = m**a**, by considering a "Newtonian sequence" thought experiment in which a force scale is *calibrated* in terms of the extension of a spring attached to a *standard body* which undergoes various constant accelerations on a horizontal runway. The inertial mass m for *any* body is then operationally defined in terms of the ratio F/a. (The Arons treatment is similar to those of e.g., the PSSC text,[4] Holton and Brush,[5] Pippard,[6] and Keller.[7]) We have devised a *real* lab version[8] of Arons's thought experiment which melds Socratic Dialogue Inducing (SDI)[9] and Tools for Scientific Thinking (TST)[10] techniques.

The present lab, *Newton's Second Law Revisited,* was tested on 91 students in a non-calculus based introductory physics course for science (but not physics) majors. These 91 volunteered to do the

*Partially sponsored by NSF Grant DUE/MDR-9253965.

lab for extra credit outside their regular lab periods and comprised 58% of a class with average Force Concept Inventory[11] scores of 35% (pretest) and 70% (posttest). [The gain of 35% is typical for low pre-test-score students in interactive engagement courses and about twice that for such students trapped in traditional TRAPT (TRAnsmission to a Passive Target) courses.[12]]

The testing was done towards the end of the semester, after students had (a) been thoroughly exposed to operational definitions of kinematic parameters, (b) considered force as defined by the extension of a spring balance, (c) completed SDI labs on all Newton's Laws (including the "Zeroth"[3,13]), (d) completed one other computer-intensive SDI-TST lab, and (e) made extensive use of Newton's Laws in problem solving. With such preparation, most students were able to satisfactorily complete the data taking and computer analysis in the first section of the lab (dealing with F = ma) in a two-hour period with little help from instructors. However, the written explanations required in the present lab (see Sec. V) posed a considerable challenge to most students and the final section of the lab concerning the comparison of gravitational and inertial mass (not treated in this paper) was satisfactorily completed by only a few students.

Apparatus

The Tools-for-Scientific-Thinking-type apparatus[10] set up is shown in Figure 1.

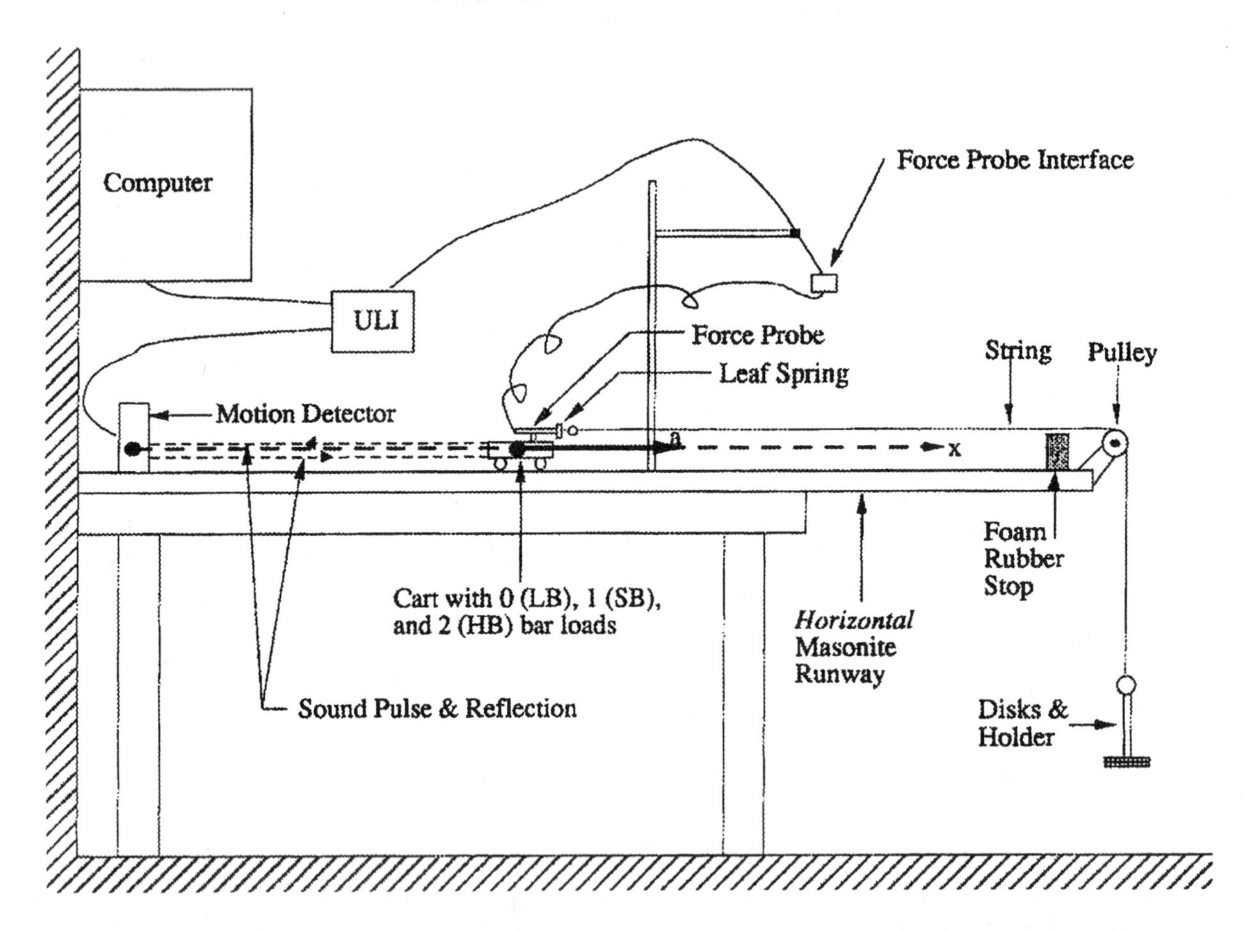

Figure 1. The apparatus for determining acceleration and Force-Probe Output (FPO) for different constant accelerations (a) of a cart on a horizontal runway. The force probe, sonic motion detector, computer tools, and software were developed at the Center for Science and Mathematics Teaching at Tufts University (see ref. 10). The computer is a Mac IIsi running MacMotion 4.0. (Almost any Mac would probably be satisfactory.) The disk holder and the disks each have a mass of 50 g. The Universal Lab Interface (ULI) is from Vernier Software. The cart and the loading bars are from PASCO.

The different accelerations are produced by varying the number of disks on the suspended holder. The cart is loaded first with one (Standard Body ≡ SB), then two (Heavy Body ≡ HB), and then zero (Light Body ≡ LB) loading bars.

The time interval for the reflection and return of the motion detector's sound-pulse yields x(t), i.e., the cart position x versus time. From x(t) the computer software calculates v(t) as $\Delta x/\Delta t$ and then a(t) as $\Delta v/\Delta t$. The x(t) data are collected at a rate of 20 points per second, and the MacMotion software data averaging of acceleration has been set to its maximum "15 point" value.

The tension T in the string deflects a leaf spring by some amount δ. This, then, causes an outward movement of a small magnet attached to the spring. Movement of the magnet results in the reduction of the magnet field applied to a semiconductor slab (visible under the spring). This, in turn, causes a decrease in a voltage ("Hall Effect") across the slab, interpreted by the (interface + computer) as proportional to a force through Hook's law $F = -k\delta$. The Hall Effect voltage as processed by the interface constitutes the Force - Probe Output.

Acceleration (a) and Force-Probe Output (FPO) are displayed in nearly real time on the computer screen. The data are graphed and analyzed by the computer software.

Force Probe Output (FPO) and Acceleration (a) as Functions of Time
Typical acceleration a(t) and FPO(t) data as shown on the screen of Figure 1 are shown in Figure 2.

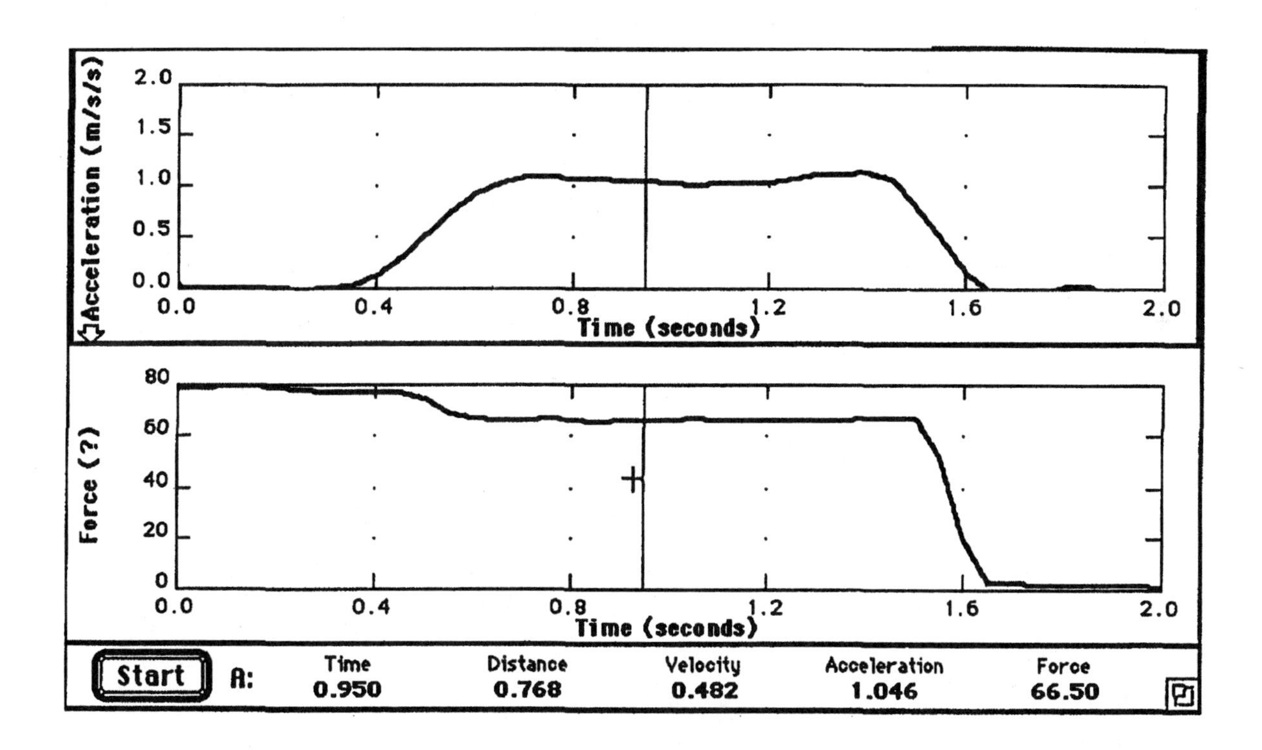

Figure 2. Typical acceleration and Force-Probe Output (FPO) data. (Note the initial reduction in the FPO as the cart goes from a stationary to an accelerating condition.) The rounding of the a(t) curve at the start and end of the run is due to the computer's data-averaging of the v(t) and a(t) data. We do not presently understand the cause of the slight dip in the plateau region of a(t) curve and other occasional eccentricities in the a(t) and FPO(t) curves.

The data of Figure 2 were obtained for the cart loaded with one bar (Standard Body) and with two disks on the disk holder. The bold numbers at the bottom of Figure 2 are values at the time setting t = 0.950 sec indicated by the vertical line. The latter can be moved by a mouse-controlled cursor arrow and is part of the "Analyze Tool" in the MacMotion program. Students are instructed to use their judgment to set the line at a time (t) so as to read off a roughly average value of acceleration (a) for the middle portion of the a(t) curve, and then record this average a and its corresponding FPO (unfortunately called "Force" by MacMotion) in a table. For the case shown, the acceleration is 1.046 m/s^2 and the FPO is 66.50. The FPO is in some arbitrary force-probe units (fpu), whose calibration is considered latter in the experiment (Sec. V, question 4b.)

At this point in the experiment a typical Socratic Dialogue Inducing sequence appears in the manual:[8]

"Note the *qualitative* variation of the distance (i.e., position), velocity, acceleration, and FPO numbers at the top of your graph (similar to Figure 2) as you scan the data by moving the vertical line along the nearly flat plateau region of the FPO vs time curve. Do the observed variations support the Aristotelian view that *velocity* is directly proportional to force? {Y, N, U, NOT}*"

In the plateau region of the FPO(t) curve, the FPO (proportional to force) is nearly constant in time, whereas scanning the data shows that the magnitude of the velocity increases rapidly with time. In discussions with Socratic instructors, students are led to state the present *scanning evidence* that v is not proportional to FPO, rather than merely parrot "F = ma" or "v is not proportional to F."

During the course of the experiment, students repeat all the above measurements for one and then no disks on the disk holder so as to obtain three points for an FPO(a) curve for the Standard Body. (Of course, it would have been desirable for students to obtain more than the bare minimum three points per line, but the requisite lab time was not available.) Then the entire sequence is repeated first for the Heavy Body (two loading bars in the cart) and then for the Light Body (no loading bars in the cart).

Force Probe Output (FPO) vs Acceleration (a) for Light, Standard, and Heavy Bodies

The data accumulated in Sec. III allow the graphing of FPO vs acceleration (a) for the three bodies as shown in Figure 3 (on page 5).

Interpretation of the Data

In an SDI manner, students are led to interpret the data by the following series of lab manual questions (here abbreviated) which demand *written* explanations and justifications, probe for conceptual understanding, and induce peer discussion and subsequent Socratic dialogue with instructors:

1. Before attempting a detailed interpretation of the FPO vs acceleration graph (Figure 3), can you give a *rough qualitative* interpretation?

2. Does your Figure 3 graph support the Aristotelian view that *velocity* is directly proportional to force?

3 ***What's* a?** Can you give an ***operational*** definition of acceleration **a**? Do you think the computer plot of a(t) in Figure 2 is consistent with your operational definition?

*{Y,N,U,NOT} ≡ {Yes, No, Uncerrtain, None of These}; According to SDI ground rules "A curly bracket {.....} indicates that you should **ENCIRCLE** O a response within the bracket and then, we **INSIST**, briefly **EXPLAIN** or **JUSTIFY** your answers in the space provided on these sheets."

4. ***What's* F?**
    a. We have tacitly assumed that frictional forces can be ignored in this experiment so that the force probe output is proportional to the *net* force acting on the carts. That friction can be ignored might seem reasonable considering the present very low-friction cart and pulley. Do the present data offer any support for this assumption?

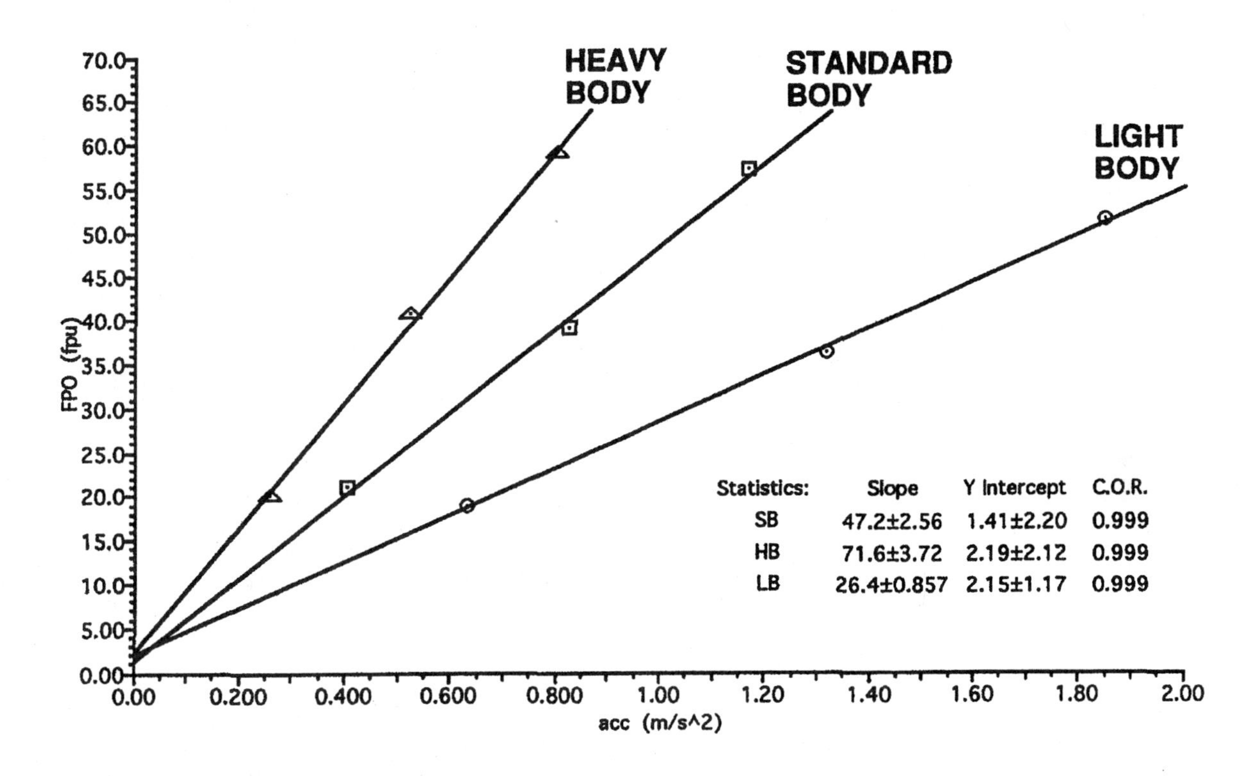

Figure 3. Force-Probe Output (FPO) vs. acceleration (a) for three bodies as obtained by feeding the data of Sec. III into Graphical Analysis, a Vernier Software program. These are the data of a four-student group, taken with little assistance from instructors on 4/27/93 during a two-hour lab period and are fairly typical of those reported by the 24 student groups that worked on this experiment. This real data graph is very similar to the thought-experiment graph of Figure 3.5.1 of ref. 2b. The "C.O.R." in the "Statistics" is the "Correlation of Regression."

    b. You have designated the cart loaded with one black bar as the "Standard Body," SB. Your graph shows Force-Probe Output (FPO) readings vs the acceleration (a) for the SB. Can you now operationally define a "Force" scale in terms of the operationally defined acceleration of the SB? [HINT: Suppose the SB had been the platinum-iridium cylinder kept at the International Bureau of Weights and Measures whose mass is by definition precisely one kilogram. What then would have been your force scale units for accelerations of 1 and 2 m/s^2?]

5. ***What's m?*** If your results are similar to those obtained by other students with this apparatus, then in Sec. IV you concluded that the data for the *three* carts (SB, HB, and LB) yield *three* straight lines with *three* different slopes for the Force Probe Output (FPO) vs. acceleration. Would you be justified in saying, then, that the F vs a slope might (with further testing) turn out to be *a unique property of every body ?*

6. ***What's "F = ma"?*** Suppose you *assume* that the F vs a slope is a unique characteristic of every body in the universe, and designate the slope as the "*inertial* mass _ m." You could then summarize the results of the present investigation by the equation "F = ma". Suppose you had done these experiments in 1665, well before Newton's work, and suppose that your "F = ma" hypothesis had been confirmed by others throughout the world.
   a. Do you think your "F = ma" relationship would have been justifiably hailed as a "new law of nature"?
   b. Do you think your "F = ma" relationship could have been justifiably dismissed on the grounds that "F = ma" is "circular"? (Here "circular" means that "F = ma" is true only because F and m are *defined* in such a way as to automatically make F = ma without yielding any new information on the behavior of nature.)

7. **Is "F = ma" the whole story?** In "6" above, suppose you are invited to lecture before the Royal Society of England on your work showing that "F = ma." You decline, stating that the road to knowledge is paved with questions and that such questions need to be settled by experiment before "F = ma" can be usefully employed. Do you have any idea as to what questions might have been on your mind (assume that you are even smarter than Newton)? [HINT: Think of the mechanics lore beyond the scalar "F=ma" that you have applied to solve mechanics problems.]

Conclusions

We have devised a laboratory experiment to help students understand the essential meaning of Newton's Second Law as a true *law of nature* rather than a circular definition of force and/or mass. Although no quantitative measure of the lab's effectiveness is yet available, judging from the dialogues, peer discussions, and written explanations, we think that at least some students were able to develop better insights on the crucial questions *What's **F**? What's m? What's **a**? What's **F**=ma?*

References

1. See, e.g., R. Weinstock, "Laws of Classical Motion: What's F? What's m? What's a?" *Am. J. Phys.* **29**, 698 (1961) and references therein; M. Hesse, "Resource Letter PhM-1 on Philosophical Foundations of Classical Mechanics," *Am. J. Phys.* **32**, 905 (1964).
2. A. B. Arons, a. *A Guide to Introductory Physics Teaching* (Wiley, 1990), Chap.4; b. ibid. p. 52-55. The Newtonian sequence first defines force in contrast to the Mach sequence (see ref. 1) which first defines inertial mass.
3. D. Hestenes, "Modeling games in the Newtonian World," *Am. J. Phys.* **60**, 732 (1992); *New Foundations for Classical Mechanics* (Klewer, 1990), Chap. 9.
4. U. Haber-Schaim, J.B. Cross, J.H. Dodge, and J.A. Walter, *PSSC Physics* (Heath, Lexington, 4th ed., 1976), Chap. 11.
5. G. Holton and S.G. Brush, *Introduction to Concepts and Theories in Physical Science* (Princeton, 2nd ed., 1985), Chap. 9.
6. A. B. Pippard, *Forces and Particles* (Wiley, 1972), Chap. 3.
7. J. B. Keller, "Newton's Second Law," *Am. J. Phys.* **55**, 1145 (1987).
8. R. Wakeland and R.R. Hake, a. SDI Lab #6, *Newton's Second Law Revisited.* This and a Teacher's Guide are available upon request. b. "A Noncircular SDI-TST-Lab Treatment of Newton's Second Law," *AAPT Announcer* **23** (2), 64 (1993).
9. R. R. Hake, "Socratic Pedagogy in the Introductory Physics Lab," *Phys. Teach.* **30**, 546 (1992).
10. R. K. Thornton and D.R. Sokoloff, "Learning motion concepts using real-time microcomputer-based laboratory tools," *Am J. Phys.* **58**, 858 (1990). R.K. Thornton, these proceedings. [Thornton-Sokoloff Tools-for-Scientific-Thinking (TST) material is incorporated into this and some other SDI labs under a cross-licensing agreement.]

11. D. Hestenes, M. Wells, and G. Swackhamer, "Force Concept Inventory," *Phys. Teach.* **30**, 141(1992).
12. R. R. Hake, unpublished data survey, NSF proposal DUE 9253965.
13. R. R. Hake and R. Wakeland, "SDI-TST Labs on 'Newton's Zeroth Law'," *AAPT Announcer* **23** (2), 72 (1993).

PHYSICS: CONCEPTS AND CONNECTIONS. A NEW TEXTBOOK FOR THE LIBERAL-ARTS PHYSICS COURSE

Art Hobson, Department of Physics
University of Arkansas

Introduction

I've just finished writing a textbook for the liberal-arts physics course, or non-scientists' physics course. It will be published in 1994 by the Macmillan Publishing Company. Here is the preface, the table of contents, and a flow chart of topics.

From the Preface

This book's central premise is that nonscientists deserve a true *liberal-arts* physics course. Thus, this is not a watered-down version of the standard technical introductory physics textbooks for scientists. Far from being a simplified version of anything, this book is designed for a course that has a cultural sophistication not found in more technical courses. It is a cultural, rather than a technical, physics textbook. It presents physics as a human endeavor that has a philosophical and societal context.

The Introductory University Physics Project, the AAAS Project 2061, and other study groups have recommended new approaches to science education and science literacy. The features in this book that reflect these recommendations include:

- Less is more. The focus is on the great ideas of physics rather than lots of details and applications. This book presents most of the main principles while deleting many narrower topics normally "covered" in introductory courses.
- Modern physics is emphasized. Fully half of this book is devoted to such post-Newtonian topics as relativity, quantum theory, nuclear physics, and quantum field theory.
- To develop scientific literacy, this book addresses the values, philosophical meaning, and cultural impact of science. It places physics within its social, historical, and philosophical context, with stress on the methods of science.
- Quantitative skills are developed, but not merely for their own sake. Non-scientists should become *numerate* as well as literate in science. The abilities to interpret graphs, to think probabilistically, and to make rough numerical estimates are important for non-scientists. On the other hand, traditional algebra-based physics problems are less important for non-scientists. Just as one can appreciate a painting without being able to paint, one can appreciate the power and beauty of a scientific idea without solving equations. Thus, this book is quantitative but non-algebraic.
- The book is built around unifying themes that provide a coherent story line. Four themes recur:

  how we know what we know in science;
  comparisons and contrasts between Newtonian and contemporary physics;
  the social context of physics;
  and the unifying concept of energy.

One justification for communicating these ideas to non-scientists is simply that, as a matter of general principle, educated people should understand the modern world as science sees it. But there is a more pressing, practical reason. In an age when science and technology are driving rapid cultural and physical changes on Earth, it is imperative that all of us, and especially our non-scientists, contribute their understanding and their perspective to helping us figure out where we are going and where we should go. Today, science is far too important to be left to scientists. I have written this book for you, the poets, teachers, historians, business people, politicians, journalists, and others who must help us find a rational and humane path through a time of rapid change and powerful technologies. We need your perspective, and your informed leadership.

I have tried to develop the scientific ideas slowly, with attention to the learning process. You will find the following learning aids in the book:

Quotations in the margins lend different perspectives, and depth, to the philosophical and societal topics. Thus, such topics are accompanied by many more quotations than are the purely scientific topics, where quotations are more sparse. Most of the quotes span a spectrum of views by scientists and nonscientists about topics, such as nuclear power, for which there are differing viewpoints, while some are scientific views that reinforce the text. Please understand that the inclusion of a particular quotation does not mean that I necessarily agree with that viewpoint, or that the scientific community necessarily agrees! Quotations represent a wide range of views.

Footnotes provide additional details for students who want them. It is a perennial problem for writers of introductory textbooks, especially ones for nonspecialists, to keep everything really correct while at the same time not burdening the student with details that might do more to confuse than to enlighten. Footnotes seemed to me to be the best way to handle such situations.

Dialogue questions in the main text of each chapter are my way of conducting a dialogue with the reader, to check and reinforce of the reader's understanding. Readers should formulate their own answer to each dialogue question as they come to it, before comparing with the answer that appears in a footnote at the bottom of the same page.

How do we know is a frequently asked question in this book, and subsections bearing this title appear regularly. It cannot be emphasized too strongly that scientists have specific observation-based reasons for their theories and principles. Most principles of physics are accompanied by a discussion of how we know.

Making estimates is one of the skills that this book seeks to develop. Examples and exercises bearing this title appear frequently. Making informed but rough estimates is one numerical technique with which educated citizens should be familiar.

Summary of ideas and terms, following each chapter, put the chapter into perspective. These summarize and clarify the main concepts, and should be helpful when studying for exams.

Review questions, about thirty, follow each chapter. Organized by sections within the chapter, they go over the main points. Answers can be found by looking them up in the appropriate section.

Home projects, a few of them, follow each chapter. There is nothing like "hands on" experience to bring out the essential experimental nature of science. These projects could be done either at home using common household items, or in physics laboratories, or as demonstrations or desktop experiments in physics lectures.

Discussion questions, a few of them at the end of each chapter. These are meant to stimulate thinking about the open-ended questions, for example values questions, that have no single correct answer. They can be used for class discussion, essays, or individual thought.

Exercises, about thirty, follow each chapter. These are designed to exercise the mind, in the same way that jogging exercises the body. They are based on the text, but they require original thought. Most are qualitative, or "conceptual." A few are quantitative, but they require no algebra. These quantitative exercises are indicated with asterisks. Answers to all odd-numbered exercises are in the back of the book.

To allow the instructor flexibility in the choice of topics, the book is purposely a little long for one semester. Chapters and sections that may be omitted without losing continuity are indicated with asterisks in the table of contents. The first chapter, about the scientific method, is one of these deletable chapters, because there is material on this topic throughout the book. Some topics, for example interdisciplinary topics such as extraterrestrial life, could be assigned for reading only. The flow chart following the table of contents shows connections between topics, and suggests different course formats.

I welcome your comments and suggestions.

Table of Contents

Chapters and sections marked by asterisks (*) can be skipped without destroying the continuity of the remaining material. For guidance in choosing and connecting topics, there is a flow chart below.

Flow Chart of Topics

For a course emphasizing modern and philosophical topics, many societal topics could be omitted. For a course emphasizing societal topics, many modern and philosophical topics could be omitted.

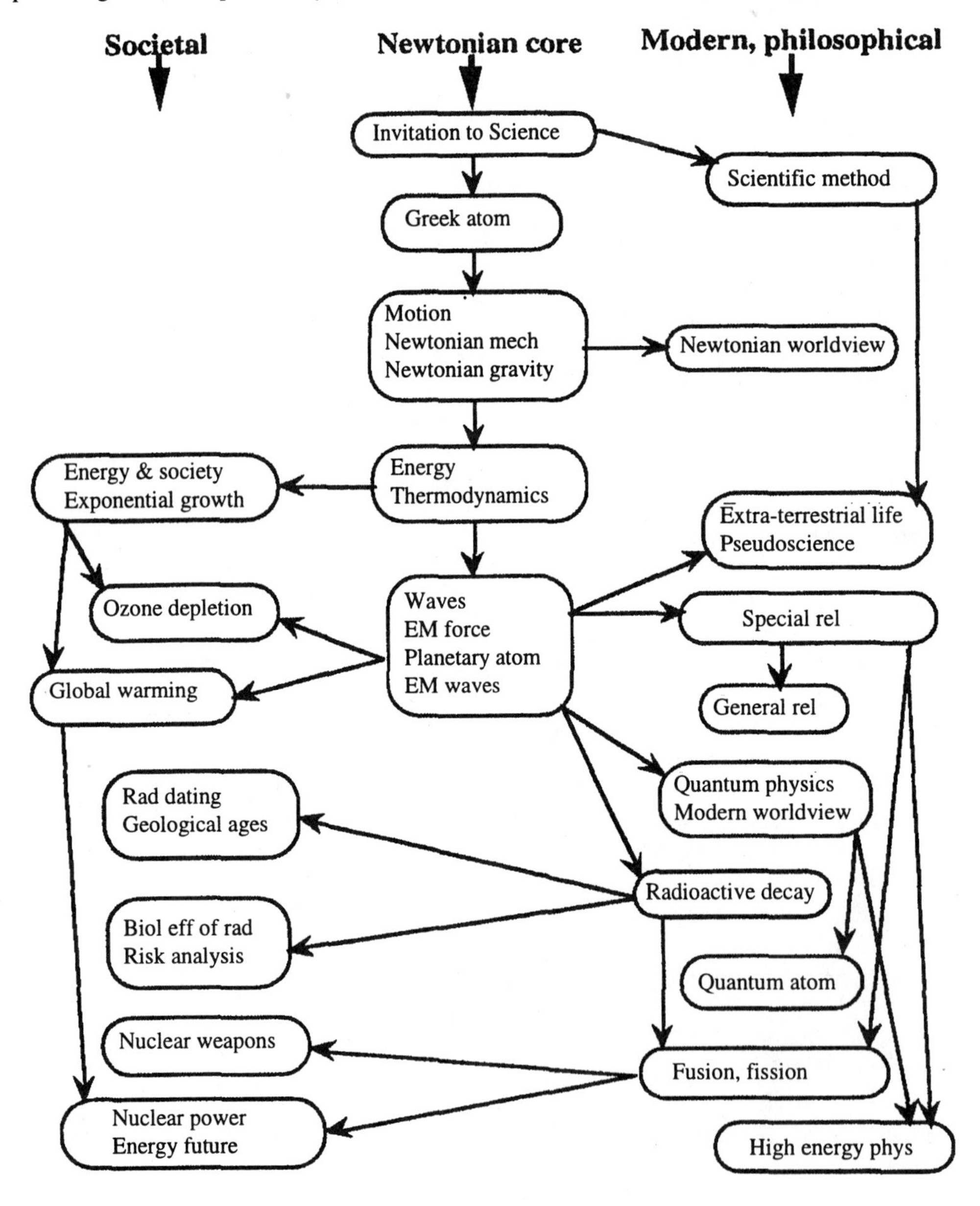

Changes in the Introductory Calculus-Based Physics Course at Iowa State University

Laurent Hodges, Department of Physics
Iowa State University

Our introductory calculus-based sequence — Physics 221 and 222 — is a two semester sequence, with both courses offered each fall, spring, and summer. Most of the 1000 or so students who take it each year are engineers, with a sprinkling of physics, chemistry, and other majors. When I first taught in this sequence in 1970, I was a fairly new assistant professor, and my courses followed the organization that had been used by my older colleagues. In 1988, after an interval of about 16 years devoted to teaching advanced courses and courses in environmental studies and solar home architecture, I once again returned to this sequence. Over the next few semesters I quickly became disenchanted — or should I say horrified? — with many aspects of our courses: their all-classical nature, the multiple-choice examinations, the overgrown formula sheets allowed at exams, the "plug-and-chug" problem assignments, the competition and grading on a curve, and the passiveness of the students in lecture.

Over the last few years I have continued to teach these courses, making some changes to address these problems. Significant changes (or even experimentation) were difficult to make on my own because two faculty members are generally involved in each course, one giving two lectures and the other giving one lecture, and all the students have the same problem assignments and examinations (given at night).

During the 1992-93 academic year, I was given permission to teach one lecture section of 221 in fall 1992 and one lecture section of 222 in spring 1993. It was my intention to decouple my section from the other two sections and teach a significantly different course, with a slightly different syllabus but entirely different problem assignments and examinations. I felt that experimenting with one-third of the students — perhaps 150-200 students — would be more manageable than experimenting with all the students, plus I did not want to force the other lecturer to teach a very different course than he was accustomed to. My 221 course was decoupled in fall 1992, but in spring 1993 the lecturer with the other two lectures wanted to recouple the two courses. This was done, permitting us to gather information from a total of 239 students who had taken the first semester in the traditional format and the second semester in my new format. What follows is a brief description of what I regard as some of the most important features of the revised course, having to do with the problem assignments, the examinations, and the lectures.

Problems

It has been traditional at Iowa State University to assign about 15-20 textbook problems each week. These are due at the weekly recitation sections, where only a few of the problems can be discussed in class. However, solutions are posted and students can also go to the course help room (staffed mainly by graduate students, but also a few faculty members) to ask questions about the problems.

We have always asked students to write up careful, well-explained solutions to the textbook problems. We expect good sketches, definitions of all symbols, use of SI units throughout, vector notation for vectors, explanations in full sentences, and so forth. However, we rarely get anything of the sort. The reason became clear one semester when I decided, instead of assigning recitation instructors to take turns writing up the solutions, to write them all myself and to write very high-quality solutions, with all the values bells and whistles we ask from the students. I discovered this took about 10 hours a week, in part because a faculty recitation instructor was very conscientious about pointing out steps and diagrams and other nice touches I should have included. If I took 10 hours a week to do this, how many students could do it in less than 20 or 30 hours? Presumably this care in writing up solutions is not expected by the textbook authors. The solutions manual for the our text has about 15 problem solutions on a page: the solutions are stripped down to the essential formulas and then numbers are plugged in.

For my course I wrote up all my own problems. There are typically only three problems assigned each week, but they have many parts to them, and extend over about 2 or 3 pages, with room for answers. The students will normally be asked to do one or more of the following, as appropriate to the problem:

- Start the problem from fundamental equations (definitions, laws, etc.), never from highly specialized results.
- Sketch the physical situation, choosing symbols for the relevant physical quantities and labeling them on the diagram, also listing the numerical values if any are given at the beginning of the problem.
- Draw free-body force diagram(s), labeling all forces and defining them, remembering to include contact forces with all objects with which the object is in contact, plus any field forces acting on the object.
- Choose a convenient coordinate system, and draw a modified free-body force diagram with forces resolved into components along the axes.
- Describe what object a force acts on, what object produces the force, and the reaction to a given force.
- Derive results as algebraic expressions and only then substitute numerical
- Read information off a graph or diagram or table.
- Plot qualitative or quantitative results on a graph, choosing appropriate scales for the axes in the latter case.
- Recognize and name the important principles of physics used in a problem and (sometimes) describe them in words and in equations. Examples of such principles are the important conservation laws, Coulomb's law, Ampere's law, Kirchhoff's loop rule, and the principle of linear superposition.
- Recognize when there is insufficient information present to answer a question, when there is superfluous information given in a problem, when there is more than one solution to a problem, and when a result is physically impossible or unreasonable.
- Determine how a change in one quantity affects the results of the problem.
- Always make proper use of units and significant figures, and write all vectors in unit vector notation consistent with the coordinate system chosen.
- Convert quantitative information or information from a graph into mathematical form.
- Name the fundamental interaction responsible for each force in the problem.
- Understand the limits of validity of an expression or approach. For example, recognize that a formula applies only to non-relativistic objects, or only to an ideal gas.
- Express physical conditions in words and in mathematical terms.
- Diagram processes like those involved in a heat engine.
- Be able to solve a simplified physical situation but also be able to explain in what ways, qualitatively and quantitatively, a more realistic physical situation would differ.
- Derive the results in more than one way, or find a way to check the results.
- Draw field lines for gravitational, electric, and magnetic fields.
- Graph results in several different ways. For example, in the case of projectile motion, be able to graph $y(x)$, $y(t)$, $x(t)$, $v_y(y)$, $a_y(x)$, etc.
- Recognize the errors in an incorrect derivations.
- Be able to identify a formula as a definition, fundamental law, phenomenological relationship, etc.

Of course, no problem assignment includes more than a few of these, but over the course of a semester all of them are encountered several times by the students in several different contexts.

Examinations

When I first taught introductory physics at ISU, the exams were written examinations, but some instructors were experimenting with multiple-choice questions. The exam usually included a short list of "useful information" such as the values of key physical quantities or a few basic equations, mainly fundamental definitions and important laws. When I returned to the course, the exams were all multiple-choice (except for an occasional instructor who preferred written examinations), and the "useful information" had evolved into an official four-page "formula sheet" handed out at the beginning of the course, a sheet which conveyed the idea that "physics is nothing but a collection of several hundred formulas, and here they are."

Now I give examinations that are completely written (except that half the final exam consists of concept-type and laboratory-based multiple-choice questions, so that we can finish grading the course before the university deadline). No formula sheet is given out, and students are only allowed to bring a writing implement, no formula sheet, no calculator. There are some numerical calculations on the exam, but they involve either estimation or simple numbers that can be determined with little mental effort, and it is the process of deriving the correct expression and substituting into it that is important and gains points, not the final numbers. By not allowing calculators I avoid the problem of the wealthier students bringing in a calculator or palmtop computer loaded with formulas. I tell the engineers in the course who express unhappiness with this policy that I don't doubt their ability to use calculators, and don't want to waste physics examination time testing that ability. The homework assignments have plenty of numerical calculations.

I make a practice on the examinations of giving small amounts of bonus points for relevant extra work on the questions. For example, on a problem which involves heat transfer from a reservoir to a system, I might ask for the heat transfers and changes in entropy of the system and reservoir, and give bonus points to anyone who then also adds up the entropy changes to show that the result is positive, in accordance with the second law of thermodynamics.

The combination of multi-part homework problems and written examination questions is intended to be reinforcing. Both emphasize to the student that it is the concepts and basic principles and laws of physics that we want them to learn. My experience with new students is that this can be difficult for them at first, but by the time the semester is one-third over they have become accustomed to it.

Lectures

My lectures are not the traditional chalkboard lectures, though I do write a good deal on the chalkboard. During the course of the lecture I try to ask the class a couple of dozen questions. Some of these are questions during a demonstration: What do they think will happen when I do such and such? A good example involves our giant yo-yo. When it is partially wound up I hold the straight vertically while supporting the yo-yo in the air and asking what will happen when I let go: will the yo-yo drop, unrolling the string, or stay motionless, or ride up the string? Everyone laughs and votes for the first alternative. They are right. Then I repeat this on a horizontal surface and the class is divided, some of them refusing to guess whether the yo-yo will unroll or roll up or slip. Then I show them that it rolls up and explain why.

During the course of the lecture, or in describing an example, I will often ask the class something rather than tell them. A plane circuit is located in a uniform magnetic field which is directed up. The magnitude of the magnetic field is decreasing. What is the direction of the induced current? I don't tell them, I give them 30 seconds to think about it and discuss it with their neighbors. When the class quiets down we vote and then I explain the direction. Usually the majority of the class will come up with the right answer, but sometimes not, and in one extraordinary case every single student in a class of nearly 200 guessed wrong.

In addition to these frequent questions during the lecture, I pass out at the beginning of the course a page of short specific questions for each lecture. The students are expected to answer these questions, usually about 6 to 8 in number, after reading the textbook assignment and thinking about them. These questions are then discussed in lecture, and only in lecture; no official solutions are posted and the recitation instructors are not responsible for them. As an example, for the lecture on heat transfer one of the questions asks the students to provide examples of heat losses from a house in winter and classify them as conduction, convection, or radiation. Questions similar to these lecture questions will show up on some of the examinations.

Student Reaction

An extensive questionnaire was passed out in May 1993, at the end of the second semester. 322 of the 360 students in the course turned it in. 83 of these students had taken the new-style course both semesters and had overwhelmingly favorable reactions to it. The other 239 students had taken a traditional first semester of the introductory course, one that assigned textbook problems and used multiple-choice exams with a four-page formula sheet, so their opinions are particularly interesting.

When asked about the problems, 77% of these 239 students definitely favored the new style of problems, 15% favored the textbook problems they had had the first semester, and the other 8% preferred a combination or had no strong opinion one way or the other.

When asked about the nature of the exams, 80% of these 239 students favored the all-written exams, 8% favored an all-multiple-choice format, and 12% favored tests that are combination of the two types of questions.

One change not mentioned earlier is that modern topics were introduced, amounting to about 15% of the lectures. When asked about the incorporation of some modern physics into the course, 62% of the 322 students favored this, 30% preferred that there be no modern physics in the course, and 8% were neutral. Feelings ran very strong on this question. Many students regarded the modern physics as the most interesting part of the course, and some wondered how you could teach an introductory physics course without including it. Some of the engineers, particularly civil and construction engineers (but not the electrical engineers), felt more classical topics would have been more useful to them, although some argued that they will encounter classical topics in plenty of detail in their engineering courses.

Since the textbook in this course was used only for reading assignments and as a reference, and not as a source of problems, students were asked the following question: Should there be a required textbook for the course, or could the lecturer hand out extensive notes and allow students to choose their own textbook to read from, as is done in Germany? The students responses (for the whole class of 322) were: 59% preferred the current scheme of having a required textbook even though no problems were assigned from it, 27% preferred the option of lecture notes and freedom to choose a textbook to read, and the other 14% preferred having a required textbook which would be the source of some or all of the problems.

Let's let a few students speak for themselves:

"This course is so much better than it used to be, even though it is tough. I learned a lot more in 222 than in 221!"

"In Physics 221, I got a D for a final grade. I barely passed any tests. I was dreading this semester of physics, but to my surprise, I really enjoyed this semester. Instead of the usual boring science classes, this one was more modern in context. As a result, I believe I will get a B in this course. The new format helped me achieve this grade. "

"This is the most interesting class I have ever had, especially the assignments and the instructor."

"If the course is taught with the same enthusiasm and effort to 'teach' rather than just 'lecture,' I think it can't really be much improved. "

"I felt thus course was well thought out. I feel the changes in the course from previous years were beneficial and made the course more interesting."

Use of Computer-based Problem-Solving Software* and Supplemetal Instruction† in Introductory Physics at Rutgers-Newark

Yuan Li & Karen G. Smith, Department of Physics
Rutgers University

Introduction

We would like to report a couple of initiatives at Rutgers-Newark to improve the environment of learning in the introductory physics courses. Although the scope is modest and similar approaches have been adopted elsewhere, what we have learned from these initiatives, such as how to relate to the students' thinking, students' acceptance (or resistance) toward computer-aids in learning, may be relevant to the current search by the physics community to find more effective means to discharge its responsibility of teaching introductory physics to a wide variety of students because of the characteristic of our institution and the composition of our students, many of them will not be a physics major.

Profile of the Students

Located in downtown Newark within the country's largest metropolis and serving a highly heterogeneous group of students many of whom are the first generation in their family having a college education, Rutgers-Newark exemplifies nation's urban campuses where most of its students not only commute but often work off-campus in non-academic-related jobs twenty hours or more per week.

Description of the Courses

At present, two introductory physics courses are offered. Elements of Physics (EP) is a calculus-based three-credit course for students in physics, mathematics, and engineering. Although all well-prepared students are encouraged to use the course to fulfill the college requirement of a laboratory science course, few elects to do so. A typical class has about forty students. The class meets twice a week for 80 minutes each period. There is no additional recitation session and the instructor is responsible for the lecture as well as problem-solving. The students take, in addition, a one-credit introductory physics laboratory which is only loosely tied to the course and which carries a separate course number. General Physics (GP) is a non-calculus-based three-credit course for students in chemistry, biological sciences, geology and computer science as well as those who wish to use it to fulfill the college requirement of an eight-credit laboratory science course when combined with the one-credit introductory physics laboratory. The class meets twice a week for 80 minutes each period. Approximately two hundred students take the course each academic year (and fifty more during the summer). Currently there are four sections in this course, taught by different instructors. As in EP, each individual instructor is responsible for all aspects in his/her section, including the selection of as well as the policy on the homework problem. Although there is a general agreement on the topics covered in the course, there is no common examination for all sections.

Description of Computer-based Problem-Solving Software

The problems used are selected from the textbook we have adopted for the course (Paul A. Tipler's "Physics for Scientists and Engineers" for EP and "College Physics" by the same author for GP) and Diploma4, a textware developed by Brownstone Research Group, is used as the shell. Many problems are explained as to

*Supported in part by a grant from the Teaching Excellence Center of Rutgers University, Newark
†Supported through a Howard Huges Medical Institute Program grant

their objectives as well as the underlying concepts and, whenever possible, are broken down into several steps to guide the students along a line of reasoning. In multiple-choice problems, wrong answers are often placed there based on common misconceptions. When students review their performance, a line of "rational" explains why that particular answer is wrong. The program runs on IBM-compatible PC's linked via Novell Netware on campus and its use is made compulsory for the students in EP and in the section of GP taught by YL. It is also made available to students from other sections as a learning-aid. One example is given in the Appendix.

Description of the Supplemental Instruction (SI)

Supplemental Instruction (SI) has been offered on a voluntary-participation basis since January 1993 to the students in EP. Based on a national model validated as an exemplary program in University of Missouri, Kansas City, MO, SI links course content and cognitive skills development through a regular schedule of out-of-class sessions. A trained student leader guides the sessions by encouraging students to actively participate as they think about, question, and confirm their understanding of the concepts of physics while learning more effective strategies of studying and learning physics. In our case, a physics major who took the course last year was chosen as the SI leader under the supervision by the Learning Resource Centers headed by KGS. To accommodate students who have different class schedules, three eighty-minute sessions are scheduled each week and the students are encouraged to attend as many as they wish (or as few as they find necessary). It is emphasized to the students that SI is not meant only for those who are not doing well in the class. The SI leader attends the class regularly and takes notes during the class.

Our Experience

Surveys were conducted at the end of the school year among the students in both courses. Although the small sample may dilute the significance of the survey, many responses from the students do show a consistency with other less tangible experiences we have had throughout the years. Some also confirm with the assertions made by experts in the field such as Lillian C. McDermott and Sheila Tobias. The following conclusions may be drawn.

1. *Inability among some of the students to bridge the conceptual gap between mathematics and its meaning in "real-life":* In her comment which appeared in the American Journal of Physics (**61**, 295, 1993), Lillian C. McDermott suggested that facility in solving standard quantitative problems is not an adequate criteria for functional understanding. Our survey supports this observation in an unexpected case. To the question, "how do you work on the assignment? Assign a percentage to each mode of working pattern", 8 students from GP (22% of the respondents) failed to recognize that the percentages must be added up to 100 percent. 2 students from EP (less than 10 % of the respondents) did the same. It is a discouraging sign on our average students who are not majoring in engineering, physics, mathematics, and computer science. It suggests the conceptual gap between mathematics (or mathematical manipulation) and its meaning in "real-life" among many students who supposedly have passed the mathematical proficiency test.

2. That *Commuting does have an effect on the study* is reflected in the general response to the question, "Would you be more interested in the SI sessions if you were living on campus and some of the SI sessions were scheduled in the evening hours?" Students who did respond to the SI sessions but attended no more than once a week stated, three to one, that they would be interested in attending the additional evening SI sessions if they were living on campus.

3. *Wide availability of computer is necessary so that the students may work at home on the computer-interactive software.* Because most of the students prefer to work at home instead of spending additional hours on campus after class, unless students can have the access to computers at home, the use of computer-interactive tutorial can be a problem. We had to print out contents of the tutorial instructions, except answers to the multiple-choice questions, and distributed to the students as guides. But its effectiveness is greatly compromised.

4. *Using SI as a possible alternative to the standard recitation session.* Not only the percentage of students (70%) participating in SI is encouraging enough to consider it as an integrated part of the course, a com-

parison of the final grade between two groups (2.44 for the SI participants vs. 3.2 for the Non-SI participants) and the percentage of students receiving a grade of D,F, or W (11% for the former vs. 37.5% for the latter) shows SI is particularly helpful to the students "in the middle". To the question, "Do you think SI should be adopted for all introductory courses in science and mathematics?", 64.7% say yes if the course does not have a recitation session, 5.9% say yes even if the course has a recitation session, and an additional 17.6% think SI should be adopted in place of the recitation session. Furthermore, because we use an undergraduate student (a sophomore physics-major) as the SI leader, not only students find a better communication with him than with a standard TA, the SI leader is also benefited by the reinforcement of what he had learned the year before.

5. *Students' acceptance (and resistance) of the computer-interactive software for homework and grading:* While it is expected that most of students (more than 3/4 of each class in the survey) want to have another learning aid at their disposal, that over sixty percent of the students wish to have (or at least to have no objection toward) mandatory implementation of computer interactive tutorial/problem-solving software suggests that the students seems to have recognized the trend of the future. But it does not mean that the students' resistance does not exist. It was evident in two occasions. In one occasion, a few students complained that they would rather do the homework problems on paper by whatever approaches they preferred rather than be guided by the step-by-step approach in the tutorial. In the second occasion when the students who came into the section in the second semester and who knew the computer-interactive tutorial as a part of homework were asked by the questions, "Do you find the detailed guide helpful to your solving the problem?" and "Do you think the computer interactive tutorial have increased your ability of doing the problem?", their response were more favorably than the overall response from the first semester when the students had not idea of the computer-interactive homework requirement in advance. Thus, the success of a program hinges as often on a rapport between the program and the students as on the "absolute" merit of the program.

Conclusion

As pointed out by Sheila Tobias in her presentation during the conference, to a heterogeneous group of students, there is no single solution to a complex problem such as the effective delivery of the introductory physics course to the students and any improvement in the environment of learning should not be restricted to innovation alone. In addition to the adoption of MBL in our introductory physics laboratory, our experimentation presents two possible avenues to improve the environment of learning. The preliminary result convinces us that computer-interactive tutorial, when properly refined to include video demonstrations and simulations as envisioned in the CUPLE project, can nurture students' interest with the satisfaction of being able to tackle a problem, albeit guided. But as one student commented in our survey, "The computer cannot take the place of an instructor in a discussion section". Our experience suggests SI seems an ideal complement.

Appendix

In this example, the idea of breaking a standard problem into several parts is emphasized:

A car making a 100-km journey covers the first 50 km at 40 km/h. How fast must it cover the second 50 km to average 50 km/h.

This problem requires a bit of logical deduction. It also points out to the definition of average velocity which depends on the initial and final position during the period of time in question.

To begin with, with an average of 50 km/h, the car makes the 100-km journey in how many hours?
(a) 2 hours; (b) 2.5 hours; (c) 2.2 hours; (d) 2.25 hours.

But at 40 km/h, how long does it take to cover the first 50 km?
(a) 1 hour; (b)1.25 hours; (c) 0.8 hour; (d) 1.2 hours.

From the answer to the question above, how much time is left to cover the second 50 km?
(a) 1 hour; (b) 1.25 hours; (c) 0.8 hour; (d) 0.75 hour.

Finally, how fast must it cover the second 50 km?
(a) 60 km/hr. (b) 75 km/hr. (c) 66.7 km/hr;(d) 65 km/hr.

IMPROVING THE PHYSICS CURRICULUM TO INCREASE ENROLLMEENT

Joan Mackin

Several innovative programs are being developed to pursue the goal of excellence in science in the United States. The America 2000 program has the goal of the United States being first in science and mathematics. New American Schools Development Corporation has funded private consortia to design and build schools that will provide better education at the same cost as today's schools. (Beardsley, 1992) The Scope, Sequence and Coordination program is being tested in six geographic areas. This program has a "core" and rejects the traditional sequence of biology, chemistry, and physics in the secondary school. Instead a time line for teaching all sciences concurrently or in an integrated manner in middle and high school is proposed. (Beardsley, 1992) Another program, Project 2061, urges the reduction in material covered, integration of the rigid subject matter boundaries, and the use of critical thinking skills. (Hofkin, 1989) Chris Whittle has taken the challenge to develop schools for profit to achieve excellence. His company earns $630,000 per day from Channel One's advertisements. (McNichol, p.4) All these programs have the same basic premise, to change our present school system radically.

Radical change is being sought to improve the science curriculum in public and private schools throughout the United States. Areas of concern include the teaching of science in the elementary schools, middle schools and high schools. One goal that was expressed was that students who graduate from secondary schools in the United States would be able to compete with students from other countries and achieve better ranking than presently attained. One must ask how we are to accomplish this if only twenty percent of high school graduates has experience in physics. One must also consider the effect of changing our present program from within by changing teaching strategies to improve students understanding of the subject.

Researchers cite that there are problems with our education system. John H. Bishop concludes that "No adolescent wants to be considered a nerd, brain, geek, grade grubber, or brown noser, yet that is what happens to students that study hard and are seen to study hard." (Beardsley, p. 103) Girls and minority groups reject math and science. When one thinks of physics there is the attitude that this course is for the "elite" and to be avoided if possible.

Must this condition exist? Approximately sixty four percent of the 1990 Avon Grove Seniors completed a course in physics. In 1991 about sixty percent of the seniors completed a first year course in physics and eight percent completed a second year course. In comparison in 1981 approximately twenty percent of the graduates completed a one year course which was the national average. During the last ten years not only has the enrollment increased, but the number of girls and minority students electing physics has increased as well. What factors motivated students to elect physics without changing the units that were studied? The analysis of these forces must begin with the changes in the science program in the high school and the middle school and with the strategies that were implemented in the physics classes.

Over the ten year period there was an evolution of three courses instead of one and the implementation of strategies that differed from the traditional ideas of teaching physics. These changes served to meet the needs of more students and to allow more understanding of the ideas presented in the courses. This revision allowed more students to study physics and emphasis was placed on students understanding the concepts and learning to use mathematics as a tool without weakening the requirements or student outcomes. Care was taken to vary the approach to the learning depending on the background of the students within each level. Students selected the course based on their background and plans for the future.

In answer to student requests and the suggestions from a Middle States Evaluation, a physics course for students with a limited background in mathematics (Algebra I and Geometry) was added. The original course required students to have a background in mathematics which included Algebra I and II as well as Geometry. The additional course allowed those students who were preparing for technical education, training programs and interested in science to take a fourth year elective. A third course, Honors Physics, was added in the 1987 school year. This course became AP Physics level B and allowed a third semester of physics to be taught to accelerated students. Students were accelerated by beginning physical science in eighth grade and by studying biology in ninth and chemistry in tenth.

Different textbooks were used for each course to suit the needs of students. The textbooks used were as follows:

> *Physics* by Giancoli for AP Physics B
> *Modern Physics* by Williams, Trinklein, Metcalfe for Physics
> *Physics Principles and Problems by Murphy,* Hollon, Zitzewitz, and Smoot for General Physics(Geometry background students)

The goals of each of the courses were based on the needs of the students for the various programs they would enter after graduation, current issues in society and technology, and the expectations of students achievement and understanding of the concepts. The teacher could use different approaches from the text or develop new ones as needed. As the course progressed, student evaluations and the continuing education of the teacher were responsible for using certain strategies to implement the curriculum and achieve student learning.

Another factor which allowed students to broaden their backgrounds was the introduction of a mandatory introduction to physical science in ninth grade or in eighth grade for accelerated students. During the first half of the year students were introduced to chemistry and during the second semester they were introduced to physics. Environmental issues were integrated into each of the semesters. Emphasis was placed on learning through laboratory and "hands on" experience.

To understand the increase in enrollment and student desire to continue the study of science, it is necessary to discuss some of the strategies that were implemented and that differed from the traditional approach. It is the belief of the author that teachers must convey that the study of science is interesting, pertinent and understandable for all students.

The first strategy that was utilized was getting to know the students. The schedule in our high school allows students to meet six periods instead of five. Class size was kept to twenty four or less whenever possible, consequently the physics teacher was responsible for at most one hundred twenty students. As a result of a survey taken, it was found that the students had the feeling that the teacher knew what they were doing during class and was interested in their individual efforts. Techniques were used to give the students common experience to use as a basis for understanding the concepts and to utilize their backgrounds and contributions to help others relate to the ideas.

Although the lecture method allows for the presentation of more facts, the students do not actively participate in the learning. Rather than lecture, a question and answer technique was used to allow discussion and building of the lesson on prior knowledge or actually modifying prior concepts. Assignments were given to prepare for class by reading the text or time was devoted to demonstrations, videos and simulations to show examples of the concepts. Student questions and answers became more meaningful and variety in the presentation added interest. The variety depends on the type of subject matter, student learning styles, the unit or concepts, the time of the year, teacher discretion or sometimes students choice.

Just as teachers need ownership in planning a curriculum, the students react in the same way when given the opportunity to create part of the work in class. One of the methods used in the physics classes to expose them to various problem solving activities is to allow them to create sample problems adding information, deciding what is to be found and then as a class offering the solution. This is done with teacher guidance and allows for analyzing parts of a natural situation. It conveys the idea of taking one aspect at a time rather than being inundated with all the factors. The students share in the solution using the concept, related formulas and one or more of the possible explanations. Using this strategy, the students not only remember the example, they also remember the concept, the units that are compatible and recognize challenges that may be added to make the problem more involved. Students question other solutions and the relevance of the information. Some become more aware of the "real world" situation and the contrived example they are formulating. Educational research shows that student's perception of control can increase their motivation and level of achievement.(Baird & White, 1982)

The use of technology is important to give the students the advantage of working with the most efficiency. Calculators are used by all students. Some are kept in the classroom for those students who will need to borrow them. Students also learn to estimate the results so that they will know if their results are valid. During class, calculators allow students to follow the computation as we discuss solutions. Programmable calculators and computers may be used in finding lab results and in preparing lab reports. Combining the skills acquired in using a program for the computer that includes word processing and spreadsheets is encouraged. One laboratory exercise is devoted to introducing the spreadsheet application for physics in addition to computer class work so that students can review and learn to use it effectively.

Every effort is made to make the course relevant to the students. Elastic collisions become more interesting if they occur on a pool table. Videos and demonstrations are effective in conveying ideas using actual examples. The decibel scale takes on new meaning when the relative intensity of rock concerts, a walkman, or the high school sound system are used for examples.

Cooperative learning takes many forms for the physics class. Students work together in lab groups sharing ideas and equipment with others at their table and with other tables. They form study groups to help each other during study halls and other times outside of class. These groups very often are formed without regard to specific classes. Study groups are often given the privilege of being able to meet in the lobby during their study hall so they can work together to understand the physics concepts. This type of interaction is encouraged since the result is that the students do understand the concepts and learn from the questions of others. There are times in class when cooperative learning is used as well. Research has shown that when students are given control over a classroom task, they become more effective and to like fellow students better. (Brady, 1987) One strategy that is used at the end of each year is a syllabus which contains three to four weeks to complete. Class discussions are noted as well as all the readings and assignments to be completed. Although the practice began to cope with the end of the year activities and demands on students, it has become a tradition since it allows students to investigate the concepts and learn from each other. Students actually seem to have a better understanding of the concepts which was lost to the approach of graduation before the strategy was used. Attendance is improved and students respond to the need for organization and for use of the study skills required.

Study skills and critical thinking skills are a valuable part of the curriculum. Emphasis is placed on understanding and applying the concepts. Some of the stress of testing is eliminated by allowing students to put formulas on 3 x 5 cards to use as an aid during the tests. Before each test, students discuss the objectives, skills and concepts they are expected to know. They also discuss the type of problem solving and offer sample test questions that might be asked. Non threatening means of having students answer these questions are used such as answering in unison or writing an answer before someone offers an answer are used to keep test stress at a minimum.

Effort is rewarded. Research has shown that students are more motivated to attempt tasks in which they believe they are competent and have control over the outcome. They are more likely to continue and be persistent in such tasks even after initial failure. (Skunk, 1984; Weisz & Stipek, 1982) Assessment is not based solely on tests. Only one-third of the assessment is compiled from test results, one third is the lab segment of the course consisting of lab work and lab reports, and one third is based on homework, classwork and class participation. Projects are completed and count as a test grade for students in the the marking period in which they are submitted. Students may complete additional projects but may only complete one per marking period(quarter). Required projects vary for each of the courses. There is also a wide variation in the format and topic choice. The presentation of project information allows the student to learn more about an area that interests the individual and to use creativity to present their findings. Students may work together on a project provided that that their plan is approved and the particular project has merit for group work.

One interesting group project that was repeated was the planning of a physics competition for all physical science and physics classes. The students of the second year class developed, planned, and held the event. They decided on responsibilities, wrote the rules, judged the qualifications of the entries and recorded data to determine a winner for a downhill race. Students fashioned a model car from a block of wood, four plastic wheels and nails which served as axles. There were limits as to size and mass. The second year physics students became experts on the operation of the track to time the entries, methods of determining that a car met the specifications and organizing the event.

Video presentations are used to show specific concepts to aid the students in their understanding. There are several sources of short video presentations that will show an idea either using humor or some type of alternate presentation that will supplement the reading of the information from the text. Research shows that students receiving instruction with humor score higher on course exams than students covering the same content without humor. (Ziv, 1988) Students look forward to Julius Sumner Miller's video demonstrations and the animated videos from TV Ontario. They can laugh when something goes wrong and become motivated to ask questions after viewing since everything is not explained. We they are absent from school, they have been known to watch the PBS channel and relate information and questions when they return to school.

Students continue to relay information that will improve the course after graduation. Usually students return during their semester break to visit high school classes. When they return, they bring the message that there is a benefit in taking the course not only for an understanding of the subject matter but for learning how to accomplish tasks and learn on their own. Graduates are always welcome and encouraged to share this message with students who are physics students or may wish to be physics students.

One concern in implementing an AP Physics course was that the techniques used for the previous physics courses could not be continued. The college board realizes that the teacher, the approach and the students involved in the program are important factors.
"This diversity of approaches should be encouraged to the high school teacher of an AP course. The success of a given program depends strongly on the interests and enthusiasm of the teacher and on the general ability and motivation of the students involved." (APCD, p.5)
This philosophy allows teachers to use different styles of teaching to achieve their goals. It allows latitude so that the college presentation, lecture, lab and section, does not have to be used. The philosophy expressed by the college board seems to be a common one for improvement of the curriculum.

At the high school level, there are other factors that aid in the success of the program. The teacher, and the teacher to teacher relationship, and the teacher to administration relationship within the school are determining factors in improving the curriculum. The support of the science department, the other departments in the school and support from the administration are necessary for the program to succeed. Schools can only improve from within if there is cooperation within the staff and administration. If an individual is alone in bringing about changes and improvement, there may be no lasting effect from these innovations.

Roland Barth proposes that schools can improve from within. He notes that the changes must occur in the relationships within the school personnel. He also states that the school must become a community of learners. Staff development takes on the meaning of assistance and encouragement in a hundred different ways. His plan includes having teachers work together and cooperate as well as the principal and/or supervisors being coaches and facilitators. School is a place where diversity is respected.

Diane Ravitch classifies what effective schools have in common as:

> "a strong academic curriculum, a principal with a vision and the courage to work for it, dedicated teachers, a commitment to learning, a mix of students from different backgrounds, and high expectations of all children." (Ravitch, p.294) She suggests taking what presently exists and developing it to its highest potential for education will be achieved by a series of small moves in the right direction. Thus education would have "better teachers, better teaching, better administration, better textbooks, better curricula and higher aspirations." (Ravitch, p.308)

The modifications to the physics course at Avon Grove certainly used teacher input for continuing improvement of the course. Changes occurred in small steps. The changes required that the teacher continue to learn new information and strategies, try new methods and have the cooperation of the students, other teachers and the administration. For the program to succeed there was a commitment of time and money for equipment, textbooks and support of the science program. The guidance department played an important role in counseling students and giving them the opportunity to take the courses. The members of the science department prepared students to progress and accept the challenge of the new course. The mathematics, business, and communications department offered students the opportunity to learn skills that would help them succeed in the course. If any one of the factors that helped the program evolve deteriorates then the program itself will struggle to survive.

The idea that a school can improve from within does need the commitment of the entire staff and community. The solution for the "crisis" in education requires an opportunity for recommitment, a respect for diversity, humor and philosophies, and a place where one can make the things happen the you believe in. (Barth, 1990)

INEXPENSIVE ELECTROSTATICS MATERIALS: PHYSICS FUN WITH A FUNDAMENTAL FORCE

Robert A. Morse,
St. Albans School, Washington, DC

The introductory physics course is criticized for being too problem-solving oriented and having too few activities that build conceptual understanding. Inexpensive disposable plastic containers provide raw materials for electrostatics equipment from electroscopes and Leyden jars to motors and generators that students can build rapidly, that perform well and that can be used in the laboratory, as seat labs in lecture, or as take home labs for homework. The ease of building and using the devices allows science and engineering students to get a 'feel' for the behavior of a fundamental force and provide experience that can be explained by models. Most of the materials can be acquired at little or no cost, so that there is no economic excuse to not do laboratory work in electrostatics. Students and teachers have responded very favorably to these activities, and it is a good workshop to do for local high school, middle school and even elementary teachers. Authors should seriously consider incorporating the use of such experiments in the study of electrostatics. The construction of many such devices has been discussed by the author in the new AAPT Publication *Teaching About Electrostatics,* a major revision of the *AAPT Electrostatics Workshop.* An example of a curriculum using similar devices is *Electrical Interactions and the Structure of Matter,* B. Sherwood, Carnegie Mellon Univ. Some of the materials and activities have been incorporated into *Electricity Visualized: The CASTLE project,* by Mel Steinberg, PASCO Scientific (1993). The workshop book also contains two sample curriculum modules for high school students which could be adapted.

The IUPP Model
Structures and Interactions

Dwight E. Neuenschwander, Department of Physics
Southern Nazarene University

Structures and Interactions is a model curriculum of the Introductory University Physics Project (see Ref. 1, and also Donald Holcomb's paper, "Curriculum Reform: IUPP, An Example" elsewhere in this volume, for a review of the IUPP). Here we present a summary of the S&I model.)

Objectives and Theme

- We seek to actively engage the student through hands-on experience in the laboratory and participation in the art of mathematical modeling.'
- Our strategy is "observe in lab, then model in lecture," instead of the traditional "explain in lecture, then confirm in lab."
- The content meets the IUPP goal of "less is more, but with 20th Century physics as a natural part of the story line."
- The content theme is "The fundamental interactions in nature produce structure. Structure tells us about the fundamental interactions" (see text contents below).

The Model's History

Structures and Interactions is one of four models whose development is sponsored by The Introductory University Physics Project;[1,2]

- IUPP-sponsored testing to date:
  - United States Military Academy, West Point, NY: 91/92 & 92/93
  - Tulane University, New Orleans, LA: 92/93 & 93/94-
- Informal testing to date:
  - Jacksonville University, Jacksonville, FL: 92/93

  Southern Nazarene University, Bethany, OK: 86/87-92/93

Communication Strategy

1. "Lab Before Lecture"

- Whenever possible, concepts are introduced through the laboratory.
- Goals, not "cookbook" instructions, are given.
- Students work in small groups of three students. Having no workbooks, the spirit of the laboratory experience is that of directed research. This builds espirit de corps in lab, and raises questions to be addressed by modeling practice in lecture.
- Whenever possible, evaluation is by an exit interview.

Discussion: Each student has an assigned task (e.g., managing equipment, handling data, communicating questions to the instructor). In the exit interview, each student is asked about the task of another student. The group either passes initially with an A, B, or C; or if necessary, they repeat the necessary portion of the experiment or analysis, and then can pass at the B or C level. Students like having more responsibility and opportunities to show creativity than is the case with fill-in-the-blank "cookbook" labs. There simply is no contest between the two approaches for stimulating student interest and discussions about conceptual and technical issues. The method was adapted from Robertson.[3] Two logistical questions frequently arise: (1) Is there a backlog with the exit interview? This is not often a serious problem, since groups finish at different times and the interviews take only a few minutes. In a pinch, we have on occasion asked the students to submit abstracts in lieu of the interview. (2) What about teaching assistants? Teaching assistants need to be briefed on the features of the S&I laboratory, as they would for any laboratory. Needed are TA's that can think on their feet and interact well with students. At Tulane and SNU, our approach has been used successfully in laboratory sections that use the services of upper-division undergraduate TA's (plus graduate TA's at Tulane).

2. Explicit Instruction in the Art of Mathematical Modeling

- Lecture time is seen as follow-up to lab and is orchestrated with it.
- To be emphasized is the view that physics is the art of reconciling two Worlds ,[4] requiring an explicit distinction to be made between them:
  - — the Physical World of real phenomena, and
  - — the Conceptual World of models.
- The craft of physics requires the development of two distinct skills:
  - — Conceptual (visualizing, manipulating ideas);
  - — Technical (problem-solving, manipulating equations).
- It is our task in "Lecture" to engage the students in the art of developing mathematical models for the phenomena studied empirically in the laboratory.

Discussion: Making explicit a sharp distinction between the Conceptual and Physical Worlds offers a window into the "game of physics" that is a more useful strategy than formula memorization, and is faithful to the way that physicists actually work. Whenever feasible, it is useful to approach a topic on two fronts: first on the Conceptual front, followed by the details of the Technical front. Engaging students outside of lab can be implemented in several ways, such as sending groups of students to the chalkboard, or by employing Eric Mazur's very effective "ConcepTest".[5]

3. Using the Computer

The S&I model is not dependent upon the computer or any other specific technology. We wanted to design something that could be as readily adapted to a class that meets under a tree, as it could to the well-equipped, well-funded micro-computer-based laboratory. At various test sites the computer has been used for data collection and analysis in lab, and for demonstrations, simulations, and numerical problem-solving in lecture and outside of class. The S&I model will, like any curriculum, benefit from the computer, but is not driven by it.

How it Works

Suppose the subject is oscillations, or perhaps one seeks an example of motion in a gravitational field, and "the pendulum" is the subject of choice:

First, in Lab:

The Instructor: "How does the period T depend on the length L?

Suggestion: Parameterize it as

$$TG = CL^n$$

then measure C and n, keeping the swing angle small."

The Students (e.g., Shawna & Carol, SNU, 1991):

$n = .52 \pm .03$ and $C = .19 \pm .05$ (cgs units).

Later, in Lecture:

Instructor & Students: *Model* the pendulum: *conceptualize* it as a particle of mass m, with a non-stretchable string of negligible mass, no air resistance, interacting with a uniform gravitational field.

Now comes the technical part: apply **F = dp/dt** to this particle, and if the swing angle is small, one finds that the model predicts

$$T = 2\pi\,(L/g)^{1/2}$$

$$n = 1/2$$

$$C = 2\pi g^{-1/2} \cong .20\text{s/cm}^{1/2}$$

Now compare the model to actual student data, and discuss it! Reflect on it! Extend it (e.g., to nonlinear and/or damped cases)! Improve the model! Perform a follow-up experiment!

Acknowledgments

"Structures and Interactions" Advisors in Development

Dwight Neuenschwander	Southern Nazarene University
John Goulden	Southern Nazarene University
David Hestenes	Arizona State University
Robert Hilborn	Amherst College
Michael Lieber	University of Arkansas
James McGuire	Tulane University
William Robertson	U.S. Space Foundation
Sallie Watkins	Univ. of Southern Colorado
Dean Zollman	Kansas State University

Test Site Faculty to Date

Al Gant (92/93)	United States Military Academy
Heidi Gruner (91/92)	United States Military Academy
James McGuire (92/93)	Tulane University
Dwight Neuenschwander	Southern Nazarene University
Paul Simony (92/93)	Jacksonville University
James Stith (91/92 & 92/93)	United States Military Academy

Liaisons with the IUPP Steering Committee

Robert Hilborn (90-92)	Amherst College
Joe Meyer (92-)	Georgia Tech

Text: Structures and Interactions in Nature.

Content Summary[6]

PART I: OVERVIEW

An overview of structures and interactions in nature, from astronomical to microscopic scales. Energy and its transformations. The art of modeling.

PART II: WAVE MODELS and SUPERPOSITION

The "shape" of the signal, or "wave function." Harmonic series, superposition, bandwidth theorems and uncertainty, diffraction and scattering.

PART III: PARTICLE MODELS and NEWTONIAN MECHANICS

Vocabularies for describing particle motion (kinematics and Newtonian relativity), the "cause and effect equation" of particle motion (or Newton's laws, in the language of "force and momentum" and "work and energy").

PART IV: MODELING THE FUNDAMENTAL INTERACTIONS

A ubiquitous case: harmonic motion. Next, the concept of a field. Modeling the gravitational interaction. Modeling the electrostatic, magnetostatic, and electromagnetic interactions. (Eschews dependence on traditional line and flux integrals. Embraces the development of Feynman's equations for the electric and magnetic fields produced by an accelerating point charge. The radiation fields, including electromagnetic waves radiated by a harmonic source, follow at once.)

PART V: A WINDOW from the MACROSCOPIC into the MICROSCOPIC WORLD

Macro-micro connections are illustrated (examples include, for instance, electrical conductivity, first in terms of the macroscopic Ohm's law, then in terms of the microscopic Drude-Lorentz model). General macroscopic constraints: thermodynamics. The microscopic view: statistical mechanics (wherein the partition function is introduced at the appropriate level, and used to derive the thermodynamic equations of state in simple models).

PART VI: ATOMIC and NUCLEAR PHYSICS

The Special theory of relativity. Quantum light (photons) and quantum mechanics (inspired by photons and deBroglie's generalization that "$p = h/\lambda$" also has meaning for particles of non-zero mass, quantization is introduced by analogy to the harmonic series. We meet the one-dimensional Schrödinger equation and the tunneling effect.) Applications of quantum mechanics to atoms and light (such as multi-electron atoms and chemical bonding, transistiors, lasers, and the thermodynamics of light in equilibrium with matter). Nuclear physics (radioactivity, nuclear models, fission, fusion, nuclear reactors and weapons). Also included is the recent saga of cold fusion, to illustrate how controversy is resolved in science. The fundamental interactions at the microscopic level of the elementary particles (using Feynman diagrams to qualitatively predict intermediate states). Ultimate structure: the universe & the Big Bang.

References

1. John Ridgen, Donald Holcomb, and Roseanne DiStefano, "The Introductory University Physics Project," *Physics Today*, April 1992, p. 32.
2. IUPP Newsletters 1-5, *American Institute of Physics*, 335 East 45th St., New York, NY 10017.
3. William Robertson, "An Alternative Approach to an Introductory Physics Laboratory," *The Physics Teacher*, September 1985, p. 348-352.
4. David Hestenes, "Modeling games in the Newtonian World," *Am. J. of Phys.* 60, 732-748 (1992).
5. Eric Mazur, presentation on peer instruction and ConcepTest, AAPT Winter Meeting, Orlando FL, January 8,1992; and "Understanding or Memorization: Are We Teaching the Right Thing?" at the Conference on the Introductory Physics Course, Rensselaer Polytechnic Institute, Troy NY, May 22, 1993.
6. Dwight E. Neuenschwander, *Structures and Interactions in Nature*

Reinforced Cluster Active Learning (RECALL) in a Large Enrollment Science Majors Course

David G. Onn, Cynthia A. Cuddy, & Barbara J. Duch, Department of Physics
University of Delaware

Abstract

Permanent learning clusters of three to five students have been instituted in a large enrollment two-semester General Physics course for science majors. Cluster responsibilities include group problem solving in large bi-weekly main class meetings, smaller weekly recitations and examinations. Consistent in-class seating facilitates student-instructor interaction in a large lecture theater. Reinforced learning occurs through pre-class review and problem solving rewarded by rapid feedback on individual and cluster quizzes. Student responses to these approaches are presented.

Introduction

Recent research in the learning of physics has shown that although a formal content-driven lecture may be the most efficient method of presenting information to a large enrollment class it is not necessarily the most efficient vehicle for effective learning to occur.[1-4] A recent effort to apply active learning principles in large accounting classes[5] encouraged us to develop a related approach, which we have termed "reinforced cluster active learning in lectures" (RECALL), in large enrollment two-semester physics courses for science majors (not Physics or Engineering).

The essential features of reinforced cluster active learning (RECALL) enter all aspects of the course including the large bi-weekly main class meetings (formerly "lectures"), recitation sessions, office hours, quizzes, examinations, out of class study and review. Student responses to this approach will be presented and student performance summarized. A parallel Honors section of this course, which shared the same main class but used more intensive group learning is described in an accompanying paper.[6] A more complete description and analysis of both approaches will be published elsewhere.[7]

Background And Objectives

At University of Delaware the General Physics course [Phys 201-2] for science majors is a two-semester sequence, requiring algebra and trigonometry. Up to 300 students from a wide range of majors enroll each semester. It has historically been taught using a bi-weekly content-driven faculty lecture for the entire class (75 mins) supplemented by weekly recitation sessions (50 mins) and laboratories (110 mins) both taught by TA's (Teaching Assistants). Recitation and laboratory sessions include between 16 and 24 students. The recitation sessions have generally consisted of problem solution presentations, lecture style, by the TA with student activity limited to short quizzes.

Early in this course students readily admit to being very anxious about physics because of the required mathematical skills, hearsay that physics is very difficult and the acknowledged competitive atmosphere due to admissions pressures for pre-professional graduate programs. They often delay the course until late in their undergraduate career because of this apprehension. Such attitudes on the part of students in turn tend to sharply reduce faculty interest and enthusiasm for teaching this course.

In light of this background our objectives in developing the RECALL approach included:

- Providing peer support groups (clusters) and competency based grading as ways of reducing anxiety, reducing the competitive atmosphere and encouraging cooperative learning
- Reducing the time devoted to content-driven lectures while increasing student cluster time for actively discussing physics concepts and problem solving in the main class
- Developing interpersonal skills and personal responsibility for learning by encouraging student pre-preparation for class reinforced by cluster review, discussion and quizzes
- Fostering instructor-student interactions both in the main class and by use of "cluster office hours"
- Developing an alternative and less stressful format for this class that other faculty may be encouraged to adopt.

From our own observation and review of student responses we find that most of these objectives were achieved, at least in part. The valuable experience gained will certainly further their more complete attainment in the future.

Features of Reinforced Cluster Active Learning (Recall)

Brief descriptions of selected special features of our RECALL approach are given below and will be described in further detail elsewhere:[7]

A. Cluster selection and consistent seating

The bi-weekly main class of up to 300 is assigned consistent seating by randomly pre-selected clusters of 3 to 5 students. The seating plan leaves every third row vacant to give the instructor direct access to every cluster. The clusters remain together for the semester. Cluster members rotate weekly through specific responsibilities (Leader, Presenter, Secretary, Recorder etc.). The roles are rotated to minimize the possibility that cluster activities would be dominated by one or two strong personalities. The 'Secretary' of the cluster is responsible for monitoring roles.

B. The reinforced cluster active learning process

RECALL uses sequenced stages designed to reinforce learning by repeated exposure to physics concepts and problem solving in a range of learning environments. The most common format for a 75-minute class is outlined below; alternate formats are described elsewhere.[7]

i) Written pre-class assignment

Before every class a written active worksheet covering assigned reading and some simpler problems *must* be completed by each individual preferably in cooperation with the cluster. This worksheet is collected early in the main class as mentioned below.

ii) Cluster review — identification of learning issues

The main class starts with each cluster verbally reviewing their pre-assigned worksheets. At this point each cluster *must* identify at least two remaining "learning issues" (i.e., something specific that they do not yet understand). These are written down by the Recorder and handed in with the completed individual pre-assigned worksheets *before* the cluster receives the "Entry Quiz".

iii) Entry quiz (individual or cluster)

This first quiz *must* be completed within a fixed time, usually 20 minutes, of the start of each main class. Each cluster has a choice of starting time since it cannot receive the Entry Quiz until the activities in (i) and (ii) are completed. At the discretion of the instructor the Entry Quiz may be taken individually or by clusters. If taken individually the solutions are then reviewed within the clusters and modification of answers permitted as a result of this discussion. Individual quizzes are rarely collected since it has been more effective to reinforce learning by providing rapid feedback through an immediate review of the quiz answers with the entire class.

When quizzes are taken by clusters one "master copy" of the quiz, a consensus solution, is handed in while each member of the cluster retains a copy. Again, rapid feedback is provided by immediate review of the quiz with the entire class. The "master copy" for each cluster is graded and contributes equally to the grades of all cluster members. During both types of quizzes TA and instructor interact with clusters through the "open rows" provided by the consistent seating described earlier.

iv) Reinforcement and extension: review of "Learning Issues"

During the Entry Quiz the "Learning Issues" provided by the clusters are collated by the instructor permitting the subsequent 30 or 40 minutes of class activities to focus on those issues identified by the class. Such focused activities stimulate student attention and participation since the class content for this segment has been student selected. The learning issues provide a springboard to more advanced material, complex problems, real-life applications and new concepts. During this time short mini-lectures are interspersed with a range of individual and cluster activities, participatory demonstrations and presentations through various media.[7]

v) Exit quiz

An "Exit Quiz", graded and returned in the next class, occupies the final 5 to 10 minutes of the main class. It is usually taken by clusters, occasionally individually. The "Exit Quiz" content is intended to reinforce learning by including questions selected from the pre-class assignment, the "Entry Quiz" (repeated questions) and new concepts just introduced.

C. Recitation sessions

Students are seated in the same clusters as during the main class, the number of clusters per section ranging from 3 to 5. Clusters spend the session discussing concepts, and developing written and oral solutions to problems under the individual 'tutorial' guidance of the TA. The TA is *not* expected to present her/his own solution on the board and conventional short quizzes are not given. Quiz activity is confined to the main class.

Up to 30 minutes of the 50 minute session is devoted to cluster discussions and formulation of consensus written solutions by the "Recorder". The "Presenters" from each cluster then present solutions to the entire section. Constructive critique by the other clusters is encouraged. 40% of the recitation grade is awarded for combined oral participation (presentation and critiquing); 60% is for the written solutions. All members of a cluster receive the same grade; a policy that enhances active participation by all cluster members.

D. Clusters in examinations, office hours, homework and review

The three "hour exams" and the final exam include a "cluster question" which may account for up to 25% of the grade on the exam. This question is answered as a collaborative effort by all members of each cluster with a single "master copy" graded and all cluster members receiving the same grade. This feature of RECALL provides students with strong peer group support at the start of the examination and builds confidence for the subsequent individual questions. Peer support is also fostered by the use of cluster office hours and strong encouragement to work homework problems and review for exams in clusters. The use of 'cluster office hours' in a large enrollment course has been particularly valuable. One or two clusters meet with the instructor at a time so that it is possible, with a reasonable time commitment on the part of the instructor, for every student to meet with the instructor before or after each examination. Further details on these aspects of RECALL will be described elsewhere.[4]

Student Responses

Student responses are described in two ways: the grades and knowledge level achieved and summaries of semester-end surveys.

A. Grades and knowledge level

RECALL has been implemented progressively over three semesters. Using identical grading scales, the same textbook and comparable examination questions there has been a progressive narrowing of the grade distribution towards the upper end. In the most recent offering (Spring 1993) there was a sharp rise in the number of A's and no D or F grades were earned.

B. Student survey responses

Student surveys, reported in detail elsewhere,[7] were given at the middle and end of the semester. For this conference paper there is space to highlight only five results.

The survey showed a significant rise in attendance with the implementation of RECALL. 95% of students were present at 90% of the main classes and 100% were present at 95% of the recitation sessions. Contrast with typical lecture attendance of 40% to 70% gives one reason for the rising grade profile with RECALL.

TABLE I summarizes student identification of the two most effective learning avenues. The three highest rated all involve considerable student activity. The lowest rated by far is "Listening to lecture by instructor". TABLE II summarizes student identification of the learning effectiveness of various activities related to the recitation sessions. Again the top three responses involve active learning while the lowest two are both passive.

Table I: Two Most Important Learning Avenues

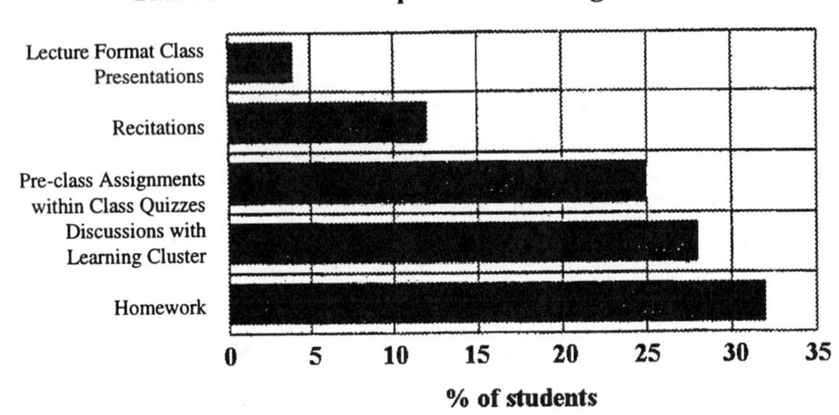

Table II: Two Most Effective Learning Avenues in Recitation

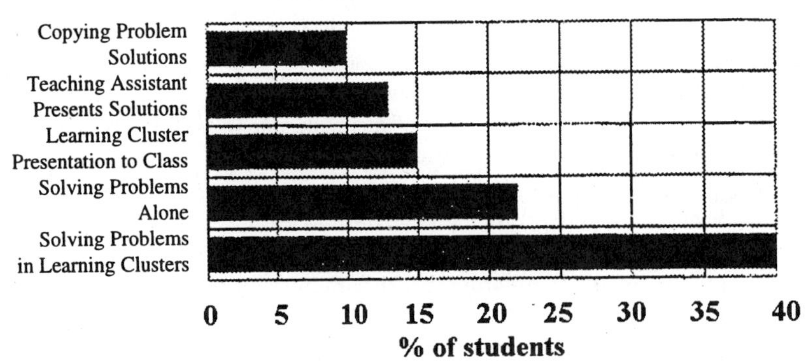

TABLE III is student response to recognition of increased personal responsibility for learning. TABLE IV shows the willingness to take another active learning course. A majority show positive responses to both, but there remains a core of about 20% unconvinced of the effectiveness of the RECALL approach. Individual student responses were also solicited and provided further insight and guidance for the future. Space precludes discussion of these comments here but they will be reported elsewhere.[4]

Table III: I have Assumed More Responsibility for My Own Learning

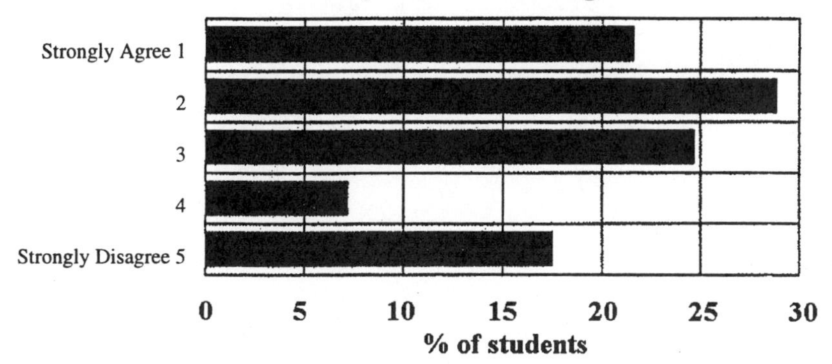

Table IV: I Would Take Another Active Team Learning Course

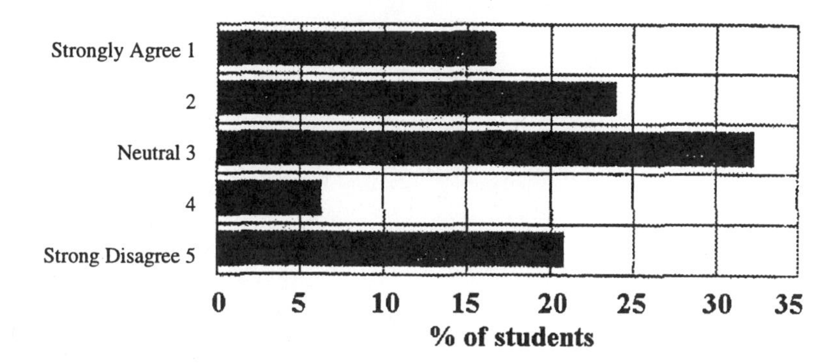

Conclusions

physics course for science majors have been described. The approach (RECALL) permits reinforced cluster active learning to occur even with large enrollments above 200. Implementation of RECALL revealed an overall positive response to active modes of learning and enhanced student awareness of personal responsibility for learning. Despite an increase in grade level, 20% of those enrolled remained unconvinced of the effectiveness of RECALL.

References

1. Charles C. Bonwell and James A. Eison; "Active Learning: Creating Excitement in the Classroom" 1991 ASHE-ERIC Higher Education Report No.1 (George Washington University, Washington D.C.), 1991
2. David W. Johnson, Roger T. Johnson and Karl A. Smith "Cooperative Learning: Increasing College Faculty Instructional Productivity" 1991 ASHE-ERIC Higher Education Report No. 4 (George Washington University, Washington D.C.), 1991
3. Sheila Tobias "They're Not Dumb, They're Different: Stalking the Second Tier" (Research Corporation, Tucson, Arizona) 1990
4. Sheila Tobias "Revitalizing Undergraduate Science: Why Some Things Work and Most Don't" (Research Corporation, Tucson, Arizona) 1992
5. Larry K. Michaelsen "Team Learning: A Comprehensive Approach for Harnessing the Power of Small Groups in Higher Education". To Improve the Academy, Vol. 11, 107 (1992)
6. Barbara J. Duch, David G. Onn and Cynthia A. Cuddy, "Problem-Based Learning in an Honors General Physics Course," Proceedings of the Conference on the Introductory Physics Course, Rensselaer Polytechnic Institute, Troy N.Y. (to be published) 1993
7. David G. Onn, Barbara J. Duch and Cynthia A. Cuddy (to be published).

Reading History: Publishing In the Digital Age

Susan Saltrick

John Wiley & Sons, Inc.

Despite my new tech persuasions, my real passion is history. Its delightfully refreshing to think about the past instead of always wondering about the future — and sometimes what has happened can help one anticipate what's to come. Take Columbus, for example.

Last year, at every turn it seemed, we were being reminded of what used to be called the discovery of America. But, of course, we now know better than to call it a discovery, since Native Americans surely have the honor of being the first to call this continent home. So we use the more accurate word — encounter — to describe events that happened some 500 years ago.

The media hoopla over Columbus brings to mind a similar "discovery" some are making in the high tech arena — I'm referring here to the discovery of content. Content has become the industry's latest love object — and is now second only to "multimedia" as the most overused — and abused — word in our vocabulary.

Hardware and software vendors alike are falling over themselves to acquire "content" — Bill Gates is acquiring a digital image bank, while his company, Microsoft, has purchased an equity position in a British publisher. IBM has established a courseware studio for the creation of very large scale multi-media projects for K-12, including one, appropriately enough, called "Columbus." Meanwhile the industry media are striving to figure out just what content is.

But is content really so mysterious a concept? Content simply means "substance" or "subject matter" — and I'd argue that it's something that traditional publishers have known about for a long time. So perhaps what we have here is not the discovery of content by the pioneers of high tech — but merely an encounter of one culture with another. Then it was the Old World running into the New, now it's New Technology bumping up against the Old.

Surveying the scene, as a publisher, I feel a bit like the Arawaks — the indigenous population of the Caribbean — when Columbus and his caravels arrived. The Europeans may have thought their arrival in the Indies a glorious affair, "the most important event in the history of the world since the crucifixion", while the Arawaks, who'd been living there for millennia, thought "there goes the neighborhood." The priests of high tech, like Miranda on Prospero's isle, may proclaim content a brave new world, while we publishers can only reply, as did Prospero, "Tis new to thee."

Publishers understand content — it is our business to do so. We have long been in the business of creating, packaging, and distributing information. These are the three critical stages leading from the creative spark of an author to a finished product in a reader's — or user's — hands. First, there is the process of authorship which is, of course, the creation of content. Secondly, decisions must be made about how that content is to be packaged — not just its physical wrapping — but, in a larger sense, what form will it take? Finally, the best means of distribution must be considered. Distribution in its broadest sense means more than just the physical means of conveying content from author to reader but also encompasses the marketing and sales of the product.

These three basic aspects of publishing are essential to any business involved in the dissemination of information — whether that information be on paper, on disk, or transmitted via network. However, great challenges lie ahead for us publishers as we move from our analog paper-based, text-driven world to the digital, electronic, and graphical world of the 21st century.

Most of you are, of course, familiar with what's known as the adoption process which is the means by which most college textbooks come to be used on a campus. Our sales force calls on you, the faculty, who the select, or "adopt" a text for use in your course. The necessary information is then relayed to the bookstores which orders and then stocks the materials on their shelves, from where students purchase them for class. Or at least that's the way it's supposed to work.

More and more these days, students and teachers are asking for more than a linear text. We are seeing the advent of a new paradigm for reading, for learning — a shift from a text-based pedagogy to graphically oriented education. Look at a new middle school social studies text — or a book form Dorling Kindersley, the publisher in which Microsoft has invested. In these books, the pictures are not just there to in support of the printed material — they are designed to carry "content" in and of themselves.

This shift from words to pictures is pervasive throughout our society — just look at the buttons on your VCR, or the levers on your dashboard — or observe the move from character-based operating systems to graphical user interfaces. Today educational materials are often collections
of multiple media — a text surrounded by print, video, and software components. Tomorrow these separate components are likely to be combined into one interactive whole. And so, to return again to our model of the publishing process, the nature of authorship must change accordingly.

Over the centuries, conventions (such as tables of contents, headers, footers, indices) and institutions (such as libraries and universities) have been established to facilitate and standardize the task of creating and interpreting text. Publishing has evolved as well. Up until this century publishers were basically printers. But over the past several decades text publishers in particular have developed an elaborate infrastructure to assist our authors in the daunting task of creating a textbook. Copy editors, marketing specialists, developmental editors, production supervisors, book designers, illustrators, photo researchers, the list goes on and on. All these individuals are there to help authors prepare their manuscript for publication. And yet, how much more support will be required in the future! The addition of each new "medium" to a work represents an order of magnitude of complexity.

Perhaps the biggest authorship challenge we face will be in the project design. Texts today are linear, proceeding from Point A to Point B. But multimedia products possess a completely different structure, one in which sequence is dependent on users interaction with the work. Today's authors, publishers, and readers are all masters of the old paradigm. As with any new technology, the first products created with it tend to mirror the technology it is supplanting. As early cars were horseless carriages, so early courseware programs were electronic page turners. Current multimedia products probably resemble their textual antecedents too greatly. We have only just begun to truly exploit the potential of this technology.

So we've seen that publishers face a number of formidable challenges as we look to the next decade and beyond. The process of creating, packaging, and disseminating information is undergoing a transformation. We will succeed? I believe at least some of us will — but none of us will do it all on our own.

The tragedy of the Arawaks is just part of the dark side of the events which began in 1492. Within fifty years of Columbus's first voyage the Arawaks have virtually ceased to exist. Oppression, forced labor, and lack of immunity to European disease decimated the population. Does the publishing industry face a similar fate? Are we headed for a collision with a technologically advanced culture, i.e., the computer industry? Perhaps — if we are slow to change, if we react rather than act, if we fail to learn from history.

We publishers must join forces with the technologists — for we understand how to develop content — they understand how to support that content with technology. Furthermore, we must integrate combined knowledge with that of our marketplace — academia — for it is there that content is both created and consumed. Together, and only together, can we begin to address the issues mentioned here today.

Five hundred and one years ago, the voyage of Columbus set into motion a chain of events that changed the course of history in may ways he could have never foreseen. Today we stand at the threshold of a new era as well. Our actions, rippling forward into the future, are changing the world as we know it. On this voyage, the seas won't always be smooth, and we're sure to encounter some new surprises on the way, but I, for one, am hopeful that the Age of Discovery is just beginning and that — working together — we can create a brave and glorious new world.

On the Exhilaration of Honors Physics

Eric Sheldon, Department of Physics
University of Massachusetts—Lowell

We delight in physics.[1] So, as we convey that joy to each new generation of students in our introductory physics courses, be it our care and delight to ensure that in their all-important exposure to *higher* education we fine-tune our presentation to harmonize with each individual's level of preparation, expectation and potential. For those students with the highest aspirations, the exhilaration of *honors physics* programs provides that compelling inducement to extend hitherto untapped capabilities and insights beyond the ordinary. A will to excel goes hand-in-hand with a readiness, an eagerness, a yearning for learning. To put it in its context,[2] *"The history of honors education is coexistent with the history of higher education. The Socratic dialogue, the Oxford tutorial, the German seminar and the Guild apprenticeship continue to serve as models for contemporary honors programs."* From academic inspiration ("drawing in") comes exhilaration ("enlivening")!

It was this viewpoint in the early 1970's that induced us, at what was then the Lowell Technological Institute (LTI), to introduce a departmental program of Honors Physics and, with the advent of the 1990's, having become The University of Massachusetts - Lowell (UML), to set about broadening this into a university-wide undertaking. Originally, at LTI the service courses in introductory physics were directed to well over 1,000 students from essentially all disciplines. Tracking into "Honors", "Advanced" and "Regular" categories was instigated in 1972, wherein a selection was made from among the uppermost 6% of the incoming freshmen for invitation to the first of what then, happily, was a 4-semester course in Honors Physics. In this program, some 60 students received 4 hours of instruction per week, enhanced with elaborate demonstrations such as had been developed at ETH-Zürich and described by Meiners,[3] with closely coordinated special *weekly* labs. The adopted text then was that by Bueche.[4] Although (declared) physics majors were accorded preference in admission to Honors Physics if they had the requisite academic credentials, it has never been the policy to include them automatically within the honors program. Rather, students have been selected (after a preliminary scrutiny of their SAT scores to ensure that, particularly in mathematics, they ranked distinctly above average and had good high-school standing among the upper 10-15%) on the basis of intellectual ability, creative potential and task commitment. Unfortunately, a combination of circumstances caused the Honors Physics Program to pass into abeyance throughout the 1980's despite its evident success. However, with the formation in 1991/92 of the Commonwealth's "UMASS" system comprising 5 campuses (at Amherst, Boston, Dartmouth, Lowell and Worcester [Medical School], of which the first three had well-established University-wide Honors Programs), the Honors Physics Program was resumed and conjoined with a special section of the mathematics course, Calculus I/II. Again, the above criteria were adopted for selection of participants (this time, we aimed for only a *single* section, comprising 21 outstanding students). In the physics course (as before, 4 hours per week, 3-credit, with special weekly 4-hour, 2-credit coordinated labs but, alas, for only 3 semesters) the text of Hudson and Nelson[5] was adopted, whereas in the mathematics course (4 hours per week, 4-credit) an experimental approach based on a preliminary text by Wattenberg[6] was employed. Extensive use is made of computers for calculation (including *Mathematica®* software) and electronic mail (in conference discussion mode). The benefits have been evident over the years, in the form of enhanced enrollment of superior students (many applied or transferred to the University and/or the Department who otherwise would not have done so), the courses have advantageously influenced regular courses and honors students have been stimulated to engage on creative projects, as instanced by the first class' design and construction of a prize-winning 12-1/2" Schmidt-Cassegrain reflector telescope, while the latest class included a student who has patented and is marketing a 3-dimensional magnetic-cube *Akimbo®* puzzle! Evidence of the acceptance of an honors "track" is provided by the fact that in the coming academic year these offerings will be augmented by honors sections in the Departments of Art, Chemistry, English and Music. During the year we also registered for institutional and regional membership in the *National Collegiate Honors Council (HCNC),*[7] who have provided invaluable documentation[2,8,9] and assistance (as have other publications[10] and organizations[11]) and in April inducted the first 110 students (*e.g.*, those earning a grade-point-average of 3.5 during their first semester or after their first year of full-time study) into the newly-formed *UMass-Lowell Freshman Honor Society*, affiliated with Alpha Lambda Delta.

The groundwork for the venture into this more broad-based honors-level instruction was established during the mid-1980's in recommendations put forward by a *Presidential Commission to Study Academic Organization at The University of Lowell* and elaborated in the deliberations of a *President's Council for Implementation*, which resolved that (among other objectives), *"A university-wide honors program be established at the University of Lowell, under the administration of an honors director.... An honors curriculum board ["Honors Council"] be appointed to review the structure of the program and approve the content and methodology of courses proposed as honors courses .... The program leading to an honors diploma [to] include a nucleus of honors core courses ...."* This was accepted by the Administration, with only minor modification but with the proviso that *"it simply will not be possible to provide these [resources] in the near future (approximately the next two years)."* There the matter rested until the assimilation of The University of Lowell into the state-wide University of Massachusetts system in 1992, when a new *Council on Undergraduate Teaching* was convened, with an *Honors Subcommittee* as one of its subsidiaries. It is through the deliberations of this 13-member Honors Subcommittee that a substantive plan for the development of a comprehensive Honors Program has been drawn up and now, having been passed by our Faculty Senate, awaits approval by our Chancellor and Trustees.

The rationale in seeking to expand the Honors Program beyond purely departmental ventures lies in the recognition that the institution *as a whole* has to affirm its commitment to such an enterprise if it is to be truly representative of the academic standards that it espouses. In setting up a comprehensive honors curriculum and advisory program, one of the important goals[10] to be borne in mind is *"an expansion of the interests of the student. The advising program not only encourages academic interests but also stimulates the student to expand his or her activities to include new advocations. Exposure to cultural events, libraries, museums, sporting events, and other examples of the rich resources on a college campus expands the student's understanding of his or her world and of the possibilities available for diverse educational experiences."*

Our tentative timetable for the growth of the honors enterprise is to devote the coming academic year to developing fresh departmental offerings (as delineated above) in two major Colleges (Arts & Sciences and Finc Arts), reaching out in 1994 to the remaining three undergraduate Colleges (Engineering, Health and Management) and, in 1995 implementing the full-scale Honors Program officially throughout the University with a view to the awarding of baccalaureate Honors Diplomas in 1999 (or possibly 2000) and joint baccalaureate/master's degrees in 2000 (or possibly 2001) as a "millennium aspiration". This is why the Honors track is envisioned as an essential ingredient in an introductory physics course, for it possesses among its most vital components: motivation, aspiration, inspiration and exhilaration.

References

1. William Shakespeare, *Macbeth* II, iii, 56. An adaptation from the full quotation, *"The labour we delight in physics pain."* Cited as a "footer" on Oxford University's Department of Nuclear Physics notepaper by Denys Wilkinson. Alas, *Macbeth*'s other quotation [V, iii, 45] is violently antithetic: *"Throw physic to the dogs; I'll none of it."* Perhaps equally unfelicitous would have been: *"Take physic, pomp; Expose thyself to feel what wretches feel."* [*King Lear* III, iv, 33]!
2. C. Grey Austin, *Honors Programs: Development, Review, and Revitalization*, Monographs in Honors Education (National Collegiate Honors Council, 1991), p.10.
3. Harry Meiners (ed.), *Physics Demonstration Experiments*, Vols. I & II (Ronald Press Co., New York, 1970); see also R.M. Sutton (ed.), *Demonstration Experiments in Physics* (McGraw-Hill, New York,

1938); Eric M. Rogers, *PHYSICS for the Inquiring Mind* (Princeton University Press, Princeton, 1960); A.B. Mlodzeevskii, *Lecture Demonstrations in Physics*, translated from Russian by Howard V. Robinson and Kira V. Robinson (American Institute of Physics, New York, 1964?), reviewed in *Am. J. Phys.* 1964, **32**, 56; Rocco C. Blasi (ed.), *Physics Fun and Demonstrations with Professor Julius Sumner Miller* (Central Scientific Company, Cenco, Chicago, 1968); G.D. Freier and F.J. Anderson, *A Demonstration Handbook for Physics* (Minneapolis, 1972); bibliographic references in *Physics Apparatus, Experiments, and Demonstrations: A Bibliographic Guide* (American Institute of Physics, New York, 1965, a publication of the Center for Educational Apparatus in Physics, obtained from University Microfilms) and *Physics Experiments and Demonstrations: Selected from the American Journal of Physics 1933-1964* (American Institute of Physics, New York, 1965, publication R-182 of the Center for Educational Apparatus in Physics, obtained from University Microfilms, Ann Arbor, Michigan).

4. Frederick J. Bueche, *Introduction to Physics for Scientists and Engineers* (McGraw-Hill, NY, 1969).
5. Alvin Hudson and Rex Nelson, *University Physics*, 2nd ed. (Saunders College Publishing, Philadelphia, New York, Chicago, San Francisco, 1990).
6. Frank Wattenberg, *Calculus in a Real and Complex World*, Preliminary Edition (Wadsworth, 1992; PWS-Kent (Wadsworth Division), 1993).
7. The National Collegiate Honors Council, established in 1966, currently has over 480 institutional members; the annual fee is $200 (institutional), $35 (individual/faculty), $15(individual/student) and $30 (regional) payable to the Executive Secretary/Treasurer, Dr. William P. Mech, Honors Program, Boise State University, 1910 University Drive, Boise, ID 83725-1125 [Tel.:(208) 385-1208; BITNET E-mail: AHPMECH@IDBSU], from whom further information and enrollment materials may be obtained.
8. Samuel Schuman, *BEGINNING IN HONORS: A Handbook* (National Collegiate Honors Council, 1989).
9. Jacqueline Reihman, Sara Varhus and William R. Whipple, *Evaluating Honors Programs: An Outcomes Approach* [National Collegiate Honors Council, 1990]. Other useful publications for smaller institutions: *Honors in the Two-Year College* (NCHC, 1983) and *Honors Programs in Smaller Colleges* (NCHC, 1988).
10. Deanell R. Tacha, "Advising and Interacting Outside the Classroom," in *Fostering Academic Excellence Through Honors Programs*, ed. by Paul Griedman and Riva C. Jenkins-Friedman (Jossey Bass, San Francisco and London, 1986), p.56.
11. *Guidelines and Information for Honors Program Faculty* (University of Georgia Honors Program, Athens, GA, 1988), kindly made available by the Honors Director, Dr. Lothar L. Tresp [(706) 542-3240].

Effectively Attacking Misconceptions In Introductory Physics Problem-Solving

Daniel M. Smith, Jr.
Northeastern University, Boston

Abstract

Observations have been made of engineering freshmen attempting to solve problems in both workshop, and individual tutorial settings. Through questions and dialogue, several recurring difficulties (consistent with physics education research) have been pinpointed in the use of vectors, kinematics equations, Newton's Laws, and energy conservation. Schemes are presented here for attacking misconceptions in kinematics and in the use of Newton's Laws, in the context of problem-solving. Also, textbooks are criticized for certain practices which may hinder learning. Grades of students attending the workshop, minorities in particular, are skewed to the higher end of the distribution compared to non-workshop participants.

Introduction

Physics education research has clearly documented the poor level of student understanding of basic physics concepts after the standard course.[1] This is a report on another effort to help students to attain a scientifically correct view of the physical world, this time by intervening in the process by which students learn to solve the standard end-of-the-chapter problems. Student tendencies to memorize algorithms for problems of a certain type are discouraged by insisting that they display, from the beginning of their solution attempt, a real understanding of a problem through drawings, and verbal and written explanations elicited by close questioning. By examining students' thinking processes in all of these ways, their erroneous notions about the physics are uncovered and, in many cases, overcome.

Physics Workshop

Freshmen engineering students in the calculus-based introductory physics course are led by an instructor in solving homework problems during Physics Workshop (PW), a one and a half hour, twice weekly, voluntarily attended meeting. (It is sponsored by NUPRIME, Northeastern University Progress in Minority Engineering, a support organization especially for African-American, Hispanic, and Native American engineering students.) The objective of PW is to transmit to minority engineering freshmen (though PW is open to anyone) the unspoken values of the physics community: (1) drawing, or picturing the physical problem, (2) describing the picture in terms of physical ideas, (3) describing physical ideas by basic equations, and (4) remembering important formulas, or equations. It is commonly observed that students value (1) finding the formula which uses the given data with only one unknown, and (2) avoiding understanding, or remembering formulas since, in many cases, formula sheets are provided along with the examinations.

A typical workshop session begins with a short, very informal introduction or review of a physical idea with emphasis on the basic equations, i.e. those which the student should remember. Students then begin work on a homework problem which has been determined by the instructor to be challenging. Only basic equations can be used in a student's attempted solutions. Close observation of each student's progress, or lack thereof reveals misconceptions and difficulties, giving the instructor the opportunity to nudge students along using questions and hints. Rarely is a solution worked out for them. Misuse of algebra is frequently corrected, as well as disorganized work. Only two or three problems of medium difficulty are completed by most students in each one and a half hour session; written solutions are distributed at week's end for problems not covered.

Kinematics

Common preconceptions and misconceptions in kinematics have been extensively documented.[2] However there is much less written about the weaknesses of textbooks in helping students to overcome their erroneous notions. These weaknesses include (1) failing to provide an adequate solution strategy for moderately difficult problems; (2) presenting so many equations that formula hunting is almost encouraged; (3) writing x, and v instead of x(t), v(t), or v(x), and improperly giving distance, $x-x_0$, primacy over position, x(t); (4) not beginning example problem solutions from basic equations; and (5) inadvertently emphasizing formulas and numerical values in worked examples rather than physical insight gained from diagrams. Perhaps weaknesses (2), (4), and (5) are inherent to any textbook and cannot be rectified, but at least instructors should be aware of them.

Young,[3] offers an excellent strategy for solving problems with constant acceleration, with one exception, which is the source of complaint (1) above. Students read that "It always helps to make a list of quantities such as x, x_0, v, v_0, and a. In general, some of these will be known and some unknown. Write down those that are known." The problem with this advice is that what is known and what is unknown is a function of time, and my observations are that this is a source of student difficulty which must be revealed, and confronted in a way later shown.

Listing five kinematic equations for motion with constant acceleration, as in Halliday and Resnick,[4] along with a "missing quantity" for each equation ($x-x_0$, v, t, a, or v_0) encourages the students' bad habit of equation hunting, noted in (2) above. Also, in common with most other texts, is the failure to make functional dependence (x(t), v(t), and v(x)) explicit which feeds into the problem discussed in the previous paragraph. Emphasizing distance, $x-x_0$, over position, x(t), is one facet of this problem highlighted by Arons.[5]

A problem-solving strategy for kinematics which allows an instructor to more easily diagnose student difficulties is as follows: (1) make a drawing of the problem, complete with a coordinate system; (2) attach a complete set of data to every significant point of the object's motion; and (3) use only the three equations (to be memorized) $x(t)=x_0+v_0t+0.5at^2$, $v(t)=v_0+at$, or $v(x)^2=v_0^2+2a(x-x_0)$ in constructing a solution. Emphasis is placed on students completing steps (1) and (2). Initially, just making appropriate substitutions into equations is sufficient progress, since the aim is to avoid having the student bury physics difficulties in the algebra.

Figure 1. Student attempt to use kinematics problem-solving strategy. "An airplane travels down the runway before taking off. If it starts from rest, moves with constant acceleration, and becomes airborne in 16.0s, what is its speed, in m/s, when it takes off?" (Ref. 3, p. 51) Note student error in a.

Figure 1 shows a student's attempt to carry out the above strategy; his confusion is apparent. This student would be questioned about his choices for acceleration, i.e. asked to justify them, and thereby led to realize his errors. (He would not be simply told what is wrong, and how to fix it.) He would be asked to remove the subscript on the time, position, and velocity since the 1 has no meaning in the context of the three equations. Though this problem could be easily solved using one of the equations in which acceleration is already eliminated, this is strongly discouraged since the aim is to emphasize that three equations are sufficient, and therefore the mental work should go into understanding the relation between the object's motion and the data.

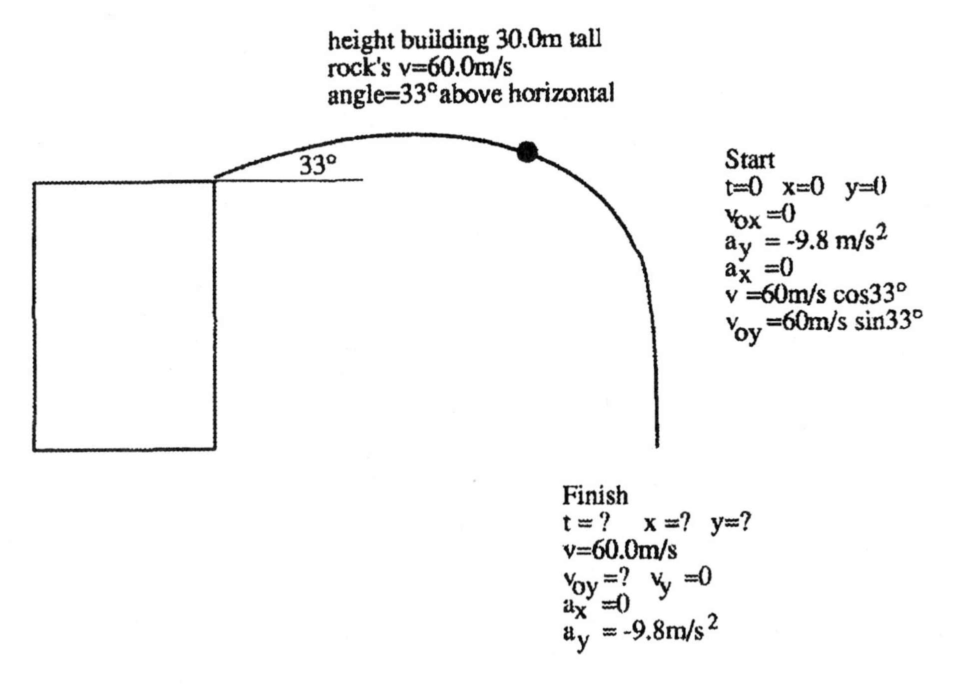

Figure 2. Another student's attempt to use same strategy. "A man stands on the roof of a building that is 30.0m tall and throws a rock with a velocity of magnitude 60.0m/s at an angle of 33.0° above the horizontal. Calculate a) the maximum height above the roof reached by the rock; b) the magnitude of the velocity of the rock just before it strikes the ground; c) the horizontal distance from the base of the building to the point where the rock strikes the ground." (Ref. 3, p. 78) Note error in y and v_y at finish.

Figure 2 shows the work of a student who has not drawn coordinate axes, nor has she attached data to the object's motion as directed. Nonetheless, two significant errors are revealed: she considers the rock's y-coordinate upon landing as unknown, and the y-component of velocity upon landing is set to 0. The latter error is common; students are asked to realize that such a velocity is appropriate at that position only *after* the object has landed.

Of course there are other problem-solving strategies for kinematics which may work just as well for some students. But this one has the advantage of demonstrating, as clearly as possible, what the student's understanding is of the relation between the actual motion of an object and the data describing that motion. Some students rebel against the prescribed regimen, preferring distance to position, or elapsed time to clock time; this is possibly a symptom of their difficulty with instantaneous quantities.

Newton's Laws

Student misunderstandings in the application of Newton's Laws have been thoroughly summarized by Arons.[6] Noted here are possible shortcomings of standard textbook treatments of problem-solving. One is that it may be counterproductive to encourage a beginning student to disassemble a multi-component mechanical system in order to understand it, given the weak physical intuition of many students. (Eventually a student must develop this skill since some problems can be solved in no other way.) What may be even worse is to have a student shrink the system to a point, as some textbooks recommend. My observations indicate that students need practice in answering such questions as "What must be the direction of the force exerted by the string, given that the mass will fall if the string is cut?", or "What would happen to the ladder if the wall gave way? Now, what does this tell you about the force exerted on the ladder by the wall?". Students pushed too soon into the abstraction of the free-body diagram risk losing contact with the physical picture which allows such questions to be answered.

Closely related to the above criticism is the fact that for many students, drawing a free-body diagram is not equivalent to isolating the system; the procedure is treated perfunctorily. For example, vectors representing "normal forces" are frequently included inappropriately in free-body diagrams, as illustrated by Figures 3 and 4. In the former case a student incorrectly shows a normal force exerted on the small mass, while in the latter a common student error is indicated by the dashed vectors.

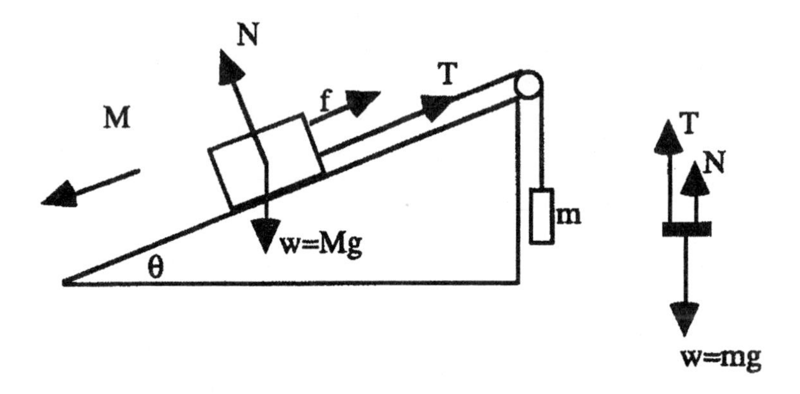

Figure 3. Student attempt to show all forces exerted on the two masses. Note incorrect "normal force" on small mass.

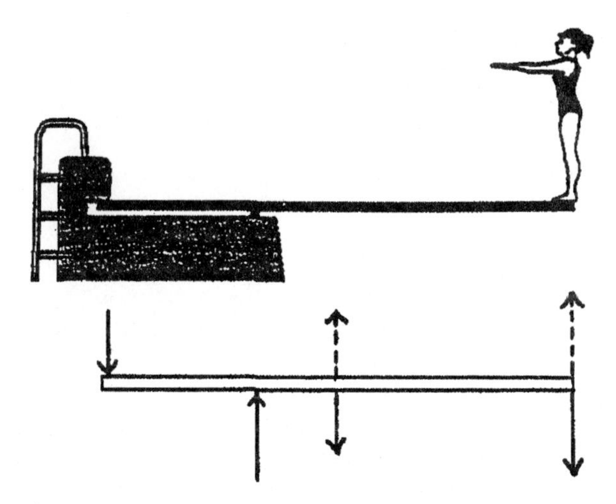

Figure 4. Dashed lines indicate common error made by students asked to draw vectors representing all forces acting on the diving board. (Ref. 3, p. 307)

Numerous observations like these convince me that there are incredibly tenacious misconceptions associated with the term "normal force" which weaken a student's problem solving ability. To counter this tendency to use the term indiscriminately (perhaps instructors should avoid it altogether), the following problem-solving strategy is recommended. A student must begin by (1) drawing a realistic sketch of the problem (not a free-body diagram) using vectors to represent all of the forces acting on the object(s) to be considered in the force equation(s); and (2) by explaining in words the cause of every force (as recommended by Arons[7]). Students must be directed away from distractions such as making a list of knowns and unknowns (unhelpful until after force equations have been written) or computing the frictional force, and directed towards these first two goals. At this point at least some of the misconceptions should be revealed, and one can begin the demonstrations and discussions which are part of the long process of overcoming them. Students should successfully complete the first two steps before (3) drawing a coordinate system, as appropriate, then (4) writing a force equation, or equations for the object(s) under consideration.

	1st Quarter		2nd Quarter		3rd Quarter		
Grade	W'shop	Non-W'shop	W'shop	Non-W'shop	W'shop	Non-W'shop	Class
	(16)	(18)	(15)	(12)	(12)	(12)*	
A	18.75%	5.56%	20.00%	8.33%	0%	0%	
B	62.5%	33.33%	40.00%	25.00%	50.00%	25.00%	1992
C	6.25%	38.89%	26.67%	16.67%	33.33%	50.00%	
D	12.5%	11.11%	6.67%	33.33%	16.67%	16.67%	
F	0%	11.11%	6.67%	16.67%	0%	8.33%	
	(18)	(18)	(17)+	(13)#	(14)	(8)	
A	5.56%	5.56%	5.88%	7.69%	7.14%	12.5%	
B	22.22%	16.67%	17.65%	23.08%	21.43%	25.00%	1993
C	44.44%	27.78%	29.41%	30.77%	50.00%	50.00%	
D	22.22%	16.67%	29.41%	15.38%	21.43%	12.5%	
F	5.56%	33.33%	17.65%	23.08%	0%	0%	
	(6)	(13)	(5)	(8)	(4)	(7)	
A	16.67%	7.69%	0%	0%	0%	14.29%	
B	50.00%	23.08%	40.00%	25.00%	0%	14.29%	1994
C	16.67%	38.46%	40.00%	12.50%	50.00%	14.29%	
D	16.67%	15.38%	20.00%	50.00%	50.00%	28.57%	
F	0%	15.38%	0%	12.50%	0%	28.57%	
	(15)	(39)	(11)	(43)			
A	6.67%	5.13%	9.09%	6.98%			
B	53.33%	30.77%	27.27%	27.91%			1997~
C	33.33%	23.08%	27.27%	32.56%			
D	0%	17.95%	36.36%	20.93%			
F	6.67%	23.08%	0%	11.63%			

*Table I. Grades of minority Workshop participants and non-participants. Number of students is given in parenthesis. * 3 new students are included. # 2 new students are included who did not continue to the 3rd quarter. + 1 new student is included. Class of '97 Workshop students are all in the same lecture section, and are compared against the rest of the students in the section (3rd Quarter data not yet available).*

Physics Workshop Benefits

The strategies above, and others, have been developed over the past few years and instituted in the Physics Workshop. Comparisons are made between the grades of Workshop participants and non-participants in Table I. One sees many fewer F's in all quarters for participants, but during the first quarter there is a significant enhancement of grades of B and above for each class except the class of 1993, where the enhancement is for grades of C and above.

References

1. A broad view of the literature is given by A. B. Arons, A Guide to Introductory Physics Teaching (Wiley, New York, 1990). See also I. A. Halloun and D. Hestenes, "The initial knowledge state of college physics students," Am. J. Phys. **53**, 1043-1055 (1985).
2. M. L. Rosenquist and L. C. McDermott, "A conceptual approach to teaching kinematics," Am. J. Phys. **55**, 407-415 (1987); Arons, op. cit., Chapt. 2.
3. H. D. Young, University Physics (Addison-Wesley, Reading, MA, 1992), 8th edition, p.41. Of course this advice is commonly found, in one form another, in many texts.
4. D. Halliday and R. Resnick, Fundamentals of Physics (Wiley, New York, 1988), 3rd extended edition, p. 20. Other texts also treat the equations in this way.
5. Arons, op. cit., p. 21.
6. Arons, op. cit., Chapt. 3; L. C. McDermott, "Research on conceptual understanding in mechanics," Phys. Today **37**, 24-32 (1984).
7. Arons, op. cit., Chapt. 3.

Halliday & Resnick's[1] PHYSICS Through the Years[2]

David L. Wallach
Penn State — McKeesport

As Halliday and Resnick's Physics evolved from edition to edition so did the figures in the book. Certain figures changed in ways which changing national priorities, others reflected the success of the text.

We are all here to reflect on the impact of Robert Resnick on the teaching of introductory physics. There is no question that Halliday & Resnick's Physics has made a profound impact on the way we go about our business of transforming high school graduates to juniors in science and engineering baccalaureate programs. It marked a change in style from the engineering approach epitomized in Sears & Zemansky's University Physics. This week we are asking questions as to whether it is time to change our heading again or stay the course. We have all seen the deterioration both in preparation and intellectual discipline of entering students and, in contrast, the increasing knowledge and skill required of our graduates. The fine series of texts begun, but not completed, by A.P. French at M.I.T. bears witness to our frustration as physics professors.

1958 **Physics,** Preliminary Edition
1960 **Physics,** 1st Edition
1966 **Physics,** 2nd Edition
1970 **Fundamentals of Physics,** 1st Edition
1978 **Physics,** 3rd Edition
1981 **Fundamentals of Physics,** 2nd Edition
1986 **Fundamentals of Physics,** 2nd Edition, Revised
1988 **Fundamentals of Physics,** 3rd Edition
1992 **Physics,** 4th Edition
1993 **Fundamentals of Physics,** 4th Edition

Figure 1. The editions of Halliday & Resnick's Physics and Fundamentals of Physics

My experience with Halliday and Resnick's Physics began in 1958 with the mimeographed, loose-leaf, preliminary edition of volume one. Figure 1. Robert Resnick and Walter Eppenstein taught our freshman class here at R.P.I. with the aid of Harry Meiner's unforgettable lecture demonstrations. A few years later I was an industrial physicist pursuing a master's degree at night at Carnegie Tech. To study for the exam my advisor said to a couple of us it would be adequate to know Halliday & Resnick and A.P. French's Introduction to Modern Physics cold and also know the content of the graduate quantum mechanics course. I made oral presentations of the first hardbound edition of Halliday and Resnick to my study partner and he did the other to me. It worked. Shortly thereafter I started teaching at Penn State's McKeesport Campus using the second hardbound edition. Over the past 25 years I have used every edition of Physics and Fundamentals of Physics.

In the course of using each edition I have noticed an interesting evolution. We all know that there are more chapters and problems as well as some improvements in the presentation of some topics. However, have you looked closely at the figures and their evolution? Let me share with you the history of just three of the figures.

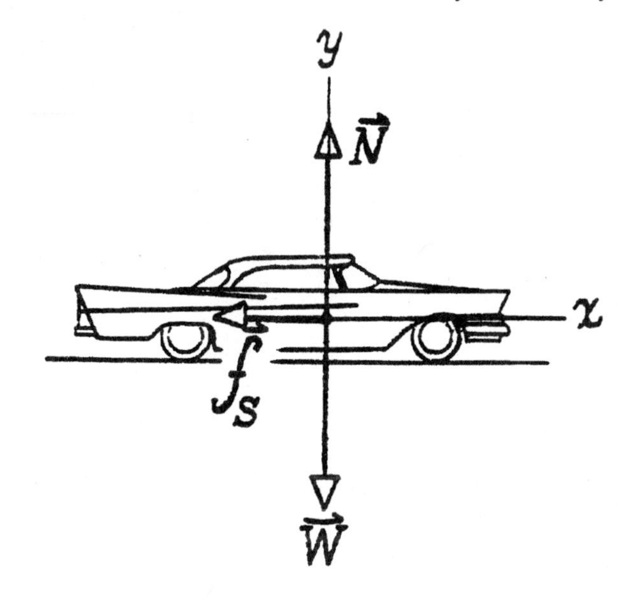

Figure 2. Physics, Preliminary Edition, Figure 6.4

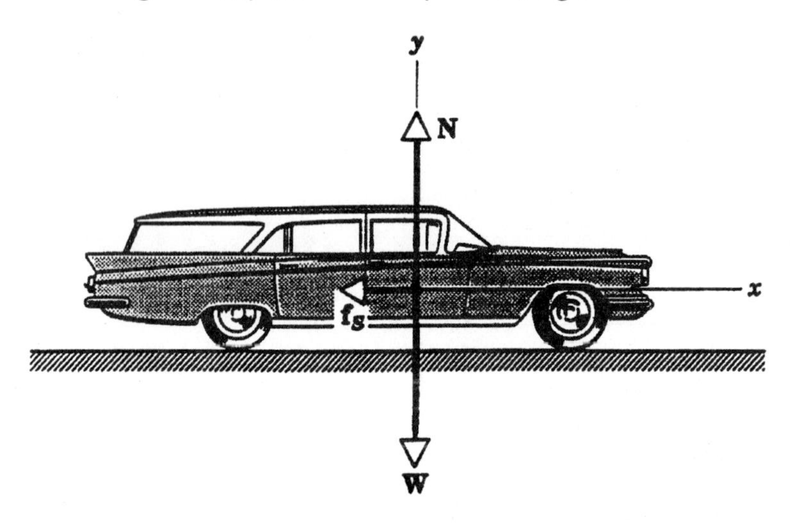

Figure 3. Physics, First Edition, Figure 6.4

One example worked out in every introductory physics text is that of a car stopping on a level road due to friction. Figure 2. In the 1958 mimeographed preliminary edition the car is a 1957 Plymouth 2-door hardtop. By the time the first hardbound edition appeared two years later it was a 1959 Buick wagon. Figure 3. Evidently this young assistant professor's family was expanding and he correctly envisioned the financial success that was just over the horizon. In the second hardbound edition, in 1966, after the unquestionably successful first edition, the car is a Lincoln Continental. Figure 4. The image is tarnished, however, because the fuel filler port is on the wrong side of the car. Of course, in the those days we did not pump our own. This figure was used in the 1978 third edition as well as the first edition of Fundamentals of Physics in 1970.

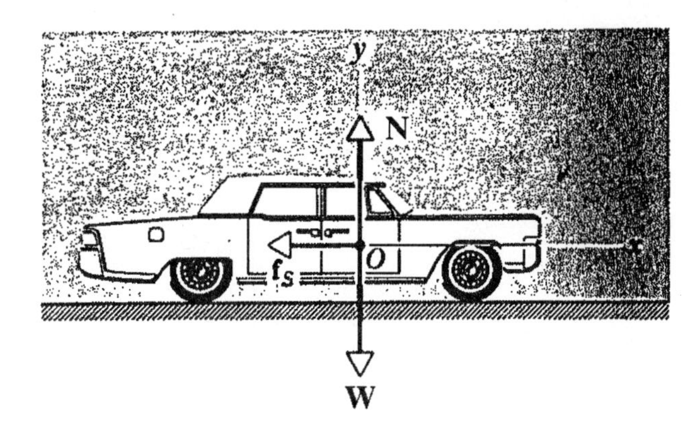

Figure 4. Physics, Second Edition, Figure 6.5

However, with the crisis brought on by what the oil companies purported to be an Arab oil embargo in 1981, the second edition of Fundamentals appeared with this fuel-efficient VW bug. Figure 5. In 1988 the third edition of Fundamentals appeared with a generic car that has defied identification by my students. Figure 6. Generics were now also popular in the supermarkets. Now financially secure, we find the authors' choice for the fourth edition of Physics in 1992 to be a Mercedes. Figure 7. We can see that retirement is on his mind this year when the fourth edition of Fundamentals was just released. The book says Jaguar, but my students say it is a Porsche. Figure 8.

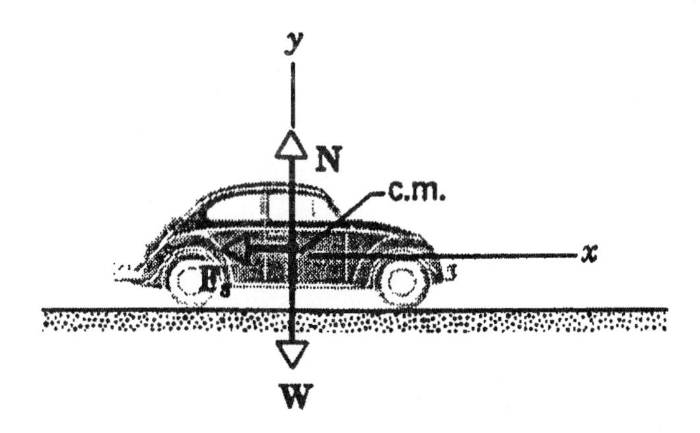

Figure 5. Fundamentals of Physics, Second Edition, Figure 6.5

Let me illustrate this evolution in the life of an author with another series of figures. We all know the example of the apparent weight of a person in an accelerating elevator. Figure 9. In the preliminary edition we the person is a beautiful young lady. Now remember the car in the corresponding edition was the '57 Plymouth two-door hardtop. Remember also that the first printed edition had a '59 Buick wagon —- suitable for a family. Here the coed has been replaced by a more mature lady in a tailored suit. Figure 10. Now, just as the Buick wagon was replaced by the Lincoln, we see that Einstein is now in the elevator. Figure 11. This figure persisted although the newest editions have slight modifications. As you can see, the force vectors have been properly placed at the coinciding centers of mass and gravity. However, my students ask me why the "P" is upward. I try to tell them that all matter, including fluids, must eventually go downward except when you are learning to change the diaper of your first son. I am glad to see that the latest figures have been modified so the discussion of fluid dynamics is begun in a later chapter. Figure 12. We should also note that Einstein, even though dead for cover 30 years, has kept up with the times and removed his tie.

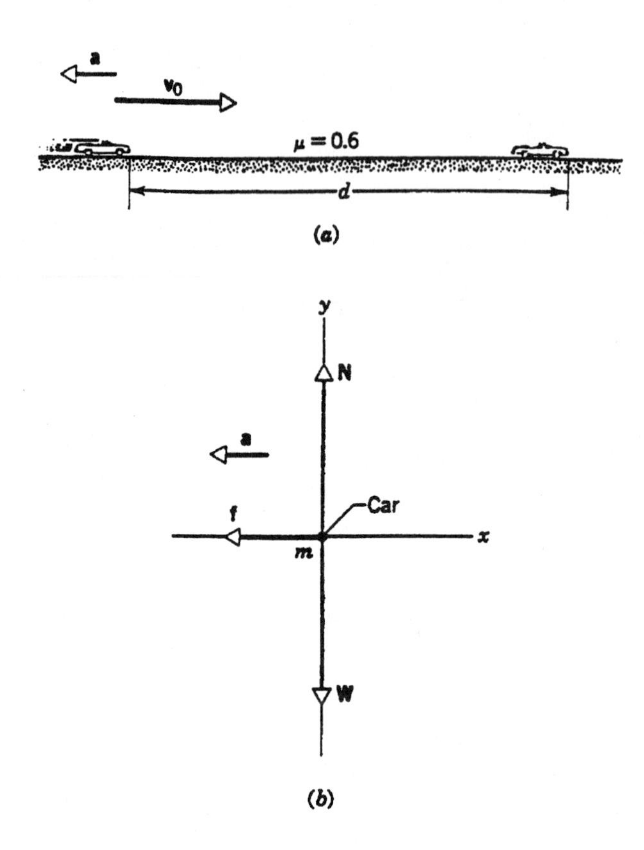

Figure 6. Fundamentals of Physics, Third Edition, Figure 6.5

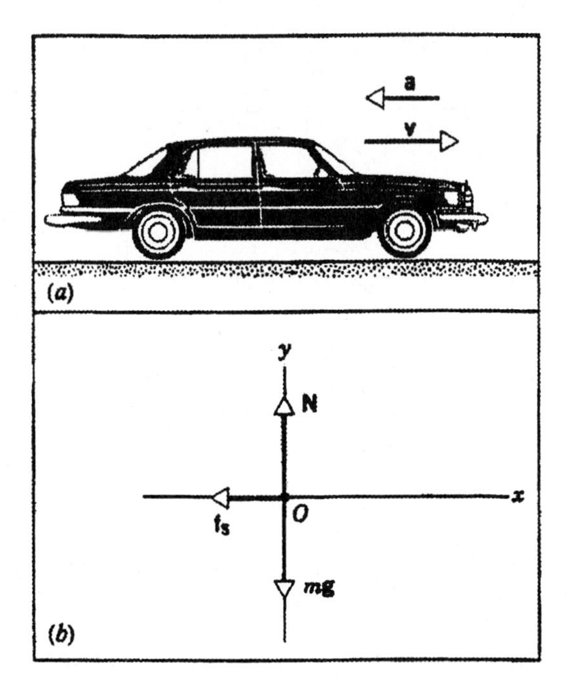

Figure 7. Physics, Fourth Edition, Figure 6.5

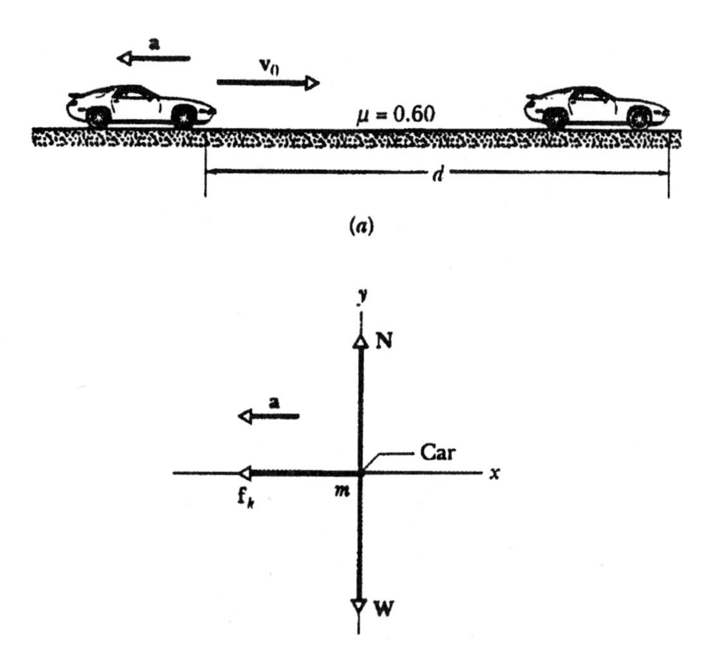

Figure 8. Fundamentals of Physics, Fourth Edition, Figure 6.5

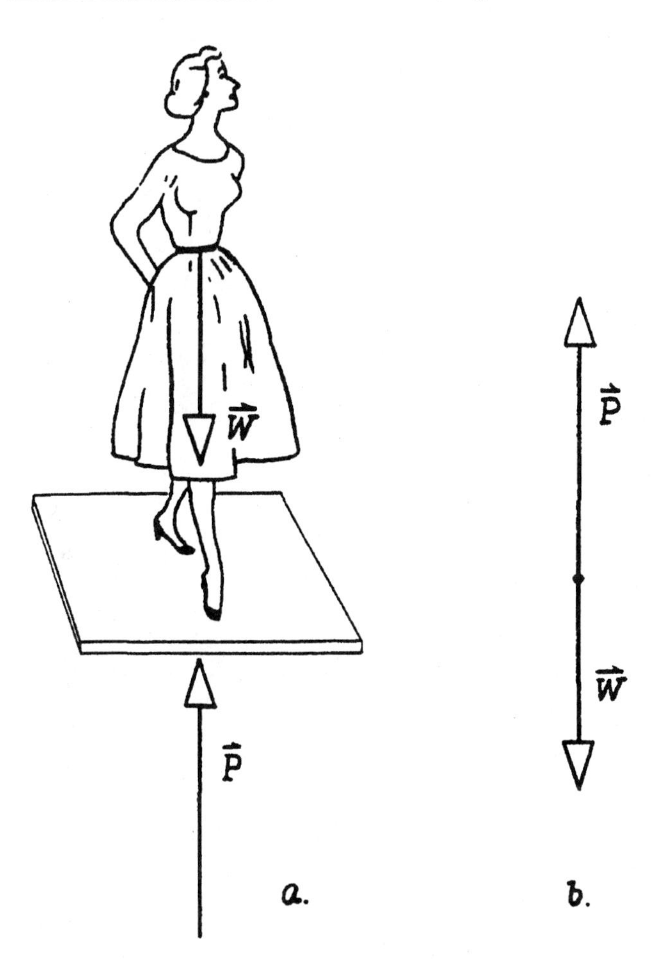

Figure 9. Physics, Preliminary Edition, Figure 5.9

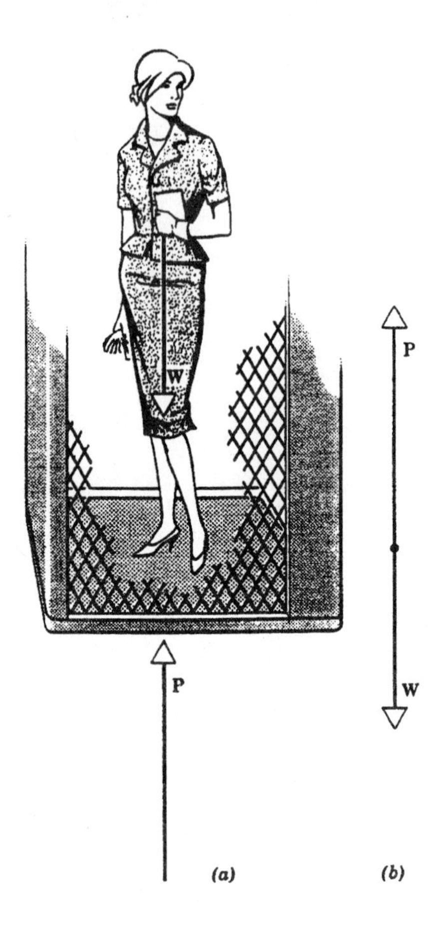

Figure 10. Physics, First Edition, Figure 5.9

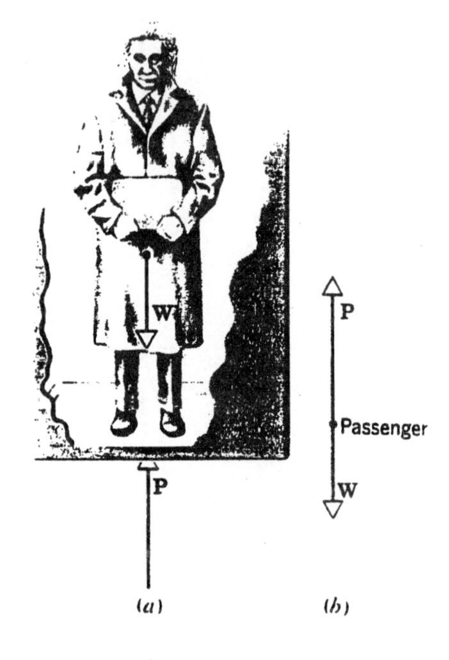

Figure 11. Physics, Second Edition, Figure 5.10

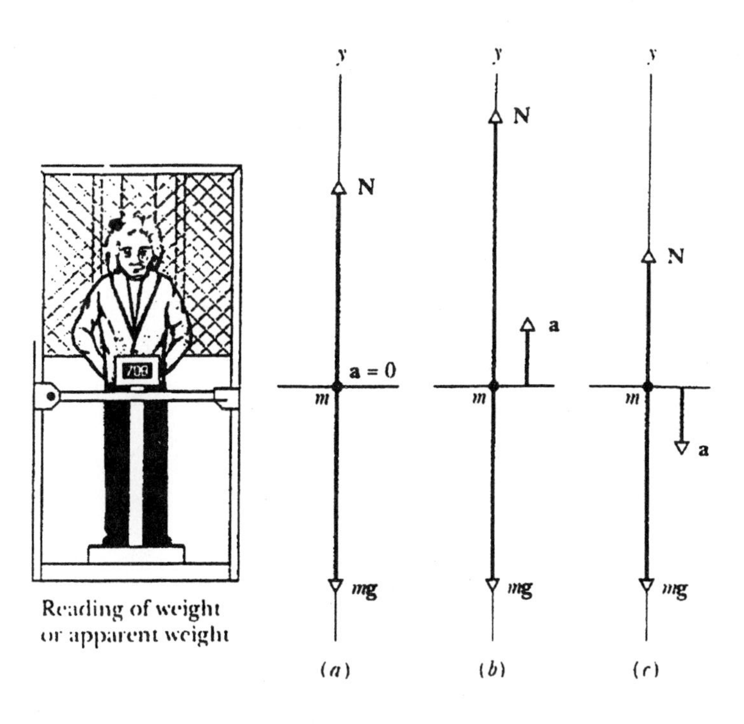

Figure 12. Fundamentals of Physics, Fourth Edition, Figure 5.30

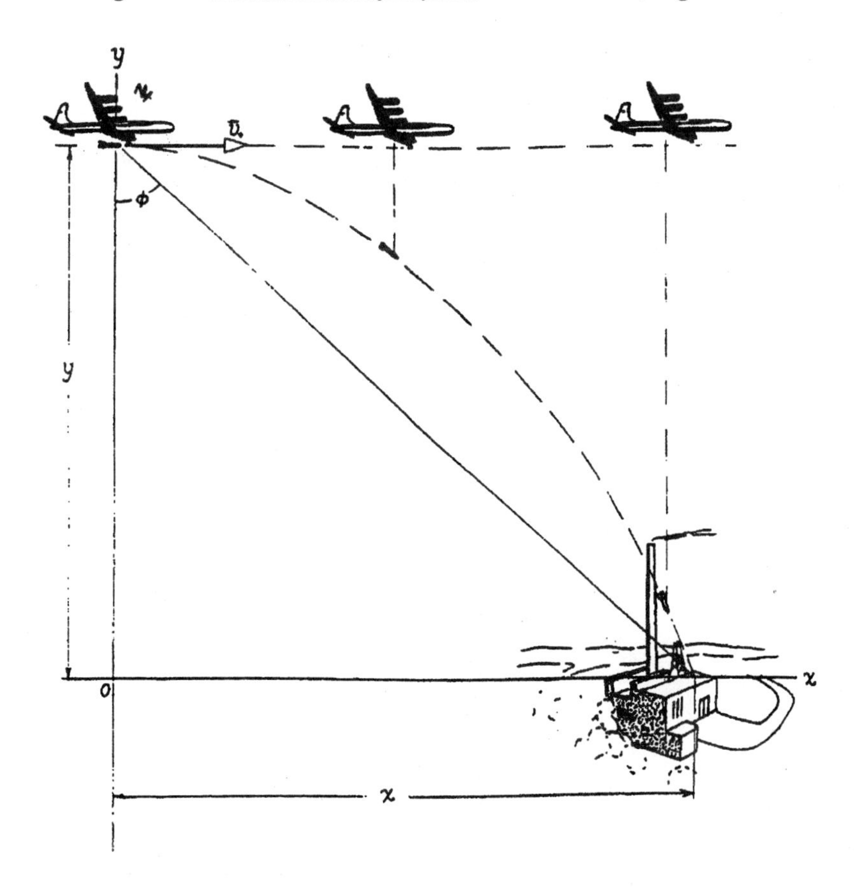

Figure 13. Physics, Preliminary Edition, Figure 4.4

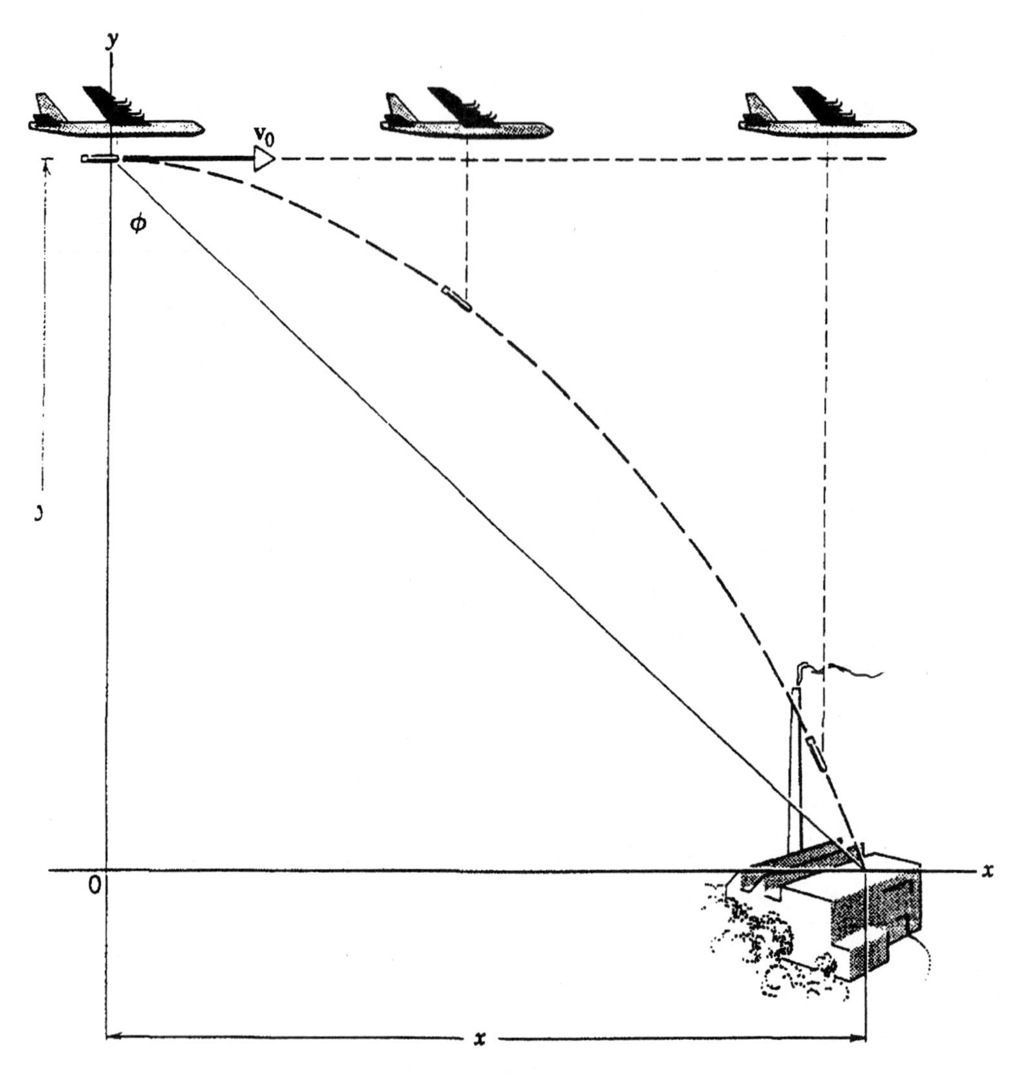

Figure 14. Physics, First Edition, Figure 4.6

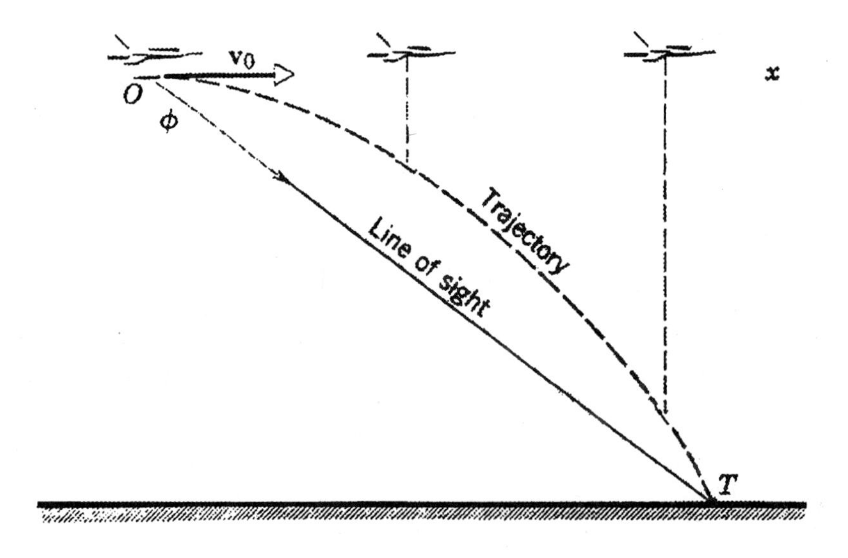

Figure 15. Physics, Second Edition, Figure 4.3

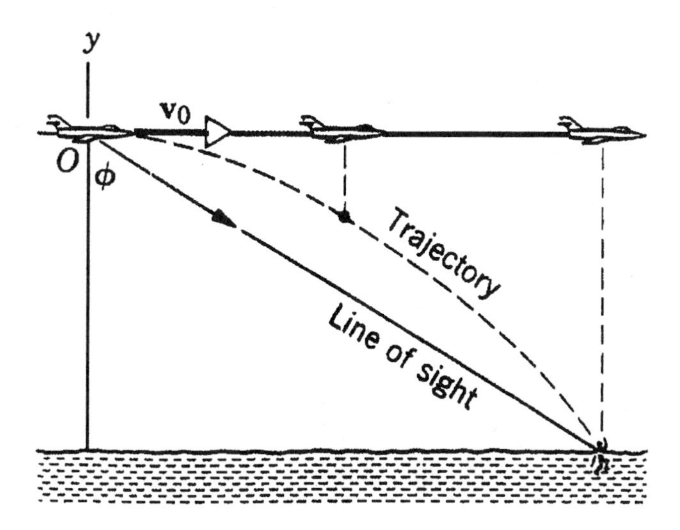

Figure 16. *Fundamentals of Physics, Second Edition, Figure 4.4*

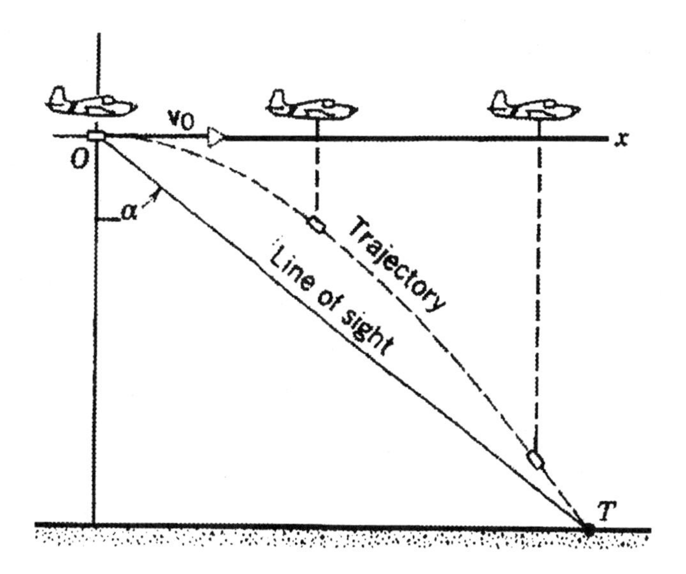

Figure 17. *Physics, Fourth Edition, Figure 4.9*

My last example gives an insight into our honored colleague's intellect. A common example in projectile motion is the calculation of the sight angle at the time of release of a bomb from an plane flying horizontally in air that miraculously offers no drag or lift to the bomb. Of course, this is a simplification for first year technical students of what the Norden bombsight did so efficiently in real life. In the preliminary edition we see an approximation to a six-engined B-25 delivering a direct hit on a well-defined power plant. Figure 13. Keeping up with world developments, the first hardbound edition has an eight-engined B-52, again with a well-detailed bomb and power plant. Figure 14. By 1966 it is a jet that closely resembles an F-4 Phantom, but neither the bomb nor target are detailed. Figure 15. We are entering the era of video warfare. Bombs and targets are dots. The figure was used again in 1970. After a decade of military agony, the figure has been transformed to that of a survival package being dropped to a person struggling in the water. Figure 16. The evolution is complete by 1992 when it is a contest to see who can drop a package closest to the target. Figure 17. Is this the maturing of a great philosopher reflecting on the latest American military venture, where the sophistication of modern technology made dropping bombs with pinpoint accuracy seem no different than playing a video game? Truly, Halliday and Resnick and their new partners, Ken Krane and Jearl Walker, have given us all a lot to think about.

Let me close with this thought:

> If you're an author of books,
> You had better beware of what cooks.
>   The figures you draw
>   You think have no flaw
> Until viewed by a fool who looks.

References

1. All editions of Physics and Fundamentals of Physics are copyrighted by John Wiley and Sons, Inc., New York, at the dates listed in Figure 1 except the 1958 edition which appears to be copyrighted by the authors.
2. Presentation at the Conference on the Introductory Physics Course, on the occasion of the retirement of Robert Resnick, Rensselaer Polytechnic Institute, Troy, NY, May 20-23, 1993.

CATAMAC AND CPIPP —

Computer Assisted Tutorials and Macintoshes and the Carleton Premedical Interactive Physics Project

P. J. S. Watson, Department of Physics
Carleton University, Canada

Although these two programs are aimed at slightly different clienteles, they share a common philosophy. The difficulties that we face at Carleton are common to almost every university in North America. We have very large first year classes, and incoming students with a huge range of abilities. Our most serious concern is neither with the very best nor the worst students, but the very large number with entry marks in the 70's and low 80's who are unable to cope with university, and see their marks drop to the low 50's.

The most serious weakness of students in science and engineering lies in their lack of basic manipulative skills in mathematics: although incoming students in principle understand calculus, they are in practice unable to apply it. We thus want to address the fairly limited task of helping students learn to solve the kind of problems that they will meet throughout the early years of university, primarily in mathematics and physics, but hopefully in a way which could be extended to other sciences, engineering and even business.

We believe our most serious failing is the lack of attention to the needs of the individual student: many more students could cope if they could be individually coached, which is simply incompatible with small numbers of faculty. There are two aspects which are almost self-evident. Firstly, it is only possible to learn how to do problems by doing problems, and most students do far too few. A reasonable ratio for, say, first year physics or calculus, might be 2-4 questions per lecture, for a total of at least 200 for the year. The less motivated students do at best about half as many. Secondly, students require help at various levels. Good students should be able to jump straight to the answer, while sometimes a detailed line-by-line solution is required. Furthermore, this feedback, if it is to be effective, should be available immediately, rather than the standard two weeks between completing an assignment and having it returned.

Hence it is essential to get into computer based learning of some kind. It is apparent that most commercially available software packages for first year university suffer from a large number of practical difficulties. They lack a consistent interface, often require specialized platforms and demand too much computer knowledge of the student. Most have a rather limited style of problems that can be solved: for example, it is trivial to write a program that will handle multiple choice, but much harder to have one that will check analytic answers, or graphs. Most seriously, however, they seem to be invariably written in such a way that they cannot be modified by the teacher, so that one is stuck with someone else's methods for solving a given problem.

The ideal system would be a flexible one, which could be modified to suit the needs of an individual teacher, and would require a minimum of computer knowledge on the part of the students. For practical reasons, it should run on Macintoshes and P.C's, and be usable by students both on networks and their own machines. It must keep track of student performance, and offer as wide a choice of styles of questions as possible. Animation at both the passive level ("press the button and watch") and the interactive level ("Sketch the following graph") should be available.

Most importantly, however, the response to the students should depend on their ability. Talented students (Tobias' "first tier"[1]) often need no help at all, and are positively insulted by being forced to proceed stepwise through a problem and repeat many very similar kinds of problem. They should not only be encouraged to proceed much faster with more challenging problems, but should be given the facilities with which to experiment. Far more common are students of some ability (the second tier) who need may repeated exposure to ideas, especially qualitative reasoning processes, as emphasized by McDermott[2] Arons,[3] and others. The third tier are students who take physics unwillingly and lack confidence in their own abilities: they will often need detailed help on a problem but in a way which builds confidence, which in practice means repeated exposure to the same kind of problem. Anyone who has ever marked an exam will know the depressing frequency with which a student will try to find a superficially appropriate formula without having the slightest idea of what principle should be applied. Hence the help should not only be available at the level of local tactics (e.g. "Gravitational Potential energy is given by U = mgh"), but at the much more general level of a quasi-expert system ("Could potential energy be used to solve this kind of problem?", "What techniques allow you to solve this equation?"). Finally there should be some hidden information back-up, with not just the physics contents of the course but also ancillary mathematical material.

Such a system is not intended to replace textbooks or lecture courses, but rather it should augment them in a form acceptable to a modern generation of students. In fact, ideally it should free the lecturer from having to devote so much time in class to the mechanics of problem-solving, and permit much more time to be spent on concepts and ideas.

We have attempted to meet all of these design goals in the CATaMac project, with, of course, varying degrees of success. It is a prototype system for first year physics which has now been used with a first year class of about 700 students. Our language of choice for the project is Hypercard (HC) for the Macs: ultimately we intend to translate this for P.C.'s by using Toolbook, but this has not been done. The main disadvantage is that the stacks tend to be large. A typical interactive tutorial would be ~ 300 kB. As an interpreted languages, HC is very slow in handling mathematics, although this is compensated for by the power of the actual commands. It is possible to compile some of the routines to XFCN's and XCMND's, with a considerable gain in speed.

At the moment we have questions which require the student to input numbers or formulae, select multiple choice with text, pictures or animation, enter points on a graph, draw vectors and analyze electrostatic fields. Formulae are entered via a palette which is specialized to each problem: this one below would allow one to enter (e.g.)

$$\propto = v_0 \cos(\theta) + \sqrt{(\mu_0^2)}.$$

It is designed to be modular, so that one could include a quite new style of problem (such as, say, one which requires spread-sheet input), and we have extended it to handle formal problems in linguistics.

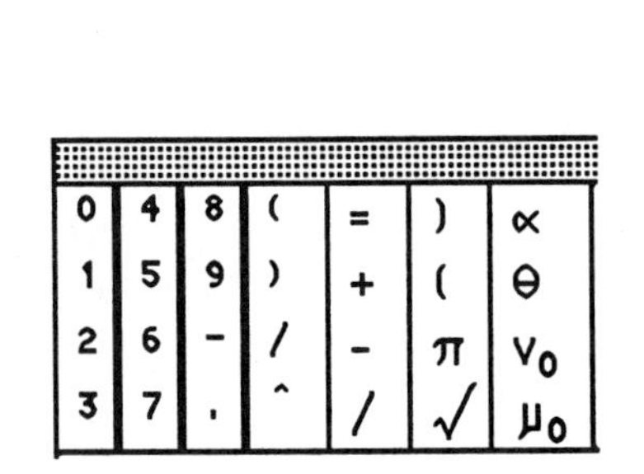

Since she is going to start moving towards the centre there must be a force towards it, and so she will hit the ground towards the centre.

It is interesting that physically impossible motions, such as the behavior of a pendulum whose string breaks at the endpoint of its motion (see above) , are almost never wrongly identified by students in the multiple choice animation questions

The current version benefits from the feedback of a complete year, and we will be making further changes this fall. It has been difficult to obtain quantitative feedback for this period, because the operation was bedeviled by a series of technical problems during the startup which were not found during the first user tests. However students were clearly in favor of keeping the tutorials, and assigning some academic weight to them (currently completion of 80% of the material counts 5% towards the course).

To address the problem of incoming students, **DVS Communications** (an information is an information technology services firm specializing in the application of digital multimedia technology) are collaborating on the development of the Carleton Premedial Interactive Physics Program. This presents basic Physics concepts and principles in a format that allows the user to work through the information and review at their own pace creating their unique best learning environment. The program is designed as a premedial review of secondary school physics prior to college or university entrance. At present the only module demonstrates the functionality in the area of interactive presentation of information without the interactive workshop component.

The program is being designed as a standalone application which can be delivered entirely electronically or in conjunction with a hard copy reference or workbook. It provides an integrated learning experience through information presentation and information utilization. The workshop option available in the options menu will be an interactive fully dynamic workspace environment. In this workshop option students can build and run physics experiments by inputting their own parameters to test the principles they are learning in the information presentation area.

Integrating sound with key text components makes the information presentation more memorable. Animation and attractive graphics are used to develop and maintain a high interest level and positive motivation ensues through work on interesting problems. The examples have been created to be gender neutral. The access mechanism allow the user to directly select from contents indexes or browse through menus. The options button allows the user to go to: 1) the main menu, 2) previous menu, 3) the workshop and 4) a demonstration of functionality, i.e.. a guided tour of the various components within the workshop. The program package will be available for both IBM and Macintosh platforms as a CD-ROM.

References

1. S. Tobias *"They're Not dumb, They're Different"* Research Corporation (1990)
2 L. McDermott, Proceedings of this conference.
3 A Arons, "*A Guide to Introductory Physics Teaching*", Wiley (1990).

MAKING WAVES:

USING A REAL APPARATUS IN CONJUNCTION WITH A SIMULATION

Charles A. Whitney & Beth Hoffman, Science Education Department
Harvard-Smithsonian Center for Astrophysics

Abstract

In this paper we describe a simulation of a weighted elastic string that provides a variety of representations (or views) of the system. We discuss alternative methods for using this program with physics students and point out some advantages of starting with a lengthy exploration of a physical device before the simulation is introduced.

The WaveMaker Simulation

The WaveMaker computer simulation represents a rubber band loaded with lead fishing weights and stretched between two fixed hooks. (Such a device can be built from supplies available in most lumber yards and hardware stores.)

The computer display consists of a menu bar, a control palette and a variety of windows, which the user can select. The main window is an animated schematic diagram of the rubber band and weights. Other windows show various representations of the motion, such as plots of position versus time, velocity versus position, etc. The screen also displays a stopwatch and the user can easily adjust the system by adding weights and changing their masses.

Why Use a Simulation and How Do We Start?

If the physical device, itself, is so easy to build and operate, what is the purpose of a simulation?

Some of the common arguments for using simulations in an introductory course are related to the power and vision that they can provide. A simulation can do things that would otherwise be too expensive or dangerous; it can let the student expand time and space and work in domains that would otherwise be inaccessible. True enough, but I think there is another argument of a rather different kind: Even when the simulation is realistic and looks like a simple, inexpensive device, it represents this device in a highly refined and abstract way. It focuses the student's attention on the concepts that can be handled mathematically. The simulation is an idealized version of reality. This is both its virtue and its weakness.

The real device, whether it is a commercial product or is constructed in the cellar, always has limitations, shortcomings, or just plain defects. The elastic may, for example, have a knot. Thus, the real device is much richer than the computer simulation, and the essence of designing the simulation is to capture the relevant features and hide or suppress the irrelevant ones. The real device, on the other hand, allows the user to decide which elements are the ones that are relevant, as oppose to those which are simply tangential to the system. The choice of "relevant" features in the simulation is guided by the conceptual goals that are held in the minds of the designer of the computer simulation. These goals are, of course, unknown to the student when the simulation is first encountered.

Where should the student start? Should the student see the apparatus first, or the computer simulation? Or to put it differently, what are the merits of various strategies for introducing the student to two types of representation?

When we began working with high school students, we were concentrating on developing the interface for the computer program, so we tended to start by turning on the computer and demonstrating what the program could do. Our next step was to lead the student through a process of validation, in the hope that the computer program would be seen as a reliable simulation of the real apparatus.

Validating the Simulation

To build up the students' confidence and interest in the simulation, we guided their work by asking questions that required prediction and verification. For example: What will happen to the period of oscillation if you pull the weight farther to the side before letting go? (In formulating these questions, great care must be taken to avoid using technical terms that may be unfamiliar. Students will be puzzled if the question is phrased in terms of "initial displacement" for example).

Occasionally the student is surprised by the behavior of the computer simulation or expresses a doubt that it is performing properly. This provides an opportunity for a key step in the process of validation. When it happens, we turn attention to the real apparatus and encourage direct comparison of the two. After a few such experiences, the student typically accepts the simulation as a substitute for the real device and talks about it as though it were the device itself. The screen objects, such as moving circles of color, are referred to by the names they are given in the real device, such as "bead" or "weight".

The exploration then continues along a path defined by questions or challenges that are presented to the student. This procedureæintroducing the simulation, then the apparatus and then going back and forthæleads to a focused and externally directed experience for the student. Depending on the student and the educational situation, this can be a useful way to proceed.

An Alternative: Hold off the Simulation

Suppose the teacher gives a physical device, such as a beaded rubber band, to small teams of students and simply asks them to explore it, without giving any further instructions.

We have found this to be a far richer and more natural way to introduce basic concepts. The device remains in center-stage, and the simulation then becomes an aid for understanding the device. Each, then, supplements and validates the other.

This approach is not only more compelling for the students, it also tends to be more satisfactory for the teacher. What happens is something like the following.

Student as a researcher

When students first see the physical device, they perceive it as a rather arcane object with a few identifiable features æ brass hooks, knots in the rubber band, fishing weights, pieces of wood. They explore it with their hands and they focus attention on what they see; they pluck it and watch it move, but they have no way of knowing what determines the motion. In fact, the motion itself is a blur and their experience is rather undifferentiated. If given the opportunity, they become engrossed in the process of making the distinctions among parts of the system, *and* among features of its behavior, such as period and amplitude. Thus, with freedom but without guidance, they take the first steps toward defining the system as a physical structure and toward defining its behavior as a dynamic structure.

Gradually, they build up a set of experiences and a vocabulary for describing what they see. They explore the device at a deeper level. Students become interested in the relationships between the physical features and its behavior. They wonder about the effect of the knot and the position of the weight; they wonder whether the way the rubber band is attached makes a difference; they will ask what happens if the weight is pulled to the side instead of straight upward. By starting with what they can see and performing a series of short experiments, they will gradually sort out the relevant and irrelevant features of the system. As they set aside some features as unimportant æ the way the rubber bands are tied to the hooks on the ends, for example æ they begin to see the device with new eyes. They begin to abstract its more important features and to perceive it as an idealized device. In this way, they prepare themselves for the simulation, which is highly idealized, no matter how "realistic" it may appear to be.

This process is rather like puzzle-solving, and it may be the closest experience to "doing science" that the teacher can provide. The physical device inspires curiosity, which is a natural precursor to experimentation.

Call in the simulation

After students have played with the physical device, they have sharpened their vision and have developed a rudimentary vocabulary for discussing the phenomena before them. The device, however, raises questions that cannot be easily answered by observation and measurement with rulers and stopwatches. For example, measuring the period by counting the oscillations while recording the seconds with a stopwatch is cumbersome and usually inaccurate. The simulation can be made slow enough for visual observation, and it includes a built-in stopwatch. The period, in this case, becomes easily measurable.

Gradually, the simulation will seem less arbitrary and less like an arcade game. To many students, it will be seen as a tool for exploring the real device that sits on the table. This was our original goal, and we now feel that the best route to this goal may be the indirect approach, which leads the student through a process of self-directed exploration to discrimination and abstraction before confrontation with the computer. If this process is successful, the simulation will serve as a convenient "abstracting" tool with which to explore the activity of the physical device; and the physical device will provide realistic and tangible evidence that supports the abstraction. Hence, the simulation and the physical system used together becomes more effective than using either one alone. *The simulation becomes the intermediary between the real object and the mathematical abstraction.*

We are working toward activities in which students approach phenomena along many avenues visualization, exploration, manipulation, measurement, and abstraction. The whole is greater than the sum of the parts when these approaches interact with each other.

Acknowledgments

The work was funded by the National Science Foundation under grant #7582 (Project InSIGHT) for which the author is principle investigator. The original concept for WaveMaker is due to Philip Sadler, Co-Investigator and Director of the Science Education Department of the Center for Astrophysics. The programming was done on an Apple Macintosh in C by Freeman S. Deutsch and the design was developed through a series of discussions with the project's advisory panel and field tests with high school students and teachers. The software is being prepared for publication in 1994.